An Introduction to Complex Analysis

Ravi P. Agarwal • Kanishka Perera
Sandra Pinelas

An Introduction to Complex Analysis

Ravi P. Agarwal
Department of Mathematics
Florida Institute of Technology
Melbourne, FL 32901, USA
agarwal@fit.edu

Kanishka Perera
Department of Mathematical Sciences
Florida Institute of Technology
Melbourne, FL 32901, USA
kperera@fit.edu

Sandra Pinelas
Department of Mathematics
Azores University, Apartado 1422
9501-801 Ponta Delgada, Portugal
sandra.pinelas@clix.pt

ISBN 978-1-4899-9716-6   ISBN 978-1-4614-0195-7 (eBook)
DOI 10.1007/978-1-4614-0195-7
Springer New York Dordrecht Heidelberg London

Mathematics Subject Classification (2010): M12074, M12007

Springer is part of Springer Science+Business Media (www.springer.com)

Dedicated to our mothers:

Godawari Agarwal, Soma Perera, and Maria Pinelas

Preface

Complex analysis is a branch of mathematics that involves functions of complex numbers. It provides an extremely powerful tool with an unexpectedly large number of applications, including in number theory, applied mathematics, physics, hydrodynamics, thermodynamics, and electrical engineering. Rapid growth in the theory of complex analysis and in its applications has resulted in continued interest in its study by students in many disciplines. This has given complex analysis a distinct place in mathematics curricula all over the world, and it is now being taught at various levels in almost every institution.

Although several excellent books on complex analysis have been written, the present rigorous and perspicuous introductory text can be used directly in class for students of applied sciences. In fact, in an effort to bring the subject to a wider audience, we provide a compact, but thorough, introduction to the subject in **An Introduction to Complex Analysis**. This book is intended for readers who have had a course in calculus, and hence it can be used for a senior undergraduate course. It should also be suitable for a beginning graduate course because in undergraduate courses students do not have any exposure to various intricate concepts, perhaps due to an inadequate level of mathematical sophistication.

The subject matter has been organized in the form of theorems and their proofs, and the presentation is rather unconventional. It comprises 50 class tested lectures that we have given mostly to math majors and engineering students at various institutions all over the globe over a period of almost 40 years. These lectures provide flexibility in the choice of material for a particular one-semester course. It is our belief that the content in a particular lecture, together with the problems therein, provides fairly adequate coverage of the topic under study.

A brief description of the topics covered in this book follows: In **Lecture 1** we first define complex numbers (imaginary numbers) and then for such numbers introduce basic operations–addition, subtraction, multiplication, division, modulus, and conjugate. We also show how the complex numbers can be represented on the xy-plane. In **Lecture 2**, we show that complex numbers can be viewed as two-dimensional vectors, which leads to the triangle inequality. We also express complex numbers in polar form. In **Lecture 3**, we first show that every complex number can be written in exponential form and then use this form to raise a rational power to a given complex number. We also extract roots of a complex number and prove that complex numbers cannot be totally ordered. In **Lecture 4**, we collect some essential definitions about sets in the complex plane. We also introduce stereographic projection and define the Riemann sphere. This ensures that in the complex plane there is only one point at infinity.

In **Lecture 5**, first we introduce a complex-valued function of a complex variable and then for such functions define the concept of limit and continuity at a point. In **Lectures 6 and 7**, we define the differentiation of complex functions. This leads to a special class of functions known as analytic functions. These functions are of great importance in theory as well as applications, and constitute a major part of complex analysis. We also develop the Cauchy-Riemann equations, which provide an easier test to verify the analyticity of a function. We also show that the real and imaginary parts of an analytic function are solutions of the Laplace equation.

In **Lectures 8 and 9**, we define the exponential function, provide some of its basic properties, and then use it to introduce complex trigonometric and hyperbolic functions. Next, we define the logarithmic function, study some of its properties, and then introduce complex powers and inverse trigonometric functions. In **Lectures 10 and 11**, we present graphical representations of some elementary functions. Specially, we study graphical representations of the Möbius transformation, the trigonometric mapping $\sin z$, and the function $z^{1/2}$.

In **Lecture 12**, we collect a few items that are used repeatedly in complex integration. We also state Jordan's Curve Theorem, which seems to be quite obvious; however, its proof is rather complicated. In **Lecture 13**, we introduce integration of complex-valued functions along a directed contour. We also prove an inequality that plays a fundamental role in our later lectures. In **Lecture 14**, we provide conditions on functions so that their contour integral is independent of the path joining the initial and terminal points. This result, in particular, helps in computing the contour integrals rather easily. In **Lecture 15**, we prove that the integral of an analytic function over a simple closed contour is zero. This is one of the fundamental theorems of complex analysis. In **Lecture 16**, we show that the integral of a given function along some given path can be replaced by the integral of the same function along a more amenable path. In **Lecture 17**, we present Cauchy's integral formula, which expresses the value of an analytic function at any point of a domain in terms of the values on the boundary of this domain. This is the most fundamental theorem of complex analysis, as it has numerous applications. In **Lecture 18**, we show that for an analytic function in a given domain all the derivatives exist and are analytic. Here we also prove Morera's Theorem and establish Cauchy's inequality for the derivatives, which plays an important role in proving Liouville's Theorem.

In **Lecture 19**, we prove the Fundamental Theorem of Algebra, which states that every nonconstant polynomial with complex coefficients has at least one zero. Here, for a given polynomial, we also provide some bounds on its zeros in terms of the coefficients. In **Lecture 20**, we prove that a function analytic in a bounded domain and continuous up to and including its boundary attains its maximum modulus on the boundary. This result has direct applications to harmonic functions.

In **Lectures 21 and 22**, we collect several results for complex sequences and series of numbers and functions. These results are needed repeatedly in later lectures. In **Lecture 23**, we introduce a power series and show how to compute its radius of convergence. We also show that within its radius of convergence a power series can be integrated and differentiated term-by-term. In **Lecture 24**, we prove Taylor's Theorem, which expands a given analytic function in an infinite power series at each of its points of analyticity. In **Lecture 25**, we expand a function that is analytic in an annulus domain. The resulting expansion, known as Laurent's series, involves positive as well as negative integral powers of $(z - z_0)$. From applications point of view, such an expansion is very useful. In **Lecture 26**, we use Taylor's series to study zeros of analytic functions. We also show that the zeros of an analytic function are isolated. In **Lecture 27**, we introduce a technique known as analytic continuation, whose principal task is to extend the domain of a given analytic function. In **Lecture 28**, we define the concept of symmetry of two points with respect to a line or a circle. We shall also prove Schwarz's Reflection Principle, which is of great practical importance for analytic continuation.

In **Lectures 29 and 30**, we define, classify, characterize singular points of complex functions, and study their behavior in the neighborhoods of singularities. We also discuss zeros and singularities of analytic functions at infinity.

The value of an iterated integral depends on the order in which the integration is performed, the difference being called the residue. In **Lecture 31**, we use Laurent's expansion to establish Cauchy's Residue Theorem, which has far-reaching applications. In particular, integrals that have a finite number of isolated singularities inside a contour can be integrated rather easily. In **Lectures 32-35**, we show how the theory of residues can be applied to compute certain types of definite as well as improper real integrals. For this, depending on the complexity of an integrand, one needs to choose a contour cleverly. In **Lecture 36**, Cauchy's Residue Theorem is further applied to find sums of certain series.

In **Lecture 37**, we prove three important results, known as the Argument Principle, Rouché's Theorem, and Hurwitz's Theorem. We also show that Rouché's Theorem provides locations of the zeros and poles of meromorphic functions. In **Lecture 38**, we further use Rouché's Theorem to investigate the behavior of the mapping f generated by an analytic function $w = f(z)$. Then we study some properties of the inverse mapping f^{-1}. We also discuss functions that map the boundaries of their domains to the boundaries of their ranges. Such results are very important for constructing solutions of Laplace's equation with boundary conditions.

In **Lecture 39**, we study conformal mappings that have the angle-preserving property, and in **Lecture 40** we employ these mappings to establish some basic properties of harmonic functions. In **Lecture 41**, we provide an explicit formula for the derivative of a conformal mapping that maps the upper half-plane onto a given bounded or unbounded polygonal region. The integration of this formula, known as the Schwarz-Christoffel transformation, is often applied in physical problems such as heat conduction, fluid mechanics, and electrostatics.

In **Lecture 42**, we introduce infinite products of complex numbers and functions and provide necessary and sufficient conditions for their convergence, whereas in **Lecture 43** we provide representations of entire functions as finite/infinite products involving their finite/infinite zeros. In **Lecture 44**, we construct a meromorphic function in the entire complex plane with preassigned poles and the corresponding principal parts.

Periodicity of analytic/meromorphic functions is examined in **Lecture 45**. Here, doubly periodic (elliptic) functions are also introduced. The Riemann zeta function is one of the most important functions of classical mathematics, with a variety of applications in analytic number theory. In **Lecture 46**, we study some of its elementary properties. **Lecture 47** is devoted to Bieberbach's conjecture (now theorem), which had been a challenge to the mathematical community for almost 68 years. A Riemann surface is an ingenious construct for visualizing a multi-valued function. These surfaces have proved to be of inestimable value, especially in the study of algebraic functions. In **Lecture 48**, we construct Riemann surfaces for some simple functions. In **Lecture 49**, we discuss the geometric and topological features of the complex plane associated with dynamical systems, whose evolution is governed by some simple iterative schemes. This work, initiated by Julia and Mandelbrot, has recently found applications in physical, engineering, medical, and aesthetic problems; specially those exhibiting chaotic behavior.

Finally, in **Lecture 50**, we give a brief history of complex numbers. The road had been very slippery, full of confusions and superstitions; however, complex numbers forced their entry into mathematics. In fact, *there is really nothing imaginary about imaginary numbers and complex about complex numbers.*

Two types of problems are included in this book, those that illustrate the general theory and others designed to fill out text material. The problems form an integral part of the book, and every reader is urged to attempt most, if not all of them. For the convenience of the reader, we have provided answers or hints to all the problems.

In writing a book of this nature, no originality can be claimed, only a humble attempt has been made to present the subject as simply, clearly, and accurately as possible. The illustrative examples are usually very simple, keeping in mind an average student.

It is earnestly hoped that **An Introduction to Complex Analysis** will serve an inquisitive reader as a starting point in this rich, vast, and ever-expanding field of knowledge.

We would like to express our appreciation to Professors Hassan Azad, Siegfried Carl, Eugene Dshalalow, Mohamed A. El-Gebeily, Kunquan Lan, Radu Precup, Patricia J.Y. Wong, Agacik Zafer, Yong Zhou, and Changrong Zhu for their suggestions and criticisms. We also thank Ms. Vaishali Damle at Springer New York for her support and cooperation.

Ravi P Agarwal
Kanishka Perera
Sandra Pinelas

Contents

Lecture 1
Complex Numbers I

We begin this lecture with the definition of complex numbers and then introduce basic operations-addition, subtraction, multiplication, and division of complex numbers. Next, we shall show how the complex numbers can be represented on the xy-plane. Finally, we shall define the modulus and conjugate of a complex number.

Throughout these lectures, the following well-known notations will be used:

$\mathbb{N} = \{1, 2, \cdots\}$, the set of all *natural numbers*;

$\mathbb{Z} = \{\cdots, -2, -1, 0, 1, 2, \cdots\}$, the set of all *integers*;

$\mathbb{Q} = \{m/n : m, n \in \mathbb{Z}, n \neq 0\}$, the set of all *rational numbers*;

$\mathbb{R} = $ the set of all *real numbers*.

A *complex number* is an expression of the form $a + ib$, where a and $b \in \mathbb{R}$, and i (sometimes j) is just a symbol.

$\mathbf{C} = \{a + ib : a, b \in \mathbb{R}\}$, the set of all *complex numbers*.

It is clear that $\mathbb{N} \subset \mathbb{Z} \subset \mathbb{Q} \subset \mathbb{R} \subset \mathbf{C}$.

For a complex number, $z = a + ib$, $\text{Re}(z) = a$ is the *real part* of z, and $\text{Im}(z) = b$ is the *imaginary part* of z. If $a = 0$, then z is said to be a *purely imaginary number*. Two complex numbers, z and w are equal; i.e., $z = w$, if and only if, $\text{Re}(z) = \text{Re}(w)$ and $\text{Im}(z) = \text{Im}(w)$. Clearly, $z = 0$ is the only number that is real as well as purely imaginary.

The following operations are defined on the complex number system:

(i). Addition: $(a + bi) + (c + di) = (a + c) + (b + d)i$.

(ii). Subtraction: $(a + bi) - (c + di) = (a - c) + (b - d)i$.

(iii). Multiplication: $(a + bi)(c + di) = (ac - bd) + (bc + ad)i$.

As in real number system, $0 = 0 + 0i$ is a complex number such that $z + 0 = z$. There is obviously a unique complex number 0 that possesses this property.

From (iii), it is clear that $i^2 = -1$, and hence, formally, $i = \sqrt{-1}$. Thus, except for zero, positive real numbers have real square roots, and negative real numbers have purely imaginary square roots.

For complex numbers z_1, z_2, z_3 we have the following easily verifiable properties:

(I).   Commutativity of addition: $z_1 + z_2 = z_2 + z_1$.

(II). Commutativity of multiplication: $z_1 z_2 = z_2 z_1$.

(III). Associativity of addition: $z_1 + (z_2 + z_3) = (z_1 + z_2) + z_3$.

(IV). Associativity of multiplication: $z_1(z_2 z_3) = (z_1 z_2)z_3$.

(V).   Distributive law: $(z_1 + z_2)z_3 = z_1 z_3 + z_2 z_3$.

As an illustration, we shall show only (I). Let $z_1 = a_1 + b_1 i$, $z_2 = a_2 + b_2 i$ then

$$
\begin{aligned}
z_1 + z_2 &= (a_1 + a_2) + (b_1 + b_2)i = (a_2 + a_1) + (b_2 + b_1)i \\
&= (a_2 + b_2 i) + (a_1 + b_1 i) = z_2 + z_1.
\end{aligned}
$$

Clearly, **C** with addition and multiplication forms a field.

We also note that, for any integer k,

$$
i^{4k} = 1, \quad i^{4k+1} = i, \quad i^{4k+2} = -1, \quad i^{4k+3} = -i.
$$

The rule for division is derived as

$$
\frac{a+bi}{c+di} = \frac{a+bi}{c+di} \cdot \frac{c-di}{c-di} = \frac{ac+bd}{c^2+d^2} + \frac{bc-ad}{c^2+d^2}i, \quad c^2 + d^2 \neq 0.
$$

Example 1.1. Find the quotient $\dfrac{(6+2i) - (1+3i)}{-1+i-2}$.

$$
\begin{aligned}
\frac{(6+2i) - (1+3i)}{-1+i-2} &= \frac{5-i}{-3+i} = \frac{(5-i)}{(-3+i)} \frac{(-3-i)}{(-3-i)} \\
&= \frac{-15 - 1 - 5i + 3i}{9+1} = -\frac{8}{5} - \frac{1}{5}i.
\end{aligned}
$$

Geometrically, we can represent complex numbers as points in the xy-plane by associating to each complex number $a + bi$ the point (a, b) in the xy-plane (also known as an Argand diagram). The plane is referred to as the *complex plane*. The x-axis is called the *real axis*, and the y-axis is called the *imaginary axis*. The number $z = 0$ corresponds to the origin of the plane. This establishes a one-to-one correspondence between the set of all complex numbers and the set of all points in the complex plane.

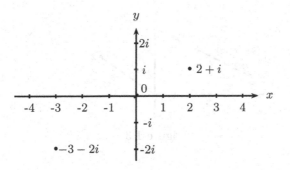

Figure 1.1

We can justify the above representation of complex numbers as follows:
Let A be a point on the real axis such that $OA = a$. Since $i \cdot i\, a = i^2\, a = -a$,
we can conclude that twice multiplication of the real number a by i amounts
to the rotation of OA through two right angles to the position OA''. Thus,
it naturally follows that the multiplication by i is equivalent to the rotation
of OA through one right angle to the position OA'. Hence, if $y'Oy$ is a
line perpendicular to the real axis $x'Ox$, then all imaginary numbers are
represented by points on $y'Oy$.

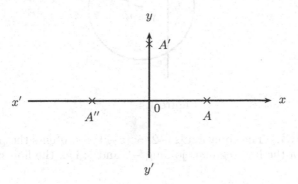

Figure 1.2

The *absolute value* or *modulus* of the number $z = a + ib$ is denoted
by $|z|$ and given by $|z| = \sqrt{a^2 + b^2}$. Since $a \le |a| = \sqrt{a^2} \le \sqrt{a^2 + b^2}$
and $b \le |b| = \sqrt{b^2} \le \sqrt{a^2 + b^2}$, it follows that $\mathrm{Re}(z) \le |\mathrm{Re}(z)| \le |z|$ and
$\mathrm{Im}(z) \le |\mathrm{Im}(z)| \le |z|$. Now, let $z_1 = a_1 + b_1 i$ and $z_2 = a_2 + b_2 i$ then

$$|z_1 - z_2| = \sqrt{(a_1 - a_2)^2 + (b_1 - b_2)^2}.$$

Hence, $|z_1 - z_2|$ is just the distance between the points z_1 and z_2. This fact
is useful in describing certain curves in the plane.

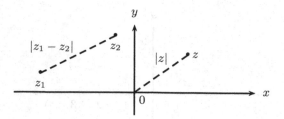

Figure 1.3

Example 1.2. The equation $|z - 1 + 3i| = 2$ represents the circle whose center is $z_0 = 1 - 3i$ and radius is $R = 2$.

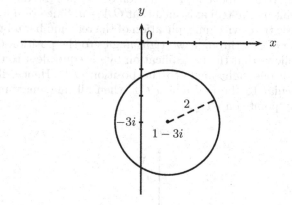

Figure 1.4

Example 1.3. The equation $|z + 2| = |z - 1|$ represents the perpendicular bisector of the line segment joining -2 and 1; i.e., the line $x = -1/2$.

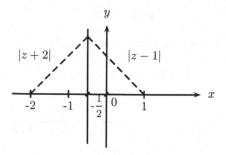

Figure 1.5

The *complex conjugate* of the number $z = a + bi$ is denoted by $\bar{z}$ and given by $\bar{z} = a - bi$. Geometrically, $\bar{z}$ is the reflection of the point z about the real axis.

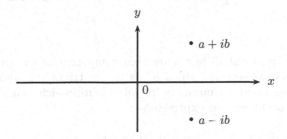

Figure 1.6

The following relations are immediate:

1. $|z_1 z_2| = |z_1||z_2|$, $\left|\dfrac{z_1}{z_2}\right| = \dfrac{|z_1|}{|z_2|}$, $(z_2 \neq 0)$.

2. $|z| \geq 0$, and $|z| = 0$, if and only if $z = 0$.

3. $z = \bar{z}$, if and only if $z \in \mathbb{R}$.

4. $z = -\bar{z}$, if and only if $z = bi$ for some $b \in \mathbb{R}$.

5. $\overline{z_1 \pm z_2} = \bar{z}_1 \pm \bar{z}_2$.

6. $\overline{z_1 z_2} = (\bar{z}_1)(\bar{z}_2)$.

7. $\overline{\left(\dfrac{z_1}{z_2}\right)} = \dfrac{\bar{z}_1}{\bar{z}_2}$, $z_2 \neq 0$.

8. $\mathrm{Re}(z) = \dfrac{z + \bar{z}}{2}$, $\mathrm{Im}(z) = \dfrac{z - \bar{z}}{2i}$.

9. $\bar{\bar{z}} = z$.

10. $|z| = |\bar{z}|$, $z\bar{z} = |z|^2$.

As an illustration, we shall show only relation 6. Let $z_1 = a_1 + b_1 i$, $z_2 = a_2 + b_2 i$. Then

$$
\begin{aligned}
\overline{z_1 z_2} &= \overline{(a_1 + b_1 i)(a_2 + b_2 i)} \\
&= \overline{(a_1 a_2 - b_1 b_2) + i(a_1 b_2 + b_1 a_2)} \\
&= (a_1 a_2 - b_1 b_2) - i(a_1 b_2 + b_1 a_2) \\
&= (a_1 - b_1 i)(a_2 - b_2 i) = (\bar{z}_1)(\bar{z}_2).
\end{aligned}
$$

Lecture 2
Complex Numbers II

In this lecture, we shall first show that complex numbers can be viewed as two-dimensional vectors, which leads to the triangle inequality. Next, we shall express complex numbers in polar form, which helps in reducing the computation in tedious expressions.

For each point (number) z in the complex plane, we can associate a vector, namely the directed line segment from the origin to the point z; i.e., $z = a + bi \longleftrightarrow \vec{v} = (a, b)$. Thus, complex numbers can also be interpreted as two-dimensional *ordered pairs*. The length of the vector associated with z is $|z|$. If $z_1 = a_1 + b_1 i \longleftrightarrow \vec{v}_1 = (a_1, b_1)$ and $z_2 = a_2 + b_2 i \longleftrightarrow \vec{v}_2 = (a_2, b_2)$, then $z_1 + z_2 \longleftrightarrow \vec{v}_1 + \vec{v}_2$.

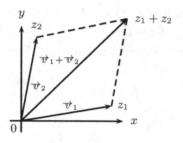

Figure 2.1

Using this correspondence and the fact that the length of any side of a triangle is less than or equal to the sum of the lengths of the two other sides, we have

$$|z_1 + z_2| \leq |z_1| + |z_2| \tag{2.1}$$

for any two complex numbers z_1 and z_2. This inequality also follows from

$$
\begin{aligned}
|z_1 + z_2|^2 &= (z_1 + z_2)\overline{(z_1 + z_2)} = (z_1 + z_2)(\overline{z}_1 + \overline{z}_2) \\
&= z_1 \overline{z}_1 + z_1 \overline{z}_2 + z_2 \overline{z}_1 + z_2 \overline{z}_2 \\
&= |z_1|^2 + (z_1 \overline{z}_2 + \overline{z_1 \overline{z}_2}) + |z_2|^2 \\
&= |z_1|^2 + 2\operatorname{Re}(z_1 \overline{z}_2) + |z_2|^2 \\
&\leq |z_1|^2 + 2|z_1 z_2| + |z_2|^2 = (|z_1| + |z_2|)^2.
\end{aligned}
$$

Applying the inequality (2.1) to the complex numbers $z_2 - z_1$ and z_1, we get

$$|z_2| = |z_2 - z_1 + z_1| \le |z_2 - z_1| + |z_1|,$$

and hence

$$|z_2| - |z_1| \le |z_2 - z_1|. \tag{2.2}$$

Similarly, we have

$$|z_1| - |z_2| \le |z_1 - z_2|. \tag{2.3}$$

Combining inequalities (2.2) and (2.3), we obtain

$$||z_1| - |z_2|| \le |z_1 - z_2|. \tag{2.4}$$

Each of the inequalities (2.1)-(2.4) will be called a *triangle inequality*. Inequality (2.4) tells us that the length of one side of a triangle is greater than or equal to the difference of the lengths of the two other sides. From (2.1) and an easy induction, we get the *generalized triangle inequality*

$$|z_1 + z_2 + \cdots + z_n| \le |z_1| + |z_2| + \cdots + |z_n|. \tag{2.5}$$

From the demonstration above, it is clear that, in (2.1), equality holds if and only if $\operatorname{Re}(z_1 \bar{z}_2) = |z_1 z_2|$; i.e., $z_1 \bar{z}_2$ is real and nonnegative. If $z_2 \ne 0$, then since $z_1 \bar{z}_2 = z_1 |z_2|^2 / z_2$, this condition is equivalent to $z_1/z_2 \ge 0$. Now we shall show that equality holds in (2.5) if and only if the ratio of any two nonzero terms is positive. For this, if equality holds in (2.5), then, since

$$
\begin{aligned}
|z_1 + z_2 + z_3 + \cdots + z_n| &= |(z_1 + z_2) + z_3 + \cdots + z_n| \\
&\le |z_1 + z_2| + |z_3| + \cdots + |z_n| \\
&\le |z_1| + |z_2| + |z_3| + \cdots + |z_n|,
\end{aligned}
$$

we must have $|z_1 + z_2| = |z_1| + |z_2|$. But, this holds only when $z_1/z_2 \ge 0$, provided $z_2 \ne 0$. Since the numbering of the terms is arbitrary, the ratio of any two nonzero terms must be positive. Conversely, suppose that the ratio of any two nonzero terms is positive. Then, if $z_1 \ne 0$, we have

$$
\begin{aligned}
|z_1 + z_2 + \cdots + z_n| &= |z_1| \left| 1 + \frac{z_2}{z_1} + \cdots + \frac{z_n}{z_1} \right| \\
&= |z_1| \left(1 + \frac{z_2}{z_1} + \cdots + \frac{z_n}{z_1} \right) \\
&= |z_1| \left(1 + \frac{|z_2|}{|z_1|} + \cdots + \frac{|z_n|}{|z_1|} \right) \\
&= |z_1| + |z_2| + \cdots + |z_n|.
\end{aligned}
$$

Example 2.1. If $|z| = 1$, then, from (2.5), it follows that

$$|z^2 + 2z + 6 + 8i| \le |z|^2 + 2|z| + |6 + 8i| = 1 + 2 + \sqrt{36 + 64} = 13.$$

Similarly, from (2.1) and (2.4), we find

$$2 \leq |z^2 - 3| \leq 4.$$

Note that the product of two complex numbers z_1 and z_2 is a new complex number that can be represented by a vector in the same plane as the vectors for z_1 and z_2. However, this product is neither the scalar (dot) nor the vector (cross) product used in ordinary vector analysis.

Now let $z = x + yi$, $r = |z| = \sqrt{x^2 + y^2}$, and θ be a number satisfying

$$\cos \theta = \frac{x}{r} \quad \text{and} \quad \sin \theta = \frac{y}{r}.$$

Then, z can be expressed in *polar (trigonometric) form* as

$$z = r(\cos \theta + i \sin \theta).$$

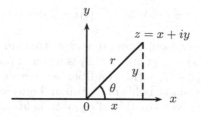

Figure 2.2

To find θ, we usually compute $\tan^{-1}(y/x)$ and adjust the quadrant problem by adding or subtracting π when appropriate. Recall that $\tan^{-1}(y/x) \in (-\pi/2, \pi/2)$.

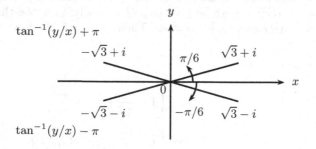

Figure 2.3

Example 2.2. Express $1 - i$ in polar form. Here $r = \sqrt{2}$ and $\theta = -\pi/4$, and hence

$$1 - i = \sqrt{2} \left[\cos \left(-\frac{\pi}{4} \right) + i \sin \left(-\frac{\pi}{4} \right) \right].$$

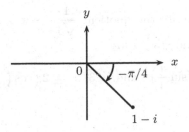

Figure 2.4

We observe that any one of the values $\theta = -(\pi/4) \pm 2n\pi$, $n = 0, 1, \cdots$, can be used here. The number θ is called an *argument* of z, and we write $\theta = \arg z$. Geometrically, $\arg z$ denotes the angle measured in radians that the vector corresponds to z makes with the positive real axis. The argument of 0 is not defined. The pair $(r, \arg z)$ is called the *polar coordinates* of the complex number z.

The *principal value* of $\arg z$, denoted by $\operatorname{Arg} z$, is defined as that unique value of $\arg z$ such that $-\pi < \arg z \le \pi$.

If we let $z_1 = r_1(\cos\theta_1 + i\sin\theta_1)$ and $z_2 = r_2(\cos\theta_2 + i\sin\theta_2)$, then

$$z_1 z_2 = r_1 r_2[(\cos\theta_1\cos\theta_2 - \sin\theta_1\sin\theta_2) + i(\sin\theta_1\cos\theta_2 + \cos\theta_1\sin\theta_2)]$$
$$= r_1 r_2[\cos(\theta_1 + \theta_2) + i\sin(\theta_1 + \theta_2)].$$

Thus, $|z_1 z_2| = |z_1||z_2|$, $\arg(z_1 z_2) = \arg z_1 + \arg z_2$.

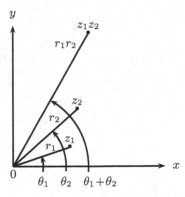

Figure 2.5

For the division, we have

$$\frac{z_1}{z_2} = \frac{r_1}{r_2}[\cos(\theta_1 - \theta_2) + i\sin(\theta_1 - \theta_2)],$$

$$\left|\frac{z_1}{z_2}\right| = \frac{|z_1|}{|z_2|}, \quad \arg\left(\frac{z_1}{z_2}\right) = \arg z_1 - \arg z_2.$$

Example 2.3. Write the quotient $\dfrac{1+i}{\sqrt{3}-i}$ in polar form. Since the polar forms of $1+i$ and $\sqrt{3}-i$ are

$$1+i = \sqrt{2}\left(\cos\frac{\pi}{4} + i\sin\frac{\pi}{4}\right) \quad \text{and} \quad \sqrt{3}-i = 2\left(\cos\left(-\frac{\pi}{6}\right) + i\sin\left(-\frac{\pi}{6}\right)\right),$$

it follows that

$$\frac{1+i}{\sqrt{3}-i} = \frac{\sqrt{2}}{2}\left\{\cos\left[\frac{\pi}{4} - \left(-\frac{\pi}{6}\right)\right] + i\sin\left[\frac{\pi}{4} - \left(-\frac{\pi}{6}\right)\right]\right\}$$

$$= \frac{\sqrt{2}}{2}\left\{\cos\left(\frac{5\pi}{12}\right) + i\sin\left(\frac{5\pi}{12}\right)\right\}.$$

Recall that, geometrically, the point $\bar{z}$ is the reflection in the real axis of the point z. Hence, $\arg \bar{z} = -\arg z$.

Lecture 3
Complex Numbers III

In this lecture, we shall first show that every complex number can be written in exponential form, and then use this form to raise a rational power to a given complex number. We shall also extract roots of a complex number. Finally, we shall prove that complex numbers cannot be ordered.

If $z = x + iy$, then e^z is defined to be the complex number

$$e^z = e^x(\cos y + i \sin y). \tag{3.1}$$

This number e^z satisfies the usual algebraic properties of the exponential function. For example,

$$e^{z_1} e^{z_2} = e^{z_1 + z_2} \quad \text{and} \quad \frac{e^{z_1}}{e^{z_2}} = e^{z_1 - z_2}.$$

In fact, if $z_1 = x_1 + iy_1$ and $z_2 = x_2 + iy_2$, then, in view of Lecture 2, we have

$$
\begin{aligned}
e^{z_1} e^{z_2} &= e^{x_1}(\cos y_1 + i \sin y_1) e^{x_2}(\cos y_2 + i \sin y_2) \\
&= e^{x_1 + x_2}(\cos(y_1 + y_2) + i \sin(y_1 + y_2)) \\
&= e^{(x_1 + x_2) + i(y_1 + y_2)} = e^{z_1 + z_2}.
\end{aligned}
$$

In particular, for $z = iy$, the definition above gives one of the most important formulas of Euler

$$e^{iy} = \cos y + i \sin y, \tag{3.2}$$

which immediately leads to the following identities:

$$\cos y = \mathrm{Re}(e^{iy}) = \frac{e^{iy} + e^{-iy}}{2}, \quad \sin y = \mathrm{Im}(e^{iy}) = \frac{e^{iy} - e^{-iy}}{2i}.$$

When $y = \pi$, formula (3.2) reduces to the amazing equality $e^{\pi i} = -1$. In this relation, the transcendental number e comes from calculus, the transcendental number π comes from geometry, and i comes from algebra, and the combination $e^{\pi i}$ gives -1, the basic unit for generating the arithmetic system for counting numbers.

Using Euler's formula, we can express a complex number $z = r(\cos \theta + i \sin \theta)$ in exponential form; i.e.,

$$z = r(\cos \theta + i \sin \theta) = re^{i\theta}. \tag{3.3}$$

11

The rules for multiplying and dividing complex numbers in exponential form are given by

$$z_1 z_2 = (r_1 e^{i\theta_1})(r_2 e^{i\theta_2}) = (r_1 r_2)e^{i(\theta_1 + \theta_2)},$$

$$\frac{z_1}{z_2} = \frac{r_1 e^{i\theta_1}}{r_2 e^{i\theta_2}} = \left(\frac{r_1}{r_2}\right)e^{i(\theta_1 - \theta_2)}.$$

Finally, the complex conjugate of the complex number $z = re^{i\theta}$ is given by $\bar{z} = re^{-i\theta}$.

Example 3.1. Compute (1). $\dfrac{1+i}{\sqrt{3}-i}$ and (2). $(1+i)^{24}$.

(1). We have $1 + i = \sqrt{2}e^{i\pi/4}$, $\sqrt{3} - i = 2e^{-i\pi/6}$, and therefore

$$\frac{1+i}{\sqrt{3}-i} = \frac{\sqrt{2}e^{i\pi/4}}{2e^{-i\pi/6}} = \frac{\sqrt{2}}{2}e^{i5\pi/12}.$$

(2). $(1+i)^{24} = (\sqrt{2}e^{i\pi/4})^{24} = 2^{12}e^{i6\pi} = 2^{12}.$

From the exponential representation of complex numbers, *De Moivre's formula*

$$(\cos\theta + i\sin\theta)^n = \cos n\theta + i\sin n\theta, \quad n = 1, 2, \cdots, \qquad (3.4)$$

follows immediately. In fact, we have

$$\begin{aligned}(\cos\theta + i\sin\theta)^n = (e^{i\theta})^n &= e^{i\theta} \cdot e^{i\theta} \cdots e^{i\theta} \\ &= e^{i\theta + i\theta + \cdots + i\theta} \\ &= e^{in\theta} = \cos n\theta + i\sin n\theta.\end{aligned}$$

From (3.4), it is immediate to deduce that

$$\left(\frac{1 + i\tan\theta}{1 - i\tan\theta}\right)^n = \frac{1 + i\tan n\theta}{1 - i\tan n\theta}.$$

Similarly, since

$$1 + \sin\theta \pm i\cos\theta = 2\cos\left(\frac{\pi}{4} - \frac{\theta}{2}\right)\left[\cos\left(\frac{\pi}{4} - \frac{\theta}{2}\right) \pm i\sin\left(\frac{\pi}{4} - \frac{\theta}{2}\right)\right],$$

it follows that

$$\left[\frac{1 + \sin\theta + i\cos\theta}{1 + \sin\theta - i\cos\theta}\right]^n = \cos\left(\frac{n\pi}{2} - n\theta\right) + i\sin\left(\frac{n\pi}{2} - n\theta\right).$$

Example 3.2. Express $\cos 3\theta$ in terms of $\cos \theta$. We have

$$
\begin{aligned}
\cos 3\theta &= \text{Re}(\cos 3\theta + i\sin 3\theta) = \text{Re}(\cos\theta + i\sin\theta)^3 \\
&= \text{Re}[\cos^3\theta + 3\cos^2\theta(i\sin\theta) + 3\cos\theta(\quad \sin^2\theta) - i\sin^3\theta] \\
&= \cos^3\theta - 3\cos\theta\sin^2\theta = 4\cos^3\theta - 3\cos\theta.
\end{aligned}
$$

Now, let $z = re^{i\theta} = r(\cos\theta + i\sin\theta)$. By using the multiplicative property of the exponential function, we get

$$
z^n = r^n e^{in\theta} \tag{3.5}
$$

for any positive integer n. If $n = -1, -2, \cdots$, we define z^n by $z^n = (z^{-1})^{-n}$. If $z = re^{i\theta}$, then $z^{-1} = e^{-i\theta}/r$. Hence,

$$
z^n = (z^{-1})^{-n} = \left[\frac{1}{r}e^{i(-\theta)}\right]^{-n} = \left(\frac{1}{r}\right)^{-n} e^{i(-n)(-\theta)} = r^n e^{in\theta}.
$$

Hence, formula (3.5) is also valid for negative integers n.

Now we shall see if (3.5) holds for $n = 1/m$. If we let

$$
\xi = \sqrt[m]{r}\, e^{i\theta/m}, \tag{3.6}
$$

then ξ certainly satisfies $\xi^m = z$. But it is well-known that the equation $\xi^m = z$ has more than one solution. To obtain all the mth roots of z, we must apply formula (3.5) to every polar representation of z. For example, let us find all the mth roots of unity. Since

$$
1 = e^{2k\pi i}, \quad k = 0, \pm 1, \pm 2, \cdots,
$$

applying formula (3.5) to every polar representation of 1, we see that the complex numbers

$$
z = e^{(2k\pi i)/m}, \quad k = 0, \pm 1, \pm 2, \cdots,
$$

are mth roots of unity. All these roots lie on the unit circle centered at the origin and are equally spaced around the circle every $2\pi/m$ radians.

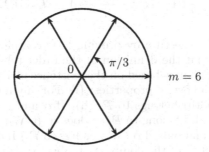

Figure 3.1

Hence, all of the distinct m roots of unity are obtained by writing

$$z = e^{(2k\pi i)/m}, \quad k = 0, 1, \cdots, m-1. \tag{3.7}$$

In the general case, the m distinct roots of a complex number $z = re^{i\theta}$ are given by

$$z^{1/m} = \sqrt[m]{r}e^{i(\theta + 2k\pi)/m}, \quad k = 0, 1, \cdots, m-1.$$

Example 3.3. Find all the cube roots of $\sqrt{2} + i\sqrt{2}$. In polar form, we have $\sqrt{2} + i\sqrt{2} = 2e^{i\pi/4}$. Hence,

$$(\sqrt{2} + i\sqrt{2})^{1/3} = \sqrt[3]{2}e^{i\left(\frac{\pi}{12} + \frac{2k\pi}{3}\right)}, \quad k = 0, 1, 2;$$

i.e.,

$$\sqrt[3]{2}\left(\cos\frac{\pi}{12} + i\sin\frac{\pi}{12}\right), \ \sqrt[3]{2}\left(\cos\frac{3\pi}{4} + i\sin\frac{3\pi}{4}\right), \ \sqrt[3]{2}\left(\cos\frac{17\pi}{12} + i\sin\frac{17\pi}{12}\right),$$

are the cube roots of $\sqrt{2} + i\sqrt{2}$.

Example 3.4. Solve the equation $(z+1)^5 = z^5$. We rewrite the equation as $\left(\dfrac{z+1}{z}\right)^5 = 1$. Hence,

$$\frac{z+1}{z} = e^{2k\pi i/5}, \quad k = 0, 1, 2, 3, 4,$$

or

$$z = \frac{1}{e^{2k\pi i/5} - 1} = -\frac{1}{2}\left(1 + i\cot\frac{\pi k}{5}\right), \quad k = 0, 1, 2, 3, 4.$$

Similarly, for any natural number n, the roots of the equation $(z+1)^n + z^n = 0$ are

$$z = -\frac{1}{2}\left(1 + i\cot\frac{\pi + 2k\pi}{n}\right), \quad k = 0, 1, \cdots, n-1.$$

We conclude this lecture by proving that complex numbers cannot be ordered. (Recall that the definition of the order relation denoted by $>$ in the real number system is based on the existence of a subset $\mathcal{P}$ (the positive reals) having the following properties: (i) For any number $\alpha \neq 0$, either α or $-\alpha$ (but not both) belongs to $\mathcal{P}$. (ii) If α and β belong to $\mathcal{P}$, so does $\alpha + \beta$. (iii) If α and β belong to $\mathcal{P}$, so does $\alpha \cdot \beta$. When such a set $\mathcal{P}$ exists, we write $\alpha > \beta$ if and only if $\alpha - \beta$ belongs to $\mathcal{P}$.) Indeed, suppose there is a nonempty subset $\mathcal{P}$ of the complex numbers satisfying (i), (ii), and (iii). Assume that $i \in \mathcal{P}$. Then, by (iii), $i^2 = -1 \in \mathcal{P}$ and $(-1)i = -i \in \mathcal{P}$. This violates (i). Similarly, (i) is violated by assuming $-i \in \mathcal{P}$. Therefore, the words positive and negative are never applied to complex numbers.

Problems

3.1. Express each of the following complex numbers in the form $x + iy$:

(a). $(\sqrt{2} - i) - i(1 - \sqrt{2}i)$, (b). $(2 - 3i)(-2 + i)$, (c). $(1 - i)(2 - i)(3 - i)$,

(d). $\dfrac{4 + 3i}{3 - 4i}$, (e). $\dfrac{1 + i}{i} + \dfrac{i}{1 - i}$, (f). $\dfrac{1 + 2i}{3 - 4i} + \dfrac{2 - i}{5i}$,

(g). $(1 + \sqrt{3}\,i)^{-10}$, (h). $(-1 + i)^7$, (i). $(1 - i)^4$.

3.2. Describe the following loci or regions:

(a). $|z - z_0| = |z - \overline{z}_0|$, where $\operatorname{Im} z_0 \neq 0$,

(b). $|z - z_0| = |z + \overline{z}_0|$, where $\operatorname{Re} z_0 \neq 0$,

(c). $|z - z_0| = |z - z_1|$, where $z_0 \neq z_1$,

(d). $|z - 1| = 1$,

(e). $|z - 2| = 2|z - 2i|$,

(f). $\left| \dfrac{z - z_0}{z - z_1} \right| = c$, where $z_0 \neq z_1$ and $c \neq 1$,

(g). $0 < \operatorname{Im} z < 2\pi$,

(h). $\dfrac{\operatorname{Re} z}{|z - 1|} > 1$, $\operatorname{Im} z < 3$,

(i). $|z - z_1| + |z - z_2| = 2a$,

(j). $az\overline{z} + kz + \overline{k}\overline{z} + d = 0$, $k \in \mathbf{C}$, $a, d \in \mathbb{R}$, and $|k|^2 > ad$.

3.3. Let $\alpha, \beta \in \mathbf{C}$. Prove that

$$|\alpha + \beta|^2 + |\alpha - \beta|^2 = 2(|\alpha|^2 + |\beta|^2),$$

and deduce that

$$|\alpha + \sqrt{\alpha^2 - \beta^2}| + |\alpha - \sqrt{\alpha^2 - \beta^2}| = |\alpha + \beta| + |\alpha - \beta|.$$

3.4. Use the properties of conjugates to show that

(a). $(\overline{z})^4 = \overline{(z^4)}$, (b). $\overline{\left(\dfrac{z_1}{z_2 z_3} \right)} = \dfrac{\overline{z}_1}{\overline{z}_2 \overline{z}_3}$.

3.5. If $|z| = 1$, then show that

$$\left| \frac{az + b}{\overline{b}z + \overline{a}} \right| = 1$$

for all complex numbers a and b.

3.6. If $|z| = 2$, use the triangle inequality to show that

$$|\text{Im}(1 - \overline{z} + z^2)| \leq 7 \quad \text{and} \quad |z^4 - 4z^2 + 3| \geq 3.$$

3.7. Prove that if $|z| = 3$, then

$$\frac{5}{13} \leq \left|\frac{2z - 1}{4 + z^2}\right| \leq \frac{7}{5}.$$

3.8. Let z and w be such that $\overline{z}w \neq 1$, $|z| \leq 1$, and $|w| \leq 1$. Prove that

$$\left|\frac{z - w}{1 - \overline{z}w}\right| \leq 1.$$

Determine when equality holds.

3.9. (a). Prove that z is either real or purely imaginary if and only if $(\overline{z})^2 = z^2$.

(b). Prove that $\sqrt{2}|z| \geq |\text{Re}\, z| + |\text{Im}\, z|$.

3.10. Show that there are complex numbers z satisfying $|z - a| + |z + a| = 2|b|$ if and only if $|a| \leq |b|$. If this condition holds, find the largest and smallest values of $|z|$.

3.11. Let $z_1, z_2, \cdots, z_n$ and $w_1, w_2, \cdots, w_n$ be complex numbers. Establish *Lagrange's identity*

$$\left|\sum_{k=1}^{n} z_k w_k\right|^2 = \left(\sum_{k=1}^{n} |z_k|^2\right)\left(\sum_{k=1}^{n} |w_k|^2\right) - \sum_{k < \ell} |z_k \overline{w}_\ell - z_\ell \overline{w}_k|^2,$$

and deduce *Cauchy's inequality*

$$\left|\sum_{k=1}^{n} z_k w_k\right|^2 \leq \left(\sum_{k=1}^{n} |z_k|^2\right)\left(\sum_{k=1}^{n} |w_k|^2\right).$$

3.12. Express the following in the form $r(\cos\theta + i\sin\theta)$, $-\pi < \theta \leq \pi$:

(a). $\dfrac{(1 - i)(\sqrt{3} + i)}{(1 + i)(\sqrt{3} - i)}$,   (b). $-8 + \dfrac{4}{i} + \dfrac{25}{3 - 4i}$.

3.13. Find the principal argument (Arg) for each of the following complex numbers:

(a). $5\left(\cos\dfrac{\pi}{8} - i\sin\dfrac{\pi}{8}\right)$,   (b). $-3 + \sqrt{3}i$,   (c). $-\dfrac{2}{1 + \sqrt{3}i}$,   (d). $(\sqrt{3} - i)^6$.

3.14. Given $z_1 z_2 \neq 0$, prove that

$$\operatorname{Re} z_1 \bar{z}_2 = |z_1||z_2| \quad \text{if and only if} \quad \operatorname{Arg} z_1 = \operatorname{Arg} z_2.$$

Hence, show that

$$|z_1 + z_2| = |z_1| + |z_2| \quad \text{if and only if} \quad \operatorname{Arg} z_1 = \operatorname{Arg} z_2.$$

3.15. What is wrong in the following?

$$1 = \sqrt{1} = \sqrt{(-1)(-1)} = \sqrt{-1}\sqrt{-1} = i\,i = -1.$$

3.16. Show that

$$\frac{(1 - i)^{49} \left(\cos \frac{\pi}{40} + i \sin \frac{\pi}{40}\right)^{10}}{(8i - 8\sqrt{3})^6} = -\sqrt{2}.$$

3.17. Let $z = re^{i\theta}$ and $w = Re^{i\phi}$, where $0 < r < R$. Show that

$$\operatorname{Re}\left(\frac{w + z}{w - z}\right) = \frac{R^2 - r^2}{R^2 - 2Rr\cos(\theta - \phi) + r^2}.$$

3.18. Solve the following equations:

(a). $z^2 = 2i$,   (b). $z^2 = 1 - \sqrt{3}i$,   (c). $z^4 = -16$,   (d). $z^4 = -8 - 8\sqrt{3}i$.

3.19. For the root of unity $z = e^{2\pi i/m}$, $m > 1$, show that

$$1 + z + z^2 + \cdots + z^{m-1} = 0.$$

3.20. Let a and b be two real constants and n be a positive integer. Prove that all roots of the equation

$$\left(\frac{1 + iz}{1 - iz}\right)^n = a + ib$$

are real if and only if $a^2 + b^2 = 1$.

3.21. A *quarternion* is an ordered pair of complex numbers; e.g., $((1, 2), (3, 4))$ and $(2+i, 1-i)$. The sum of quaternions (A, B) and (C, D) is defined as $(A + C, B + D)$. Thus, $((1, 2), (3, 4)) + ((5, 6), (7, 8)) = ((6, 8), (10, 12))$ and $(1 - i, 4 + i) + (7 + 2i, -5 + i) = (8 + i, -1 + 2i)$. Similarly, the scalar multiplication by a complex number A of a quaternion (B, C) is defined by the quadternion (AB, AC). Show that the addition and scalar multiplication of quaternions satisfy all the properties of addition and multiplication of real numbers.

3.22. Observe that:

(a). If $x = 0$ and $y > 0$ $(y < 0)$, then $\operatorname{Arg} z = \pi/2$ $(-\pi/2)$.

(b). If $x > 0$, then $\operatorname{Arg} z = \tan^{-1}(y/x) \in (-\pi/2, \pi/2)$.

(c). If $x < 0$ and $y > 0$ $(y < 0)$, then $\operatorname{Arg} z = \tan^{-1}(y/x) + \pi$ $(\tan^{-1}(y/x) - \pi)$.

(d). $\operatorname{Arg}(z_1 z_2) = \operatorname{Arg} z_1 + \operatorname{Arg} z_2 + 2m\pi$ for some integer m. This m is uniquely chosen so that the LHS $\in (-\pi, \pi]$. In particular, let $z_1 = -1$, $z_2 = -1$, so that $\operatorname{Arg} z_1 = \operatorname{Arg} z_2 = \pi$ and $\operatorname{Arg}(z_1 z_2) = \operatorname{Arg}(1) = 0$. Thus the relation holds with $m = -1$.

(e). $\operatorname{Arg}(z_1/z_2) = \operatorname{Arg} z_1 - \operatorname{Arg} z_2 + 2m\pi$ for some integer m. This m is uniquely chosen so that the LHS $\in (-\pi, \pi]$.

Answers or Hints

3.1. (a). $-2i$, (b). $-1 + 8i$, (c). $-10i$, (d). i, (e). $(1 - i)/2$, (f). $-2/5$, (g). $2^{-11}(-1 + \sqrt{3}i)$, (h). $-8(1 + i)$, (i). -4.

3.2. (a). Real axis, (b). imaginary axis, (c). perpendicular bisector (passing through the origin) of the line segment joining the points z_0 and z_1, (d). circle center $z = 1$, radius 1; i.e., $(x - 1)^2 + y^2 = 1$, (e). circle center $(-2/3, 8/3)$, radius $\sqrt{32}/3$, (f). circle, (g). $0 < y < 2\pi$, infinite strip, (h). region interior to parabola $y^2 = 2(x - 1/2)$ but below the line $y = 3$, (i). ellipse with foci at z_1, z_2 and major axis $2a$ (j). circle.

3.3. Use $|z|^2 = z\bar{z}$.

3.4. (a). $\overline{z^4} = \overline{zzzz} = \bar{z}\,\bar{z}\,\bar{z}\,\bar{z} = (\bar{z})^4$, (b). $\overline{\left(\dfrac{z_1}{z_2 z_3}\right)} = \dfrac{\bar{z}_1}{\overline{z_2 z_3}} = \dfrac{\bar{z}_1}{\bar{z}_2 \bar{z}_3}$.

3.5. If $|z| = 1$, then $\bar{z} = z^{-1}$.

3.6. $\left|\operatorname{Im}\left(1 - \bar{z} + z^2\right)\right| \leq |1 - \bar{z} + z^2| \leq |1| + |\bar{z}| + |z^2| \leq 7$, $|z^4 - 4z^2 + 3| = |z^2 - 3||z^2 - 1| \geq (|z^2| - 3)(|z^2| - 1)$.

3.7. We have

$$\left|\frac{2z - 1}{4 + z^2}\right| \leq \frac{2|z| + 1}{|4 - |z|^2|} = \frac{2 \cdot 3 + 1}{|4 - 3^2|} = \frac{7}{5}$$

and

$$\left|\frac{2z - 1}{4 + z^2}\right| \geq \frac{|2|z| - 1|}{|4 + |z|^2|} = \frac{2 \cdot 3 - 1}{4 + 3^2} = \frac{5}{13}.$$

3.8. We shall prove that $|1 - \bar{z}w| \geq |z - w|$. We have $|1 - \bar{z}w|^2 - |z - w|^2 = (1 - \bar{z}w)(1 - z\bar{w}) - (z - w)(\bar{z} - \bar{w}) = 1 - z\bar{w} - \bar{z}w + \bar{z}wz\bar{w} - z\bar{z} + z\bar{w} + w\bar{z} - w\bar{w} = 1 - |z|^2 - |w|^2 + |z|^2|w|^2 = (1 - |z|^2)(1 - |w|^2) \geq 0$ since $|z| \leq 1$ and $|w| \leq 1$. Equality holds when $|z| = |w| = 1$.

3.9. (a). $(\bar{z})^2 = z^2$ iff $z^2 - (\bar{z})^2 = 0$ iff $(z + \bar{z})(z - \bar{z}) = 0$ iff either $2\operatorname{Re}(z) = z + \bar{z} = 0$ or $2i\operatorname{Im}(z) = z - \bar{z} = 0$ iff z is purely imaginary or z is real. (b). Write $z = x + iy$. Consider $2|z|^2 - (|\operatorname{Re} z| + |\operatorname{Im} z|)^2 = 2(x^2 + y^2) - (|x| + |y|)^2 = 2x^2 + 2y^2 - (x^2 + y^2 + 2|x||y|) = x^2 + y^2 - 2|x||y| = (|x| - |y|)^2 \geq 0$.

3.10. Use the triangle inequality.

3.11. We have

$$
\left| \sum_{k=1}^{n} z_k w_k \right|^2 = \left(\sum_{k=1}^{n} z_k w_k \right) \left(\sum_{\ell=1}^{n} \overline{z_\ell} \overline{w_\ell} \right) = \sum_{k=1}^{n} |z_k|^2 |w_k|^2 + \sum_{k \neq \ell} z_k w_k \overline{z_\ell} \overline{w_\ell}
$$

$$
= \left(\sum_{k=1}^{n} |z_k|^2 \right) \left(\sum_{k=1}^{n} |w_k|^2 \right) - \sum_{k \neq \ell} |z_k|^2 |w_\ell|^2 + \sum_{k \neq \ell} z_k w_k \overline{z_\ell} \overline{w_\ell}
$$

$$
= \left(\sum_{k=1}^{n} |z_k|^2 \right) \left(\sum_{k=1}^{n} |w_k|^2 \right) - \sum_{k < \ell} |z_k \overline{w_\ell} - z_\ell \overline{w_k}|^2.
$$

3.12. (a). $\cos(-\pi/6) + i\sin(-\pi/6)$, (b). $5(\cos\pi + i\sin\pi)$.

3.13. (a). $-\pi/8$, (b). $5\pi/6$, (c). $2\pi/3$, (d). π.

3.14. Let $z_1 = r_1 e^{i\theta_1}$, $z_2 = r_2 e^{i\theta_2}$. Then, $z_1 \overline{z_2} = r_1 r_2 e^{i(\theta_1 - \theta_2)}$. $\mathrm{Re}(z_1 \overline{z_2}) = r_1 r_2 \cos(\theta_1 - \theta_2) = r_1 r_2$ if and only if $\theta_1 - \theta_2 = 2k\pi$, $k \in \mathbf{Z}$. Thus, if and only if $\mathrm{Arg}\, z_1 \text{-} \mathrm{Arg}\, z_2 = 2k\pi$, $k \in \mathbf{Z}$. But for $-\pi < \mathrm{Arg}\, z_1$, $\mathrm{Arg}\, z_2 \leq \pi$, the only possibility is $\mathrm{Arg}\, z_1 = \mathrm{Arg}\, z_2$. Conversely, if $\mathrm{Arg}\, z_1 = \mathrm{Arg}\, z_2$, then $\mathrm{Re}\,(z_1 \overline{z_2}) = r_1 r_2 = |z_1||z_2|$. Now, $|z_1 + z_2| = |z_1| + |z_2| \iff z_1 \overline{z_1} + z_2 \overline{z_2} + z_1 \overline{z_2} + z_2 \overline{z_1} = |z_1|^2 + |z_2|^2 + 2|z_1||z_2| \iff z_1 \overline{z_2} + z_2 \overline{z_1} = 2|z_1||z_2| \iff \mathrm{Re}(z_1 \overline{z_2} + z_2 \overline{z_1}) = \mathrm{Re}(z_1 \overline{z_2}) + \mathrm{Re}(z_2 \overline{z_1}) = 2|z_1||z_2| \iff \mathrm{Re}(z_1 \overline{z_2}) = |z_1||z_2|$ and $\mathrm{Re}(\overline{z_1} z_2) = |z_1||z_2| \iff \mathrm{Arg}\,(z_1) = \mathrm{Arg}\,(z_2)$.

3.15. If a is a positive real number, then $\sqrt{a}$ denotes the positive square root of a. However, if w is a complex number, what is the meaning of $\sqrt{w}$? Let us try to find a reasonable definition of $\sqrt{w}$. We know that the equation $z^2 = w$ has two solutions, namely $z = \pm\sqrt{|w|}e^{i(\mathrm{Arg}\, w)/2}$. If we want $\sqrt{-1} = i$, then we need to define $\sqrt{w} = \sqrt{|w|}e^{i(\mathrm{Arg}\, w)/2}$. However, with this definition, the expression $\sqrt{w}\sqrt{w} = \sqrt{w^2}$ will not hold in general. In particular, this does not hold for $w = -1$.

3.16. Use $1 - i = \sqrt{2}\left[\cos\left(-\frac{\pi}{4}\right) + i\sin\left(-\frac{\pi}{4}\right)\right]$ and $8i - 8\sqrt{3} = 16\left[\cos\frac{5\pi}{6} + i\sin\frac{5\pi}{6}\right]$.

3.17. Use $|w - z|^2 = (w - z)(\overline{w} - \overline{z})$.

3.18. (a). $z^2 = 2i = 2e^{i\pi/2}$, $z = \sqrt{2}e^{i\pi/4}$, $\sqrt{2}\exp\left[\frac{i}{2}\left(\frac{\pi}{2} + 2\pi\right)\right]$,

(b). $z^2 = 1 - \sqrt{3}i = 2e^{-i\pi/3}$, $z = \sqrt{2}e^{-i\pi/6}$, $\sqrt{2}e^{i5\pi/6}$,

(c). $z^4 = -16 = 2^4 e^{i\pi}$, $z = 2\exp\left[i\left(\frac{\pi + 2k\pi}{4}\right)\right]$, $k = 0, 1, 2, 3$,

(d). $z^4 = -8 - 8\sqrt{3}i = 16e^{i4\pi/3}$, $z = 2\exp\left[\frac{i}{4}\left(\frac{4\pi}{3} + 2k\pi\right)\right]$, $k = 0, 1, 2, 3$.

3.19. Multiply $1 + z + z^2 + \cdots + z^{m-1}$ by $1 - z$.

3.20. Suppose all the roots are real. Let $z = x$ be a real root. Then $a + ib = \left(\frac{1+ix}{1-ix}\right)^n$ implies that $|a + ib|^2 = \left|\frac{1+ix}{1-ix}\right|^{2n} = \left(\frac{1+x^2}{1+x^2}\right)^n = 1$, and hence $a^2 + b^2 = 1$. Conversely, suppose $a^2 + b^2 = 1$. Let $z = x + iy$ be a root. Then we have $1 = a^2 + b^2 = |a + ib|^2 = \left|\frac{(1-y)+ix}{(1+y)-ix}\right|^{2n} = \left(\frac{(1-y)^2+x^2}{(1+y)^2+x^2}\right)^n$, and hence $(1 + y)^2 + x^2 = (1 - y)^2 + x^2$, which implies that $y = 0$.

3.21. Verify directly.

Lecture 4
Set Theory
in the Complex Plane

In this lecture, we collect some essential definitions about sets in the complex plane. These definitions will be used throughout without further mention.

The set S of all points that satisfy the inequality $|z - z_0| < \epsilon$, where ϵ is a positive real number, is called an *open disk* centered at z_0 with radius ϵ and denoted as $B(z_0, \epsilon)$. It is also called the *ϵ-neighborhood* of z_0, or simply a neighborhood of z_0. In Figure 4.1, the dashed boundary curve means that the boundary points do not belong to the set. The neighborhood $|z| < 1$ is called the *open unit disk*.

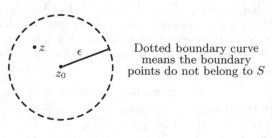

Dotted boundary curve
means the boundary
points do not belong to S

Figure 4.1

A point z_0 that lies in the set S is called an *interior point* of S if there is a neighborhood of z_0 that is completely contained in S.

Example 4.1. Every point z in an open disk $B(z_0, \epsilon)$ is an interior point.

Example 4.2. If S is the right half-plane $\text{Re}(z) > 0$ and $z_0 = 0.01$, then z_0 is an interior point of S.

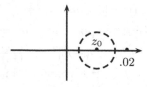

Figure 4.2

Example 4.3. If $S = \{z : |z| \leq 1\}$, then every complex number z such that $|z| = 1$ is not an interior point, whereas every complex number z such that $|z| < 1$ is an interior point.

If every point of a set S is an interior point of S, we say that S is an *open set*. Note that the empty set and the set of all complex numbers are open, whereas a finite set of points is not open.

It is often convenient to add the element ∞ to $\mathbf{C}$. The enlarged set $\mathbf{C} \cup \{\infty\}$ is called the *extended complex plane*. Unlike the extended real line, there is no $-\infty$. For this, we identify the complex plane with the xy-plane of $\mathbb{R}^3$, let S denote the sphere with radius 1 centered at the origin of $\mathbb{R}^3$, and call the point $N = (0, 0, 1)$ on the sphere the north pole. Now, from a point P in the complex plane, we draw a line through N. Then, the point P is mapped to the point P' on the surface of S, where this line intersects the sphere. This is clearly a one-to-one and onto (bijective) correspondence between points on S and the extended complex plane. In fact, the open disk $B(0, 1)$ is mapped onto the southern hemisphere, the circle $|z| = 1$ onto the equator, the exterior $|z| > 1$ onto the northern hemisphere, and the north pole N corresponds to ∞. Here, S is called the *Riemann sphere* and the correspondence is called a *stereographic projection* (see Figure 4.3). Thus, the sets of the form $\{z : |z - z_0| > r > 0\}$ are open and called *neighborhoods of* ∞. In what follows we shall make the following conventions: $z_1 + \infty = \infty + z_1 = \infty$ for all $z_1 \in \mathbf{C}$, $z_2 \times \infty = \infty \times z_2 = \infty$ for all $z_2 \in \mathbf{C}$ but $z_2 \neq 0$, $z_1/0 = \infty$ for all $z_1 \neq 0$, and $z_2/\infty = 0$ for $z_2 \neq \infty$.

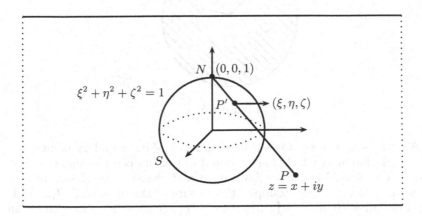

Figure 4.3

A point z_0 is called an *exterior point* of S if there is some neighborhood of z_0 that does not contain any points of S. A point z_0 is said to be a *boundary point* of a set S if every neighborhood of z_0 contains at least one point of S and at least one point not in S. Thus, a boundary point is neither an interior point nor an exterior point. The set of all boundary points of S denoted as ∂S is called the *boundary* or *frontier* of S. In Figure 4.4, the solid boundary curve means the boundary points belong to S.

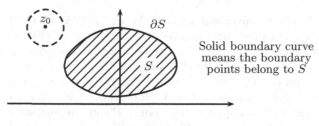

Solid boundary curve means the boundary points belong to S

Figure 4.4

Example 4.4. Let $0 < \rho_1 < \rho_2$ and $S = \{z : \rho_1 < |z| \leq \rho_2\}$. Clearly, the *circular annulus* S is neither open nor closed. The boundary of S is the set $\{z : |z| = \rho_2\} \cup \{z : |z| = \rho_1\}$.

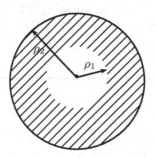

Figure 4.5

A set S is said to be *closed* if it contains all of its boundary points; i.e., $\partial S \subseteq S$. It follows that S is open if and only if its complement $\mathbf{C} - S$ is closed. The sets $\mathbf{C}$ and $\emptyset$ are both open and closed. The *closure* of S is the set $\overline{S} = S \cup \partial S$. For example, the closure of the open disk $B(z_0, r)$ is the closed disk $\overline{B}(z_0, r) = \{z : |z - z_0| \leq r\}$. A point z^* is said to be an *accumulation point* (*limit point*) of the set S if every neighborhood of z^* contains infinitely many points of the set S. It follows that a set S is closed if it contains all its accumulation points. A set of points S is said to be *bounded* if there exists a positive real number R such that $|z| < R$ for every z in S. An *unbounded* set is one that is not bounded.

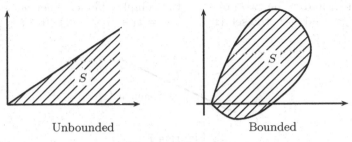

Unbounded   Bounded

Figure 4.6

Let S be a subset of complex numbers. The *diameter* of S, denoted as diam S, is defined as

$$\text{diam } S = \sup_{z,w \in S} |z - w|.$$

Clearly, S is bounded if and only if diam $S < \infty$. The following result, known as the *Nested Closed Sets Theorem*, is very useful.

Theorem 4.1 (Cantor). Suppose that $S_1, S_2, \cdots$ is a sequence of nonempty closed subsets of $\mathbf{C}$ satisfying

1. $S_n \supset S_{n+1}$, $n = 1, 2, \cdots$,
2. diam $S_n \to 0$ as $n \to \infty$.

Then, $\bigcap_{n=1}^{\infty} S_n$ contains precisely one point.

Theorem 4.1 is often used to prove the following well-known result.

Theorem 4.2 (Bolzano-Weierstrass). If S is an infinite bounded set of complex numbers, then S has at least one accumulation point.

A set is called *compact* if it is closed and bounded. Clearly, all closed disks $\overline{B}(z_0, r)$ are compact, whereas every open disk $B(z_0, r)$ is not compact. For compact sets, the following result is fundamental.

Theorem 4.3. Let S be a compact set and $r > 0$. Then, there exists a finite number of open disks of radius r whose union contains S.

Let $S \subset \mathbf{C}$ and $\{S_\alpha : \alpha \in \Lambda\}$ be a family of open subsets of $\mathbf{C}$, where Λ is any indexing set. If $S \subseteq \bigcup_{\alpha \in \Lambda} S_\alpha$, we say that the family $\{S_\alpha : \alpha \in \Lambda\}$ *covers* S. If $\Lambda' \subset \Lambda$, we call the family $\{S_\alpha : \alpha \in \Lambda'\}$ a *subfamily*, and if it covers S, we call it a *subcovering* of S.

Theorem 4.4. Let $S \subset \mathbf{C}$ be a compact set, and let $\{S_\alpha : \alpha \in \Lambda\}$ be an open covering of S. Then, there exists a finite subcovering; i.e., a finite number of open sets $S_1, \cdots, S_n$ whose union covers S. Conversely, if every open covering of S has a finite subcovering, then S is compact.

Let z_1 and z_2 be two points in the complex plane. The *line segment* ℓ joining z_1 and z_2 is the set $\{w \in \mathbf{C} : w = z_1 + t(z_2 - z_1),\ 0 \le t \le 1\}$.

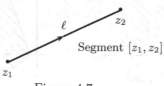

Figure 4.7

Now let $z_1, z_2, \cdots, z_{n+1}$ be $n+1$ points in the complex plane. For each $k = 1, 2, \cdots, n$, let ℓ_k denote the line segment joining z_k to z_{k+1}. Then the successive line segments $\ell_1, \ell_2, \cdots, \ell_n$ form a continuous chain known as a *polygonal path* joining z_1 to z_{n+1}.

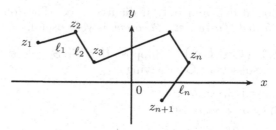

Figure 4.8

An open set S is said to be *connected* if every pair of points z_1, z_2 in S can be joined by a polygonal path that lies entirely in S. The polygonal path may contain line segments that are either horizontal or vertical. An open connected set is called a *domain*. Clearly, all open disks are domains. If S is a domain and $S = A \cup B$, where A and B are open and disjoint; i.e., $A \cap B = \emptyset$, then either $A = \emptyset$ or $B = \emptyset$. A domain together with some, none, or all of its boundary points is called a *region*.

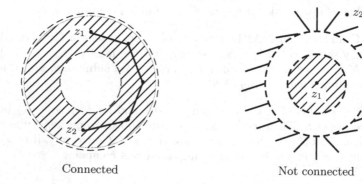

Connected   Not connected

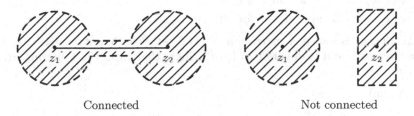

Connected	Not connected

Figure 4.9

A set S is said to be *convex* if each pair of points P and Q can be joined by a line segment PQ such that every point in the line segment also lies in S. For example, open disks and closed disks are convex; however, the union of two intersecting discs, while neither lies inside the other, is not convex. Clearly, every convex set is necessarily connected. Furthermore, it follows that the intersection of two or more convex sets is also convex.

Problems

4.1. Shade the following regions and determine whether they are open and connected:

(a). $\{z \in \mathbf{C} : -\pi/3 \le \arg z < \pi/2\}$,

(b). $\{z \in \mathbf{C} : |z - 1| < |z + 1|\}$,

(c). $\{z \in \mathbf{C} : |z - 1| + |z - i| < 2\sqrt{2}\}$,

(d). $\{z \in \mathbf{C} : 1/2 < |z - 1| < \sqrt{2}\} \bigcup \{z \in \mathbf{C} : 1/2 < |z + 1| < \sqrt{2}\}$.

4.2. Let S be the open set consisting of all points z such that $|z| < 1$ or $|z - 2| < 1$. Show that S is not connected.

4.3. Show that:

(a). If $S_1, \cdots, S_n$ are open sets in $\mathbf{C}$, then so is $\bigcap_{k=1}^{n} S_k$.

(b). If $\{S_\alpha : \alpha \in \Lambda\}$ is a collection of open sets in $\mathbf{C}$, where Λ is any indexing set, then $S = \bigcup_{\alpha \in \Lambda} S_\alpha$ is also open.

(c). The intersection of an arbitrary family of open sets in $\mathbf{C}$ need not be open.

4.4. Let S be a nonempty set. Suppose that to each ordered pair $(x, y) \in S \times S$ a nonnegative real number $d(x, y)$ is assigned that satisfies the following conditions:

(i). $d(x, y) \ge 0$ and $d(x, y) = 0$ if and only if $x = y$, (ii). $d(x, y) = d(y, x)$ for all $x, y \in S$,

(iii). $d(x, z) \leq d(x, y) + d(y, z)$ for all $x, y, z \in S$.

Then, $d(x, y)$ is called a *metric* on S. The set S with metric d is called a *metric space* and is denoted as (S, d). Show that in $\mathbf{C}$ the following are metrics:

(a). $d(z, w) = |z - w|$,

(b). $d(z, w) = \dfrac{|z - w|}{1 + |z - w|}$,

(c). $d(z, w) = \begin{cases} 0 \text{ if } z = w \\ 1 \text{ if } z \neq w. \end{cases}$

4.5. Let the point $z = x + iy$ correspond to the point (ξ, η, ζ) on the Riemann sphere (see Figure 4.3). Show that

$$\xi = \frac{2\,\mathrm{Re}\,z}{|z|^2 + 1}, \quad \eta = \frac{2\,\mathrm{Im}\,z}{|z|^2 + 1}, \quad \zeta = \frac{|z|^2 - 1}{|z|^2 + 1},$$

and

$$\mathrm{Re}\,z = \frac{\xi}{1 - \zeta}, \quad \mathrm{Im} = \frac{\eta}{1 - \zeta}.$$

4.6. Show that if z_1 and z_2 are finite points in the complex plane $\mathbf{C}$, then the distance between their stereographic projection is given by

$$d(z_1, z_2) = \frac{2|z_1 - z_2|}{\sqrt{1 + |z_1|^2}\sqrt{1 + |z_2|^2}}.$$

This distance is called the *spherical distance* or *chordal distance* between z_1 and z_2. Also, show that if $z_2 = \infty$, then the corresponding distance is given by

$$d(z_1, \infty) = \frac{2}{\sqrt{1 + |z_1|^2}}.$$

Answers or Hints

4.1.

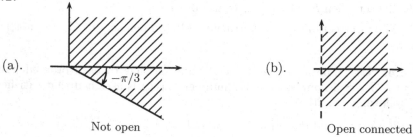

(a).   (b).

Not open   Open connected

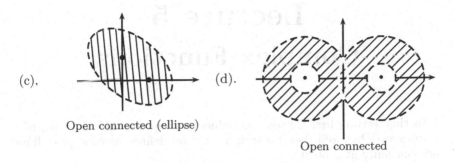

(c).

Open connected (ellipse)

(d).

Open connected

4.2. Points 0 and 2 cannot be connected by a polygonal line.

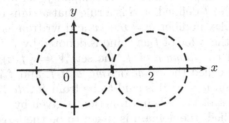

4.3. (a). Let $w \in \cap_{k=1}^{n} S_k$. Then $w \in S_k$, $k = 1, \cdots, n$. Since each S_k is open, there is an $r_k > 0$ such that $\{z : |z - w| < r_k\} \subset S_k$, $k = 1, \cdots, n$. Let $r = \min\{r_1, \cdots, r_n\}$. Then $\{z : |z - w| < r\} \subseteq \{z : |z - w| < r_k\} \subset S_k$, $k = 1, \cdots, n$. Thus, $\{z : |z - w| < r\} \subset \cap_{k=1}^{n} S_k$.
(b). Use the property of open sets.
(c). $\cap_{n=1}^{\infty} \{z : |z| < 1/n\} = \{0\}$.
4.4. (a). Verify directly. (b). For $a, b, c \geq 0$ and $c \leq a + b$, use $\frac{c}{1+c} \leq \frac{a}{1+a} + \frac{b}{1+b}$. (c). Verify directly.
4.5. The straight line passing through $(x, y, 0)$ and $(0, 0, 1)$ in parametric form is $(tx, ty, 1 - t)$. This line also passes through the point (ξ, η, ζ) on the Riemann sphere, provided $t^2 x^2 + t^2 y^2 + (1 - t)^2 = 1$. This gives $t = 0$ and $t = 2/(x^2 + y^2 + 1)$. The value $t = 0$ gives the north pole, whereas $t = 2/(x^2 + y^2 + 1)$ gives $(\xi, \eta, \zeta) = \left(\frac{2x}{x^2+y^2+1}, \frac{2y}{x^2+y^2+1}, \frac{x^2+y^2-1}{x^2+y^2+1}\right)$. From this, it also follows that $|z|^2 + 1 = \frac{2}{1-\zeta}$.
4.6. If (ξ_1, η_1, ζ_1) and (ξ_2, η_2, ζ_2) are the points on S corresponding to z_1 and z_2, then $d(z_1, z_2) = [(\xi_1 - \xi_2)^2 + (\eta_1 - \eta_2)^2 + (\zeta_1 - \zeta_2)^2]^{1/2} = [2 - 2(\xi_1\xi_2 + \eta_1\eta_2 + \zeta_1\zeta_2)]^{1/2}$. Now use Problem 4.5. If $z_2 = \infty$, then again from Problem 4.5, we have $d(z_1, \infty) = [\xi_1^2 + \eta_1^2 + (\zeta_1 - 1)^2]^{1/2} = [2 - 2\zeta_1]^{1/2} = [2 - 2(|z_1|^2 - 1)/(|z_1|^2 + 1)]^{1/2} = 2/(|z_1|^2 + 1)^{1/2}$.

Lecture 5
Complex Functions

In this lecture, first we shall introduce a complex-valued function of a complex variable, and then for such a function define the concept of limit and continuity at a point.

Let S be a set of complex numbers. A *complex function* (complex-valued of a complex variable) f defined on S is a rule that assigns to each $z = x + iy$ in S a unique complex number $w = u + iv$ and written as $f : S \to \mathbf{C}$. The number w is called the value of f at z and is denoted by $f(z)$; i.e., $w = f(z)$. The set S is called the *domain* of f, the set $W = \{f(z) : z \in S\}$, often denoted as $f(S)$, is called the *range* or *image* of f, and f is said to map S *onto* W. The function $w = f(z)$ is said to be from S *into* W if the range of S under f is a subset of W. When a function is given by a formula and the domain is not specified, the domain is taken to be the largest set on which the formula is defined. A function f is called *one-to-one* (or *univalent*, or *injective*) on a set S if the equation $f(z_1) = f(z_2)$, where z_1 and z_2 are in S, implies that $z_1 = z_2$. The function $f(z) = iz$ is one-to-one, but $f(z) = z^2$ is not one-to-one since $f(i) = f(-i) = -1$. A one-to-one and onto function is called *bijective*. We shall also consider multi-valued functions: a *multi-valued function* is a rule that assigns a finite or infinite non-empty subset of $\mathbf{C}$ for each element of its domain S. In Lecture 2, we have already seen that the function $f(z) = \arg z$ is multi-valued.

As every complex number z is characterized by a pair of real numbers x and y, a complex function f of the complex variable z can be specified by two real functions $u = u(x, y)$ and $v = v(x, y)$. It is customary to write $w = f(z) = u(x, y) + iv(x, y)$. The functions u and v, respectively, are called the real and imaginary parts of f. The common domain of the functions u and v corresponds to the domain of the function f.

Example 5.1. For the function $w = f(z) = 3z^2 + 7z$, we have

$$f(x + iy) = 3(x + iy)^2 + 7(x + iy) = (3x^2 - 3y^2 + 7x) + i(6xy + 7y),$$

and hence $u = 3x^2 - 3y^2 + 7x$ and $v = 6xy + 7y$. Similarly, for the function $w = f(z) = |z|^2$, we find

$$f(x + iy) = |x + iy|^2 = x^2 + y^2,$$

and hence $u = x^2 + y^2$ and $v = 0$. Thus, this function is a real-valued function of a complex variable. Clearly, the domain of both of these functions is

C. For the function $w = f(z) = z/|z|$, the domain is $\mathbf{C}\backslash\{0\}$, and its range is $|z| = 1$.

Example 5.2. The *complex exponential function* $f(z) = e^z$ is defined by the formula (3.1). Clearly, for this function, $u = e^x \cos y$ and $v = e^x \sin y$, which are defined for all $(x, y) \in \mathbb{R}^2$. Thus, for the function e^z the domain is $\mathbf{C}$. The exponential function provides a basic tool for the application of complex variables to electrical circuits, control systems, wave propagation, and time-invariant physical systems.

Recall that a vector-valued function of two real variables $\mathbf{F}(x, y) = (P(x, y), Q(x, y))$ is also called a *two-dimensional vector filed*. Using the standard orthogonal unit basis vectors $\mathbf{i}$ and $\mathbf{j}$, we can express this vector field as $\mathbf{F}(x, y) = P(x, y)\mathbf{i} + Q(x, y)\mathbf{j}$. There is a natural way to represent this vector field with a complex function $f(z)$. In fact, we can use the functions P and Q as the real and imaginary parts of f, in which case we say that the complex function $f(z) = P(x, y) + iQ(x, y)$ is the *complex representation* of the vector field $\mathbf{F}(x, y) = P(x, y)\mathbf{i} + Q(x, y)\mathbf{j}$. Conversely, any complex function $f(z) = u(x, y) + iv(x, y)$ has an associated vector field $\mathbf{F}(x, y) = u(x, y)\mathbf{i} + v(x, y)\mathbf{j}$. From this point of view, both $\mathbf{F}(x, y) = P(x, y)\mathbf{i} + Q(x, y)\mathbf{j}$ and $f(z) = u(x, y) + iv(x, y)$ can be called vector fields. This interpretation is often used to study various applications of complex functions in applied mathematical problems.

Let f be a function defined in some neighborhood of z_0, with the possible exception of the point z_0 itself. We say that the *limit* of $f(z)$ as z approaches z_0 (independent of the path) is the number w_0 if $|f(z) - w_0| \to 0$ as $|z - z_0| \to 0$ and we write $\lim_{z \to z_0} f(z) = w_0$. Hence, $f(z)$ can be made arbitrarily close to w_0 if we choose z sufficiently close to z_0. Equivalently, we say that w_0 is the limit of f as z approaches z_0 if, for any given $\epsilon > 0$, there exists a $\delta > 0$ such that

$$0 < |z - z_0| < \delta \implies |f(z) - w_0| < \epsilon.$$

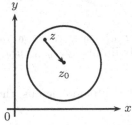 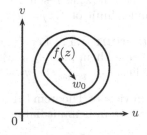

Figure 5.1

Example 5.3. By definition, we shall show that (i) $\lim_{z \to 1 - i}(\bar{z}^2 - 2) = -2 + 2i$ and (ii) $\lim_{z \to 1 - i} |\bar{z}^2 - 2| = \sqrt{8}$.

(i). Given any $\epsilon > 0$, we have

$$
\begin{aligned}
|\bar{z}^2 - 2 - (-2 + 2i)| &= |\bar{z}^2 - 2i| = |\overline{z^2 + 2i}| = |z^2 + 2i| \\
&= |z - (1 - i)||z + (1 - i)| \\
&\leq |z - (1 - i)|(|z - (1 - i)| + 2|1 - i|) \\
&\leq |z - (1 - i)|(1 + 2\sqrt{2}) \quad \text{if} \quad |z - (1 - i)| < 1 \\
&< \epsilon \quad \text{if} \quad |z - (1 - i)| < \min\left\{1, \frac{\epsilon}{1 + 2\sqrt{2}}\right\}.
\end{aligned}
$$

(ii). Given any $\epsilon > 0$, from (i) we have

$$
\begin{aligned}
||\bar{z}^2 - 2| - \sqrt{8}| &= ||\bar{z}^2 - 2| - |-2 + 2i|| \\
&\leq |\bar{z}^2 - 2 - (-2 + 2i)| \\
&< \epsilon \quad \text{if} \quad |z - (1 - i)| < \min\left\{1, \frac{\epsilon}{1 + 2\sqrt{2}}\right\}.
\end{aligned}
$$

Example 5.4. (i). Clearly, $\lim_{z \to z_0} z = z_0$. (ii). From the inequalities

$$
\begin{aligned}
|\mathrm{Re}(z - z_0)| &\leq [(\mathrm{Re}(z - z_0))^2 + (\mathrm{Im}(z - z_0))^2]^{1/2} = |z - z_0|, \\
|\mathrm{Im}(z - z_0)| &\leq |z - z_0|,
\end{aligned}
$$

it follows that $\lim_{z \to z_0} \mathrm{Re}\, z = \mathrm{Re}\, z_0$, $\lim_{z \to z_0} \mathrm{Im}\, z = \mathrm{Im}\, z_0$.

Example 5.5. $\lim_{z \to 0}(z/\bar{z})$ does not exist. Indeed, we have

$$
\lim_{\substack{z \to 0 \\ \text{along } x\text{-axis}}} \frac{z}{\bar{z}} = \lim_{x \to 0} \frac{x + i0}{x - i0} = 1,
$$

$$
\lim_{\substack{z \to 0 \\ \text{along } y\text{-axis}}} \frac{z}{\bar{z}} = \lim_{y \to 0} \frac{0 + iy}{0 - iy} = -1.
$$

The following result relates real limits of $u(x, y)$ and $v(x, y)$ with the complex limit of $f(z) = u(x, y) + iv(x, y)$.

Theorem 5.1. Let $f(z) = u(x, y) + iv(x, y)$, $z_0 = x_0 + iy_0$, and $w_0 = u_0 + iv_0$. Then, $\lim_{z \to z_0} f(z) = w_0$ if and only if $\lim_{x \to x_0,\, y \to y_0} u(x, y) = u_0$ and $\lim_{x \to x_0,\, y \to y_0} v(x, y) = v_0$.

In view of Theorem 5.1 and the standard results in calculus, the following theorem is immediate.

Theorem 5.2. If $\lim_{z \to z_0} f(z) = A$ and $\lim_{z \to z_0} g(z) = B$, then
(i) $\lim_{z \to z_0}(f(z) \pm g(z)) = A \pm B$, (ii) $\lim_{z \to z_0} f(z)g(z) = AB$, and
(iii) $\lim_{z \to z_0} \dfrac{f(z)}{g(z)} = \dfrac{A}{B}$ if $B \neq 0$.

For the *composition* of two functions f and g denoted and defined as $(f \circ g)(z) = f(g(z))$, we have the following result.

Theorem 5.3. If $\lim_{z \to z_0} g(z) = w_0$ and $\lim_{w \to w_0} f(w) = A$, then

$$\lim_{z \to z_0} f(g(z)) = A = f\left(\lim_{z \to z_0} g(z)\right).$$

Now we shall define limits that involve ∞. For this, we note that $z \to \infty$ means $|z| \to \infty$, and similarly, $f(z) \to \infty$ means $|f(z)| \to \infty$.

The statement $\lim_{z \to z_0} f(z) = \infty$ means that for any $M > 0$ there is a $\delta > 0$ such that $0 < |z - z_0| < \delta$ implies $|f(z)| > M$ and is equivalent to $\lim_{z \to z_0} 1/f(z) = 0$.

The statement $\lim_{z \to \infty} f(z) = w_0$ means that for any $\epsilon > 0$ there is an $R > 0$ such that $|z| > R$ implies $|f(z) - w_0| < \epsilon$, and is equivalent to $\lim_{z \to 0} f(1/z) = w_0$.

The statement $\lim_{z \to \infty} f(z) = \infty$ means that for any $M > 0$ there is an $R > 0$ such that $|z| > R$ implies $|f(z)| > M$.

Example 5.6. Since

$$\frac{2z + 3}{3z + 2} = \frac{2 + 3/z}{3 + 2/z},$$

$\lim_{z \to \infty}(2z + 3)/(3z + 2) = 2/3$. Similarly, $\lim_{z \to \infty}(2z + 3)/(3z^2 + 2) = 0$ and $\lim_{z \to \infty}(2z^2 + 3)/(3z + 2) = \infty$.

Let f be a function defined in a neighborhood of z_0. Then, f is *continuous* at z_0 if $\lim_{z \to z_0} f(z) = f(z_0)$. Equivalently, f is *continuous* at z_0 if for any given $\epsilon > 0$, there exists a $\delta > 0$ such that

$$|z - z_0| < \delta \implies |f(z) - f(z_0)| < \epsilon.$$

A function f is said to be continuous on a set S if it is continuous at each point of S.

Example 5.7. The functions $f(z) = \operatorname{Re}(z)$ and $g(z) = \operatorname{Im}(z)$ are continuous for all z.

Example 5.8. The function $f(z) = |z|$ is continuous for all z. For this, let z_0 be given. Then

$$\lim_{z \to z_0} |z| = \lim_{z \to z_0} \sqrt{(\operatorname{Re} z)^2 + (\operatorname{Im} z)^2} = \sqrt{(\operatorname{Re} z_0)^2 + (\operatorname{Im} z_0)^2} = |z_0|.$$

Hence, $f(z)$ is continuous at z_0. Since z_0 is arbitrary, we conclude that $f(z)$ is continuous for all z.

It follows from Theorem 5.1 that a function $f(z) = u(x,y) + iv(x,y)$ of a complex variable is continuous at a point $z_0 = x_0 + iy_0$ if and only if $u(x,y)$ and $v(x,y)$ are continuous at (x_0, y_0).

Example 5.9. The exponential function $f(z) = e^z$ is continuous on the whole complex plane since $e^x \cos y$ and $e^x \sin y$ both are continuous for all $(x,y) \in \mathbb{R}^2$.

The following result is an immediate consequence of Theorem 5.2.

Theorem 5.4. If $f(z)$ and $g(z)$ are continuous at z_0, then so are (i) $f(z) \pm g(z)$, (ii) $f(z)g(z)$, and (iii) $f(z)/g(z)$ provided $g(z_0) \neq 0$.

Now let $f : S \to W$, $S_1 \subset S$, and $W_1 \subset W$. The *inverse image* denoted as $f^{-1}(W_1)$ consists of all $z \in S$ such that $f(z) \in W_1$. It follows that $f(f^{-1}(W_1)) \subset W_1$ and $f^{-1}(f(S_1)) \supset S_1$. By definition, in terms of inverse image continuous functions can be characterized as follows: A function is continuous if and only if the inverse image of every open set is open. Similarly, a function is continuous if and only if the inverse image of every closed set is closed.

For continuous functions we also have the following result.

Theorem 5.5. Let $f : S \to \mathbf{C}$ be continuous. Then,

(i).   a compact set of S is mapped onto a compact set in $f(S)$, and

(ii).   a connected set of S is mapped onto a connected set of $f(S)$.

It is easy to see that the constant function and the function $f(z) = z$ are continuous on the whole plane. Thus, from Theorem 5.4, we deduce that the *polynomial functions*; i.e., functions of the form

$$P(z) = a_0 + a_1 z + a_2 z^2 + \cdots + a_n z^n, \tag{5.1}$$

where a_i, $0 \leq i \leq n$ are constants, are also continuous on the whole plane. *Rational functions* in z, which are defined as quotients of polynomials; i.e.,

$$\frac{P(z)}{Q(z)} = \frac{a_0 + a_1 z + \cdots + a_n z^n}{b_0 + b_1 z + \cdots + b_m z^m}, \tag{5.2}$$

are therefore continuous at each point where the denominator does not vanish.

Example 5.10. We shall find the limits as $z \to 2i$ of the functions

$$f_1(z) = z^2 - 2z + 1, \quad f_2(z) = \frac{z + 2i}{z}, \quad f_3(z) = \frac{z^2 + 4}{z(z - 2i)}.$$

Since $f_1(z)$ and $f_2(z)$ are continuous at $z = 2i$, we have $\lim_{z \to 2i} f_1(z) = f_1(2i) = -3 - 4i$, $\lim_{z \to 2i} f_2(z) = f_2(2i) = 2$. Since $f_3(z)$ is not defined at

$z = 2i$, it is not continuous. However, for $z \neq 2i$ and $z \neq 0$, we have

$$f_3(z) = \frac{(z + 2i)(z - 2i)}{z(z - 2i)} = \frac{z + 2i}{z} = f_2(z)$$

and so $\lim_{z \to 2i} f_3(z) = \lim_{z \to 2i} f_2(z) = 2$. Thus, the discontinuity of $f_3(z)$ at $z = 2i$ can be removed by setting $f_2(2i) = 2$. The function $f_3(z)$ is said to have a *removable discontinuity* at $z = 2i$.

Problems

5.1. For each of the following functions, describe the domain of definition that is understood:

(a). $f(z) = \dfrac{z}{z^2 + 3}$,   (b). $f(z) = \dfrac{z}{z + \bar{z}}$,   (c). $f(z) = \dfrac{1}{1 - |z|^2}$.

5.2. (a). Write the function $f(z) = z^3 + 2z + 1$ in the form $f(z) = u(x, y) + iv(x, y)$.

(b). Suppose that $f(z) = x^2 - y^2 - 2y + i(2x - 2xy)$. Express $f(z)$ in terms of z.

5.3. Show that when a limit of a function $f(z)$ exists at a point z_0, it is unique.

5.4. Use the definition of limit to prove that:

(a). $\lim\limits_{z \to z_0} (z^2 + 5) = z_0^2 + 5$,   (b). $\lim\limits_{z \to 1-i} \bar{z}^2 = (1 + i)^2$,

(c). $\lim\limits_{z \to z_0} \bar{z} = \bar{z}_0$,   (d). $\lim\limits_{z \to 2-i} (2z + 1) = 5 - 2i$.

5.5. Find each of the following limits:

(a). $\lim\limits_{z \to 2+3i} (z - 5i)^2$,   (b). $\lim\limits_{z \to 2} \dfrac{z^2 + 3}{iz}$,   (c). $\lim\limits_{z \to 3i} \dfrac{z^2 + 9}{z - 3i}$,

(d). $\lim\limits_{z \to i} \dfrac{z^2 + 1}{z^4 - 1}$,   (e). $\lim\limits_{z \to \infty} \dfrac{z^2 + 1}{z^2 + z + 1 - i}$,   (f). $\lim\limits_{z \to \infty} \dfrac{z^3 + 3iz^2 + 7}{z^2 - i}$.

5.6. Prove that:

(a). $\lim\limits_{z \to 0} \left(\dfrac{z}{\bar{z}}\right)^2$ does not exist,   (b). $\lim\limits_{z \to 0} \dfrac{\bar{z}^2}{z} = 0$.

5.7. Show that if $\lim\limits_{z \to z_0} f(z) = 0$ and there exists a positive number M such that $|g(z)| \leq M$ for all z in some neighborhood of z_0, then $\lim\limits_{z \to z_0} f(z)g(z) = 0$. Use this result to show that $\lim_{z \to 0} z e^{i/|z|} = 0$.

5.8. Show that if $\lim_{z \to z_0} f(z) = w_0$, then $\lim_{z \to z_0} |f(z)| = |w_0|$.

5.9. Suppose that f is continuous at z_0 and g is continuous at $w_0 = f(z_0)$. Prove that the composite function $g \circ f$ is continuous at z_0.

5.10. Discuss the continuity of the function

$$f(z) = \begin{cases} \dfrac{z^3 - 1}{z - 1}, & |z| \neq 1 \\ 3, & |z| = 1 \end{cases}$$

at the points 1, -1, i, and $-i$.

5.11. Prove that the function $f(z) = \mathrm{Arg}(z)$ is discontinuous at each point on the nonpositive real axis.

5.12 (Cauchy's Criterion). Show that $\lim_{z \to z_0} f(z) = w_0$ if and only if for a given $\epsilon > 0$ there exists a $\delta > 0$ such that for any z, z' satisfying $|z - z_0| < \delta$, $|z' - z_0| < \delta$, the inequality $|f(z) - f(z')| < \epsilon$ holds.

5.13. Prove Theorem 5.5.

5.14. The function $f : S \to \mathbf{C}$ is said to be *uniformly continuous* on S if for every given $\epsilon > 0$ there exists a $\delta = \delta(\epsilon) > 0$ such that $|f(z_1) - f(z_2)| < \epsilon$ for all $z_1, z_2 \in S$ with $|z_1 - z_2| < \delta$. Show that on a compact set every continuous function is uniformly continuous.

Answers or Hints

5.1. (a). $z^2 \neq -3 \iff z \neq \pm\sqrt{3}i$, (b). $z + \overline{z} \neq 0 \iff z$ is not purely imaginary; i.e., $\mathrm{Re}(z) \neq 0$, (c). $|z|^2 \neq 1 \iff |z| \neq 1$.

5.2. (a). $(x+iy)^3 + 2(x+iy) + 1 = (x^3 - 3xy^2 + 2x + 1) + i(3x^2y - y^3 + 2y)$. (b). Use $x = (z + \overline{z})/2$, $y = (z - \overline{z})/2i$ to obtain $f(z) = \overline{z}^2 + 2iz$.

5.3. Suppose that $\lim_{z \to z_0} f(z) = w_0$ and $\lim_{z \to z_0} f(z) = w_1$. Then, for any positive number ϵ, there are positive numbers δ_0 and δ_1 such that $|f(z) - w_0| < \epsilon$ whenever $0 < |z - z_0| < \delta_0$ and $|f(z) - w_1| < \epsilon$ whenever $0 < |z - z_0| < \delta_1$. So, if $0 < |z - z_0| < \delta = \min\{\delta_0, \delta_1\}$, then $|w_0 - w_1| = |-(f(z) - w_0) + (f(z) - w_1)| \leq |f(z) - w_0| + |f(z) - w_1| < 2\epsilon$; i.e., $|w_0 - w_1| < 2\epsilon$. But, ϵ can be chosen arbitrarily small. Hence, $w_0 - w_1 = 0$, or $w_0 = w_1$.

5.4. (a). $|z^2 + 5 - (z_0^2 + 5)| = |z - z_0||z + z_0| \leq |z - z_0|(|z - z_0| + 2|z_0|)$
$\leq (1 + 2|z_0|)|z - z_0|$ if $|z - z_0| < 1$
$\quad < \epsilon$ if $0 < |z - z_0| < \min\left\{\dfrac{\epsilon}{1 + 2|z_0|}, 1\right\}$,
(b). $|\overline{z}^2 - (1 + i)^2| = |z^2 - (1 - i)^2| \leq |z - (1 - i)||z + (1 - i)|$
$\leq |z - (1 - i)|(|z - (1 - i)| + 2|1 - i|) < 5|z - (1 - i)|$ if $|z - (1 - i)| < 1$

$< \epsilon$ if $|z - (1 - i)| < \min\{1, \epsilon/5\}$,

(c). $|\bar{z} - \bar{z}_0| = |z - z_0| < \epsilon$ if $|z - z_0| < \epsilon$,

(d). $|2z + 1 - (5 - 2i)| = |2z - (4 - 2i)| = 2|z - (2 - i)| < \epsilon$ if $|z - (2 - i)| < \epsilon/2$.

5.5. (a). $-8i$, (b). $-7i/2$, (c). $6i$, (d). $1/2$, (e). 1, (f). ∞.

5.6. (a). $\lim_{z \to 0, z=x}(z/\bar{z})^2 = 1$, $\lim_{z \to 0, y=x}(z/\bar{z})^2 = -1$.

(b). Let $\epsilon > 0$. Choose $\delta = \epsilon$. Then, $0 < |z - 0| < \delta$ implies $|\bar{z}^2/z - 0| = |z| < \epsilon$.

5.7. Since $\lim_{z \to z_0} f(z) = 0$, given any $\epsilon > 0$, there exists $\delta > 0$ such that $|f(z) - 0| < \epsilon/M$ whenever $|z - z_0| < \delta$. Thus, $|f(z)g(z) - 0| = |f(z)||g(z)| \le M|f(z)| < \epsilon$ if $|z - z_0| < \delta$.

5.8. Use the fact $||f(z)| - |w_0|| \le |f(z) - w_0|$.

5.9. Let $\epsilon > 0$. Since g is continuous at w_0, there exists a $\delta_1 > 0$ such that $|w - w_0| < \delta_1$ implies that $|g(w) - g(w_0)| < \epsilon$. Now, f is continuous at z_0, so there exists a $\delta_2 > 0$ such that $|z - z_0| < \delta_2$ implies $|f(z) - f(z_0)| < \delta_1$. Combining these, we find that $|z - z_0| < \delta_2$ implies $|f(z) - f(z_0)| < \delta_1$, which in turn implies $|(g \circ f)(z) - (g \circ f)(z_0)| = |g[f(z)] - g[f(z_0)]| < \epsilon$.

5.10. Continuous at 1, discontinuous at $-1, i, -i$.

5.11. f is not continuous at z_0 if there exists $\epsilon_0 > 0$ with the following property: For every $\delta > 0$, there exists z_δ such that $|z_\delta - z_0| < \delta$ and $|f(z_\delta) - f(z_0)| \ge \epsilon_0$. Now let $z_0 = x_0 < 0$. Take $\epsilon_0 = 3\pi/2$. For each $\delta > 0$, let $z_\delta = x_0 - i(\delta/2)$. Then, $|z_\delta - z_0| = |i\delta/2| = \delta/2 < \delta$, $-\pi < f(z_\delta) < -\pi/2$, $f(z_0) = \pi$, so $|f(z_\delta) - f(z_0)| > 3\pi/2 = \epsilon_0$, and f is not continuous at z_0. Thus, f is not continuous at every point on the negative real axis. It is also not continuous at $z = 0$ because it is not defined there.

5.12. If f is continuous at z_0, then given $\epsilon > 0$ there exists a $\delta > 0$ such that $|z_1 - z_0| < \delta/2 \Rightarrow |f(z_1) - f(z_0)| < \epsilon/2$ and $|z_2 - z_0| < \delta/2 \Rightarrow |f(z_2) - f(z_0)| < \epsilon/2$. But then $|z_1 - z_2| \le |z_1 - z_0| + |z_0 - z_2| < \delta \Rightarrow |f(z_1) - f(z_2)| \le |f(z_1) - f(z_0)| + |f(z_2) - f(z_0)| < \epsilon$. For the converse, we assume that $0 < |z - z_0| < \delta$, $0 < |z' - z_0| < \delta$; otherwise, we can take $z' = z_0$ and then there is nothing to prove. Let $z_n \to z_0$, $z_n \ne z_0$, and $\epsilon > 0$. There is a $\delta > 0$ such that $0 < |z - z_0| < \delta$, $|z' - z_0| < \delta$ implies $|f(z) - f(z')| < \epsilon$, and there is an N such that $n \ge N$ implies $0 < |z_n - z_0| < \delta$. Then, for $m, n \ge N$, we have $|f(z_m) - f(z_n)| < \epsilon$. So, $w_0 = \lim_{n \to \infty} f(z_n)$ exists. To see that $\lim_{z \to z_0} f(z) = w_0$, take a $\delta_1 > 0$ such that $0 < |z - z_0| < \delta_1$, $0 < |z' - z_0| < \delta_1$ implies $|f(z) - f(z')| < \epsilon/2$, and an N_1 such that $n \ge N_1$ implies $0 < |z_n - z_0| < \delta_1$ and $|f(z_n) - w_0| < \epsilon/2$. Then, $0 < |z - z_0| < \delta_1$ implies $|f(z) - w_0| \le |f(z) - f(z_{N_1})| + |f(z_{N_1}) - w_0| < \epsilon$.

5.13. (i). Suppose that $f : U \to \mathbf{C}$ is continuous and U is compact. Consider a covering of $f(U)$ to be open sets V. The inverse images $f^{-1}(V)$ are open and form a covering of U. Since U is compact, by Theorem 4.4 we can select a finite subcovering such that $U \subset f^{-1}(V_1) \cup \cdots \cup f^{-1}(V_n)$. It follows that $f(U) \subset V_1 \cup \cdots \cup V_n$, which in view of Theorem 4.4 implies that $f(U)$ is compact. (ii). Suppose that $f : U \to \mathbf{C}$ is continuous and U is connected. If $f(U) = A \cup B$ where A and B are open and disjoint, then $U = f^{-1}(A) \cup f^{-1}(B)$, which is a union of disjoint and open sets. Since U

is connected, either $f^{-1}(A) = \emptyset$ or $f^{-1}(B) = \emptyset$, and hence either $A = \emptyset$ or $B = \emptyset$. This implies that $f(U)$ is connected.

5.14. Use Theorem 4.4.

Lecture 6
Analytic Functions I

In this lecture, using the fundamental notion of limit, we shall define the differentiation of complex functions. This leads to a special class of functions known as analytic functions. These functions are of great importance in theory as well as applications, and constitute a major part of complex analysis. We shall also develop the Cauchy-Riemann equations which provide an easier test to verify the analyticity of a function.

Let f be a function defined in a neighborhood of a point z_0. Then, the *derivative* of f at z_0 is given by

$$\frac{df}{dz}(z_0) = f'(z_0) = \lim_{\Delta z \to 0} \frac{f(z_0 + \Delta z) - f(z_0)}{\Delta z}, \qquad (6.1)$$

provided this limit exists. Such a function f is said to be *differentiable* at the point z_0. Alternatively, f is differentiable at z_0 if and only if it can be written as

$$f(z) = f(z_0) + A(z - z_0) + \eta(z)(z - z_0); \qquad (6.2)$$

here, $A = f'(z_0)$ and $\eta(z) \to 0$ as $z \to z_0$. Clearly, in (6.1), Δz can go to zero in infinite different ways.

Example 6.1. Show that, for any positive integer n, $\dfrac{d}{dz}z^n = nz^{n-1}$. Using the binomial formula, we find

$$\frac{(z + \Delta z)^n - z^n}{\Delta z} = \frac{\binom{n}{1}z^{n-1}\Delta z + \binom{n}{2}z^{n-2}(\Delta z)^2 + \cdots + \binom{n}{n}(\Delta z)^n}{\Delta z}$$

$$= nz^{n-1} + \frac{n(n-1)}{2}z^{n-2}\Delta z + \cdots + (\Delta z)^{n-1}.$$

Thus,

$$\frac{d}{dz}z^n = \lim_{\Delta z \to 0} \frac{(z + \Delta z)^n - z^n}{\Delta z} = nz^{n-1}.$$

Example 6.2. Clearly, the function $f(z) = \overline{z}$ is continuous for all z. We shall show that it is nowhere differentiable. Since

$$\frac{f(z_0 + \Delta z) - f(z_0)}{\Delta z} = \frac{\overline{(z_0 + \Delta z)} - \overline{z}_0}{\Delta z} = \frac{\overline{\Delta z}}{\Delta z}.$$

If Δz is real, then $\Delta z = \overline{\Delta z}$ and the difference quotient is 1. If Δz is purely imaginary, then $\Delta z = -\overline{\Delta z}$ and the quotient is -1. Hence, the limit does not exist as $\Delta z \to 0$. Thus, $\overline{z}$ is not differentiable. In real analysis, construction of functions that are continuous everywhere but differentiable nowhere is hard.

The proof of the following results is almost the same as in calculus.

Theorem 6.1. If f and g are differentiable at a point z_0, then

(i). $(f \pm g)'(z_0) = f'(z_0) \pm g'(z_0)$,

(ii). $(cf)'(z_0) = cf'(z_0)$ (c is a constant),

(iii). $(fg)'(z_0) = f(z_0)g'(z_0) + f'(z_0)g(z_0)$,

(iv). $\left(\dfrac{f}{g}\right)'(z_0) = \dfrac{g(z_0)f'(z_0) - f(z_0)g'(z_0)}{(g(z_0))^2}$ if $g(z_0) \neq 0$, and

(v). $(f \circ g)'(z_0) = f'(g(z_0))g'(z_0)$, provided f is differentiable at $g(z_0)$.

Theorem 6.2. If f is differentiable at a point z_0, then f is continuous at z_0.

A function f of a complex variable is said to be *analytic* (or *holomorphic*, or *regular*) in an open set S if it has a derivative at every point of S. If S is not an open set, then we say f is *analytic in S* if f is analytic in an open set containing S. We call f *analytic at the point z_0* if f is analytic in some neighborhood of z_0. It is important to note that while differentiability is defined at a point, analyticity is defined on an open set. If a function f is analytic on the whole complex plane, then it is said to be *entire*.

For example, all polynomial functions of z are entire. For the rational function $f(z) = P(z)/Q(z)$, where $P(z)$ and $Q(z)$ are polynomials, let $\{\alpha_1, \alpha_2, \cdots, \alpha_r\}$ be the roots of $Q(z)$. By the quotient rule, $f'(z)$ exists for all $z \in S = \mathbf{C} - \{\alpha_1, \alpha_2, \cdots, \alpha_r\}$. Since S is open, f is analytic in S.

If the functions f and g are analytic in a set S, then in view of Theorem 6.1, the sum $f(z) + g(z)$, difference $f(z) - g(z)$, and product $f(z)g(z)$ are analytic in S. The quotient $f(z)/g(z)$ is analytic provided $g(z) \neq 0$ in S.

The proof of the following result is also similar to that of real-valued functions.

Theorem 6.3 (L'Hôpital's Rule). Suppose f and g are analytic functions at a point z_0 and $f(z_0) = g(z_0) = 0$, but $g'(z_0) \neq 0$. Then,

$$\lim_{z \to z_0} \frac{f(z)}{g(z)} = \frac{f'(z_0)}{g'(z_0)}.$$

Example 6.3. Consider the functions $f(z) = z^{14} + 1$ and $g(z) = z^7 + i$.

Clearly, $f(i) = g(i) = 0$, $g'(i) = 7(i)^6 = -7 \neq 0$, and hence

$$\lim_{z \to i} \frac{z^{14} + 1}{z^7 + i} = \lim_{z \to i} \frac{14z^{13}}{7z^6} = \lim_{z \to i} 2z^7 = -2i.$$

If the function $f(z) = u(x, y) + iv(x, y)$ is differentiable at $z_0 = x_0 + iy_0$, then the limit (6.1) exists and can be computed by allowing $\Delta z = (\Delta x + i\Delta y)$ to approach zero from any convenient direction in the complex plane. If it approaches zero horizontally, then $\Delta z = \Delta x$ and we obtain

$$f'(z_0)$$
$$= \lim_{\Delta x \to 0} \frac{1}{\Delta x}[u(x_0 + \Delta x, y_0) + iv(x_0 + \Delta x, y_0) - u(x_0, y_0) - iv(x_0, y_0)]$$
$$= \lim_{\Delta x \to 0} \left[\frac{u(x_0 + \Delta x, y_0) - u(x_0, y_0)}{\Delta x} \right] + i \lim_{\Delta x \to 0} \left[\frac{v(x_0 + \Delta x, y_0) - v(x_0, y_0)}{\Delta x} \right].$$

Since the limits of the bracketed expressions are just the partial derivatives of u and v with respect to x, we deduce that

$$f'(z_0) = \frac{\partial u}{\partial x}(x_0, y_0) + i\frac{\partial v}{\partial x}(x_0, y_0). \tag{6.3}$$

On the other hand, if Δz approaches zero vertically, then $\Delta z = i\Delta y$ and we have

$$f'(z_0)$$
$$= \lim_{\Delta y \to 0} \left[\frac{u(x_0, y_0 + \Delta y) - u(x_0, y_0)}{i\Delta y} \right] + i \lim_{\Delta y \to 0} \left[\frac{v(x_0, y_0 + \Delta y) - v(x_0, y_0)}{i\Delta y} \right],$$

and hence

$$f'(z_0) = -i\frac{\partial u}{\partial y}(x_0, y_0) + \frac{\partial v}{\partial y}(x_0, y_0). \tag{6.4}$$

Equating the real and imaginary parts of (6.3) and (6.4), we see that the equations

$$\frac{\partial u}{\partial x} = \frac{\partial v}{\partial y}, \quad \frac{\partial u}{\partial y} = -\frac{\partial v}{\partial x}, \tag{6.5}$$

must be satisfied at $z_0 = x_0 + iy_0$. These equations are called the *Cauchy-Riemann equations*; however, D'Alembert had stated the equations earlier in the eighteenth century.

Theorem 6.4. A *necessary* condition for a function $f(z) = u(x, y) + iv(x, y)$ to be differentiable at a point z_0 is that the Cauchy-Riemann equations hold at z_0. Consequently, if f is analytic in an open set S, then the Cauchy-Riemann equations must be satisfied at every point of S.

Example 6.4. The function $f(z) = (x^2 + y) + i(y^2 - x)$ is not analytic at any point. Since $u(x, y) = x^2 + y$ and $v(x, y) = y^2 - x$, we have

$$\frac{\partial u}{\partial x} = 2x, \quad \frac{\partial u}{\partial y} = 1, \quad \frac{\partial v}{\partial x} = -1, \quad \frac{\partial v}{\partial y} = 2y.$$

Hence, the Cauchy-Riemann equations are simultaneously satisfied only on the line $x = y$ and therefore in no open disk. Thus, by the theorem above, the function $f(z)$ is nowhere analytic.

Example 6.5. The function $f(z) = \operatorname{Re} z$ is not analytic at any point. Here $u(x, y) = x$ and $v(x, y) = 0$, and so $\dfrac{\partial u}{\partial x} = 1$, $\dfrac{\partial u}{\partial y} = 0$, $\dfrac{\partial v}{\partial x} = 0$, $\dfrac{\partial v}{\partial y} = 0$.

Example 6.6. The function $f(z) = \overline{z}$ is not analytic at any point. Here $u(x, y) = x$ and $v(x, y) = -y$, and so $\dfrac{\partial u}{\partial x} = 1$, $\dfrac{\partial u}{\partial y} = 0$, $\dfrac{\partial v}{\partial x} = 0$, $\dfrac{\partial v}{\partial y} = -1$.

If f is differentiable at z_0, then Theorem 6.2 ensures that f is continuous at z_0. However, the following example shows that the converse is not true.

Example 6.7. The function $f(z) = |z|^2 = x^2 + y^2$ is continuous everywhere but not differentiable at all points $z \neq 0$.

Example 6.7 also shows that, even if u and v have continuous partial derivatives, f need not be differentiable.

Example 6.8. Let $f = u + iv$, where $u = \dfrac{xy}{x^2 + y^2}$ for $(x, y) \neq (0, 0)$ and $u(0, 0) = 0$, $v(x, y) = 0$ for all (x, y). Clearly, at $(0, 0)$, $\dfrac{\partial u}{\partial x} = \dfrac{\partial u}{\partial y} = \dfrac{\partial v}{\partial x} = \dfrac{\partial v}{\partial y} = 0$; i.e., all the partial derivatives exist and satisfy the Cauchy-Riemann equations. However, u is not continuous at $(0, 0)$, and hence f is not differentiable at $(0, 0)$. Thus, even if the function f satisfies the Cauchy-Riemann equations at a point z_0, it need not be differentiable at z_0.

In spite of the two examples above, we have the following result.

Theorem 6.5 (*Sufficient Conditions for Differentiability*). Let $f(z) = u(x, y) + iv(x, y)$ be defined in some open set S containing the point z_0. If the first order partial derivatives of u and v exist in S, are continuous at z_0, and satisfy the Cauchy-Riemann equations at z_0, then f is differentiable at z_0. Moreover,

$$
\begin{aligned}
f'(z_0) &= \frac{\partial u}{\partial x}(x_0, y_0) + i\frac{\partial v}{\partial x}(x_0, y_0) \\
&= \frac{\partial v}{\partial y}(x_0, y_0) - i\frac{\partial u}{\partial y}(x_0, y_0).
\end{aligned}
$$

Consequently, if the first-order partial derivatives are continuous and satisfy the Cauchy-Riemann equations at all points of S, then f is analytic in S. We also note that if f is differentiable only at finitely many points, then it is nowhere analytic.

Lecture 7
Analytic Functions II

In this lecture, we shall first prove Theorem 6.5 and then through simple examples demonstrate how easily this result can be used to check the analyticity of functions. We shall also show that the real and imaginary parts of an analytic function are solutions of the Laplace equation.

Proof of Theorem 6.5. From calculus, the increments of the functions $u(x, y)$ and $v(x, y)$ in the neighborhood of the point (x_0, y_0) can be written as

$$u(x_0 + \Delta x, y_0 + \Delta y) - u(x_0, y_0) = u_x(x_0, y_0)\Delta x + u_y(x_0, y_0)\Delta y + \eta_1(x, y)$$

and

$$v(x_0 + \Delta x, y_0 + \Delta y) - v(x_0, y_0) = v_x(x_0, y_0)\Delta x + v_y(x_0, y_0)\Delta y + \eta_2(x, y),$$

where

$$\lim_{|\Delta z| \to 0} \frac{\eta_1(x, y)}{|\Delta z|} = 0 \quad \text{and} \quad \lim_{|\Delta z| \to 0} \frac{\eta_2(x, y)}{|\Delta z|} = 0.$$

Thus, in view of the Cauchy-Riemann conditions (6.5), it follows that

$$
\begin{aligned}
\frac{f(z_0 + \Delta z) - f(z_0)}{\Delta z} &= u_x(x_0, y_0)\frac{\Delta x + i\Delta y}{\Delta x + i\Delta y} + v_x(x_0, y_0)\frac{i\Delta x - \Delta y}{\Delta x + i\Delta y} \\
&\quad + \frac{\eta_1(x, y) + i\eta_2(x, y)}{\Delta x + i\Delta y} \\
&= [u_x(x_0, y_0) + iv_x(x_0, y_0)] + \frac{\eta(z)}{\Delta z},
\end{aligned}
$$

where $\eta(z) = \eta_1(x, y) + i\eta_2(x, y)$. Now, taking the limit as $\Delta z \to 0$ on both sides and using the fact that $\eta(z)/\Delta z \to 0$ as $\Delta z \to 0$, we obtain

$$f'(z_0) = u_x(x_0, y_0) + iv_x(x_0, y_0). \quad \blacksquare$$

Combining Theorems 6.4 and 6.5, we find that a necessary and sufficient condition for the analyticity of a function $f(z) = u(x, y) + iv(x, y)$ in a domain S is the existence of the continuous partial derivatives u_x, u_y, v_x, and v_y, which satisfy the Cauchy-Riemann conditions (6.5). From this it immediately follows that if $f(z)$ is analytic in a domain S, then the function $g(z) = \overline{f(z)}$ is not analytic in S.

Example 7.1. Consider the exponential function $f(z) = e^z = e^x(\cos y + i \sin y)$. Then, $u(x,y) = e^x \cos y$, $v(x,y) = e^x \sin y$, and

$$\frac{\partial u}{\partial x} = \frac{\partial v}{\partial y} = e^x \cos y, \quad \frac{\partial u}{\partial y} = \frac{\partial v}{\partial x} = -e^x \sin y$$

everywhere, and these derivatives are everywhere continuous. Hence, $f'(z)$ exists and

$$f'(z) = \frac{\partial u}{\partial x} + i\frac{\partial v}{\partial x} = e^x \cos y + ie^x \sin y = e^z = f(z).$$

Example 7.2. The function $f(z) = z^3 = x^3 - 3xy^2 + i(3x^2y - y^3)$ is an entire function and $f'(z) = 3z^2$.

Example 7.3. Consider the function $f(z) = x^2 + y + i(2y - x)$. We have $u(x,y) = x^2 + y$, $v(x,y) = 2y - x$, and $u_x = 2x$, $u_y = 1$, $v_x = -1$, $v_y = 2$. Thus, the Cauchy-Riemann equations are satisfied when $x = 1$. Since all partial derivatives of f are continuous, we conclude that $f'(z)$ exists only on the line $x = 1$ and

$$f'(1 + iy) = \frac{\partial u}{\partial x}(1,y) + i\frac{\partial v}{\partial x}(1,y) = 2 - i.$$

Example 7.4. Let $f(z) = \bar{z}e^{-|z|^2}$. Determine the points at which $f'(z)$ exists, and find $f'(z)$ at these points. Since $f(z) = (x - iy)e^{-(x^2+y^2)}$, $u(x,y) = xe^{-(x^2+y^2)}$, $v(x,y) = -ye^{-(x^2+y^2)}$, and

$$\frac{\partial u}{\partial x} = e^{-(x^2+y^2)} - 2x^2e^{-(x^2+y^2)}, \quad \frac{\partial u}{\partial y} = -2xye^{-(x^2+y^2)},$$

$$\frac{\partial v}{\partial x} = 2xye^{-(x^2+y^2)}, \quad \frac{\partial v}{\partial y} = -e^{-(x^2+y^2)} + 2y^2e^{-(x^2+y^2)}.$$

Thus, $\dfrac{\partial u}{\partial y} = -\dfrac{\partial v}{\partial x}$ is always satisfied, and $\dfrac{\partial u}{\partial x} = \dfrac{\partial v}{\partial y}$ holds, if and only if

$$2e^{-(x^2+y^2)} - 2x^2e^{-(x^2+y^2)} - 2y^2e^{-(x^2+y^2)} = 0,$$

or

$$2e^{-(x^2+y^2)}(1 - x^2 - y^2) = 0,$$

or $x^2 + y^2 = 1$. Since all the partial derivatives of f are continuous, we conclude that $f'(z)$ exists on the unit circle $|z| = 1$. Furthermore, on $|z| = 1$,

$$\begin{aligned}
f'(x + iy) &= \frac{\partial u}{\partial x}(x,y) + i\frac{\partial v}{\partial x}(x,y) \\
&= e^{-(x^2+y^2)} - 2x^2e^{-(x^2+y^2)} + 2ixye^{-(x^2+y^2)} \\
&= e^{-1}(1 - 2x^2 + 2xyi).
\end{aligned}$$

Now recall the following result from calculus.

Theorem 7.1.   Suppose $\phi(x, y)$ is a real-valued function defined in a domain S. If $\phi_x = \phi_y = 0$ at all points in S, then ϕ is a constant in S.

Analogously, for an analytic function, we have the following theorem.

Theorem 7.2.   If $f(z)$ is analytic in a domain S and if $f'(z) = 0$ everywhere in S, then $f(z)$ is a constant in S.

Proof.   Since $f'(z) = 0$ in S, all first-order partial derivatives of u and v vanish in S; i.e., $u_x = u_y = v_x = v_y = 0$. Now, since S is connected, we have $u = $ a constant and $v = $ a constant in S. Consequently, $f = u + iv$ is also a constant in S.   ■

Remark 7.1.   The connectedness property of S is essential. In fact, if $f(z)$ is defined by $f(z) = \begin{cases} 1 & \text{if } |z| < 1 \\ 2 & \text{if } |z| > 2, \end{cases}$ then f is analytic and $f'(z) = 0$ on its domain of definition, but f is not a constant.

Theorem 7.3.   If f is analytic in a domain S and if $|f|$ is constant there, then f is constant.

Proof.   As usual, let $f(z) = u(x, y) + iv(x, y)$. If $|f| = 0$, then $f = 0$. Otherwise, $|f|^2 = u^2 + v^2 \equiv c \neq 0$. Taking the partial derivatives with respect to x and y, we have

$$uu_x + vv_x = 0 \quad \text{and} \quad uu_y + vv_y = 0.$$

Using the Cauchy-Riemann equations, we find

$$uu_x - vu_y = 0 \quad \text{and} \quad vu_x + uu_y = 0$$

so that

$$(u^2 + v^2)u_x = 0 \quad \text{and} \quad (u^2 + v^2)u_y = 0,$$

and hence $u_x = u_y = 0$. Similarly, we have $v_x = v_y = 0$. Thus, f is a constant.   ■

Next, let $f(z) = u(x, y) + iv(x, y)$ be an analytic function in a domain S, so that the Cauchy-Riemann equations $u_x = v_y$ and $u_y = -v_x$ are satisfied. Differentiating both sides of these equations with respect to x (assuming that the functions u and v are twice continuously differentiable, although we shall see in Lecture 18 that this assumption is superfluous), we get

$$\frac{\partial^2 u}{\partial x^2} = \frac{\partial^2 v}{\partial x \partial y} \quad \text{and} \quad \frac{\partial^2 u}{\partial x \partial y} = -\frac{\partial^2 v}{\partial x^2}.$$

Similarly, differentiation with respect to y yields

$$\frac{\partial^2 u}{\partial y \partial x} = \frac{\partial^2 v}{\partial y^2} \quad \text{and} \quad \frac{\partial^2 u}{\partial y^2} = -\frac{\partial^2 v}{\partial y \partial x}.$$

Hence, it follows that

$$\frac{\partial^2 u}{\partial x^2} + \frac{\partial^2 u}{\partial y^2} = 0 \qquad (7.1)$$

and

$$\frac{\partial^2 v}{\partial x^2} + \frac{\partial^2 v}{\partial y^2} = 0. \qquad (7.2)$$

The partial differential equation (7.1) ((7.2)) is called the *Laplace equation*. It occurs in the study of problems dealing with electric and magnetic fields, stationary states, hydrodynamics, diffusion, and so on.

A real-valued function $\phi(x, y)$ is said to be *harmonic* in a domain S if all its second-order partial derivatives are continuous in S and it satisfies $\phi_{xx} + \phi_{yy} = 0$ at each point of S.

Theorem 7.4. If $f(z) = u(x, y) + iv(x, y)$ is analytic in a domain S, then each of the functions $u(x, y)$ and $v(x, y)$ is harmonic in S.

Example 7.5. Does there exist an analytic function on the complex plane whose real part is given by $u(x, y) = 3x^2 + xy + y^2$? Clearly, $u_{xx} = 6$, $u_{yy} = 2$, and hence $u_{xx} + u_{yy} \neq 0$; i.e., u is not harmonic. Thus, no such analytic function exists.

Let $u(x, y)$ and $v(x, y)$ be two functions harmonic in a domain S that satisfy the Cauchy-Riemann equations at every point of S. Then, $u(x, y)$ and $v(x, y)$ are called *harmonic conjugates* of each other. Knowing one of them, we can reconstruct the other to within an arbitrary constant.

Example 7.6. Construct an analytic function whose real part is $u(x, y) = x^3 - 3xy^2 + 7y$. Since $u_{xx} + u_{yy} = 6x - 6x = 0$, u is harmonic in the whole plane. We have to find a function $v(x, y)$ so that u and v satisfy the Cauchy-Riemann equations; i.e.,

$$v_y = u_x = 3x^2 - 3y^2 \qquad (7.3)$$

and

$$v_x = -u_y = 6xy - 7. \qquad (7.4)$$

Integrating (7.3) with respect to y, we get

$$v(x, y) = 3x^2 y - y^3 + \psi(x).$$

Substituting this expression into (7.4), we obtain

$$v_x = 6xy + \psi'(x) = 6xy - 7,$$

and hence $\psi'(x) = -7$, which implies that $\psi(x) = -7x + a$, where a is some constant. It follows that $v(x, y) = 3x^2 y - y^3 - 7x + a$. Thus, the required analytic function is

$$f(z) = x^3 - 3xy^2 + 7y + i(3x^2 y - y^3 - 7x + a).$$

Example 7.7. Find an analytic function f whose imaginary part is given by $e^{-y}\sin x$. Let $v(x,y) = e^{-y}\sin x$. Then it is easy to check that $v_{xx} + v_{yy} = 0$. We have to find a function $u(x,y)$ such that

$$u_x = v_y = -e^{-y}\sin x, \tag{7.5}$$

$$u_y = -v_x = -e^{-y}\cos x. \tag{7.6}$$

From (7.5), we get $u(x,y) = e^{-y}\cos x + \phi(y)$. Substituting this expression in (7.6), we obtain

$$-e^{-y}\sin x + \phi'(y) = -e^{-y}\sin x.$$

Hence, $\phi'(y) = 0$; i.e., $\phi(y) = c$ for some constant c. Thus, $u(x,y) = e^{-y}\cos x + c$ and

$$f(z) = e^{-y}\cos x + c + ie^{-y}\sin x = e^{-y+ix} + c.$$

Problems

7.1. Find $f'(z)$ when

(a). $f(z) = \dfrac{z-1}{2z+1}$ $(z \neq -1/2)$,   (b). $f(z) = e^{z^3}$,

(c). $f(z) = \dfrac{(1+z^2)^4}{z^4}$ $(z \neq 0)$,   (d). $f(z) = z^3 + z$.

7.2. Use the definition to find $f'(z)$ when

(a). $f(z) = \dfrac{1}{z}$ $(z \neq 0)$,   (b). $f(z) = z^2 - z$.

7.3. Show that

(a). $f(z) = x - iy^2$ is differentiable only at $y = -1/2$ and $f'(z) = 1$,
(b). $f(z) = x^2 + iy^2$ is differentiable only when $x = y$ and $f'(z) = 2x$,
(c). $f(z) = yx + iy^2$ is differentiable only at $x = y = 0$ and $f'(z) = 0$,
(d). $f(z) = x^3 + i(1-y)^3$ is differentiable only at $x = 0$, $y = 1$ and $f'(z) = 0$.

7.4. For each of the following functions, determine the set of points at which it is (i) differentiable and (ii) analytic. Find the derivative where it exists.

(a). $f(z) = (x^3 + 3xy^2 - 3x) + i(y^3 + 3x^2y - 3y)$,
(b). $f(z) = 6\bar{z}^2 - 2\bar{z} - 4i|z|^2$,
(c). $f(z) = (3x^2 + 2x - 3y^2 - 1) + i(6xy + 2y)$,

(d). $f(z) = \dfrac{2z^2 + 6}{z(z^2 + 4)}$,

(e). $f(z) = e^{y^2 - x^2}(\cos(2xy) - i\sin(2xy))$.

7.5. Find a, b, c so that the function $w = (ay^3 + ix^3) + xy(bx + icy)$ is analytic. If $z = x + iy$, express dw/dz in the form $\phi(z)$.

7.6. Show that there are no analytic functions of the form $f = u + iv$ with $u = x^2 + y^2$.

7.7. Let $f(z)$ be an analytic function in a domain S. Show that

$$\left(\frac{\partial^2}{\partial x^2} + \frac{\partial^2}{\partial y^2}\right)|f(z)|^2 = 4|f'(z)|^2.$$

7.8. Let $f : \mathbf{C} \to \mathbf{C}$ be defined by

$$f(z) = \begin{cases} \dfrac{\bar{z}^2}{z} & \text{if } z \neq 0 \\ 0 & \text{if } z = 0. \end{cases}$$

(a). Verify that the Cauchy-Riemann equations for f are satisfied at $z = 0$.

(b). Show that $f'(0)$ does not exist.

This problem shows that satisfaction of the Cauchy-Riemann equations at a point alone is not enough to ensure that the function is differentiable there.

7.9. Let D and S be domains, and let $f : D \to \mathbf{C}$ and $g : S \to \mathbf{C}$ be analytic functions such that $f(D) \subseteq S$. Show that $g \circ f : D \to \mathbf{C}$ is analytic.

7.10. In polar coordinates $x = r\cos\theta$, $y = r\sin\theta$, the function $f(z) = u(r, \theta) + iv(r, \theta)$. Show that the Cauchy-Riemann conditions can be written as

$$\frac{\partial u}{\partial r} = \frac{1}{r}\frac{\partial v}{\partial \theta}, \qquad \frac{1}{r}\frac{\partial u}{\partial \theta} = -\frac{\partial v}{\partial r}, \tag{7.7}$$

and

$$f'(z) = e^{-i\theta}(u_r + iv_r). \tag{7.8}$$

In particular, show that $f(z) = \sqrt{r}e^{i\theta/2}$ is differentiable at all z except $z = 0$ and $f'(z) = (1/2\sqrt{r})e^{-i\theta/2}$.

7.11. Show that the Cauchy-Riemann conditions are equivalent to $w_{\bar{z}} = 0$. Hence, deduce that the function $w = f(z) = \bar{z}e^{-|z|^2}$ is not analytic.

7.12. Let $w = f(z)$ be analytic in a neighborhood of z_0, and $w_0 = f(z_0)$, $f'(z_0) \neq 0$. Show that f defines a one-to-one mapping of a neighborhood of z_0 onto a neighborhood of w_0.

7.13. Use L'Hôpital's Rule to find the following limits:

(a). $\displaystyle\lim_{z\to i}\frac{z^7+i}{z^{14}+1}$,   (b). $\displaystyle\lim_{z\to 3i}\frac{z^4-81}{z^2+9}$,   (c). $\displaystyle\lim_{1+i\sqrt{3}}\frac{z^6-64}{z^3+8}$.

7.14. Show that if $f = u+iv$ is analytic in a region S and u is a constant function (i.e., independent of x and y), then f is a constant.

7.15. Show that if $h : \mathbb{R}^2 \to \mathbb{R}$ and $f = 2h^3 + ih$ is an entire function, then h is a constant.

7.16. Suppose f is a real-valued function defined in a domain $S \subseteq \mathbf{C}$. If f is complex differentiable at $z_0 \in S$, show that $f'(z_0) = 0$.

7.17. Suppose that $f = u + iv$ is analytic in a rectangle with sides parallel to the coordinate axes and satisfies the relation $u_x + v_y = 0$ for all x and y. Show that there exist a real constant c and a complex constant d such that $f(z) = -icz + d$.

7.18. Show that the functions

(a). $u(x,y) = e^{-x}\sin y$,   (b). $v(x,y) = \cos x \cosh y$,

are harmonic, and find the corresponding analytic function $u + iv$ in each case.

7.19. Suppose that u is harmonic in a domain S. Show that:

(a). If v is a harmonic conjugate of u, then $-u$ is a harmonic conjugate of v.

(b). If v_1 and v_2 are harmonic conjugates of u, then v_1 and v_2 differ by a real constant.

(c). If v is a harmonic conjugate of u, then v is also a harmonic conjugate of $u + c$, where c is any real constant.

7.20. Show that a necessary and sufficient condition for a function $f(z) = u(x,y) + iv(x,y)$ to be analytic in a domain S is that its real part $u(x,y)$ and imaginary part $v(x,y)$ be conjugate harmonic functions in S.

7.21. Let $f(z) = u(r,\theta) + iv(r,\theta)$ be analytic in a domain S that does not include the point $z = 0$. Use (7.7) to show that both u and v satisfy the Laplace equation in polar coordinates

$$r^2\frac{\partial^2\phi}{\partial r^2} + r\frac{\partial\phi}{\partial r} + \frac{\partial^2\phi}{\partial\theta^2} = 0.$$

7.22. Use $f(t) = \dfrac{t}{t^2+1} = \text{Re}\left(\dfrac{1}{t-i}\right)$ to show that

$$f^{(n)}(t) = \frac{(-1)^n n!(n+1)!}{(t^2+1)^{n+1}} \sum_{k=0}^{[(n+1)/2]} \frac{(-1)^k t^{n+1-2k}}{(2k)!(n+1-2k)!}.$$

7.23 (a). If $f(x+iy) = Re^{i\phi}$ is analytic, show that $\dfrac{\partial R}{\partial x} = R\dfrac{\partial \phi}{\partial y}$.

(b). Consider the two annulus domains $D_a = \{z : a < |z| < 1\}$ and $D_b = \{z : b < |z| < 1\}$. Define $f : D_a \to D_b$ by

$$f\left(re^{i\theta}\right) = \left[\left(\frac{1-b}{1-a}\right)r + \frac{b-a}{1-a}\right]e^{i\theta}.$$

Show that (i) f is bijective and (ii) f is analytic if and only if $a = b$.

7.24. Let $f(z) = u(x,y) + iv(x,y)$ be an analytic function. Show that level curves of the family $u(x,y) = c$ are orthogonal to the level curves of the family $v(x,y) = d$; i.e., the intersection of a member of one family with that of another takes place at a $90°$ angle, except possibly at a point where $f'(z) = 0$. Verify this result for the function $f(z) = z^2$.

Answers or Hints

7.1. (a). $3/(2z+1)^2$, (b). $3z^2 e^{z^3}$, (c). $4(1+z^2)^3(z^2-1)/z^5$, (d). $3z^2+1$.

7.2. (a). $f'(z) = \lim_{\Delta z \to 0} \dfrac{\frac{1}{z+\Delta z} - \frac{1}{z}}{\Delta z} = \lim_{\Delta z \to 0} \dfrac{z-(z+\Delta z)}{(\Delta z)(z+\Delta z)z} = -\dfrac{1}{z^2}$,

(b). $f'(z) = \lim_{\Delta z \to 0} \dfrac{(z+\Delta z)^2-(z+\Delta z)-(z^2-z)}{\Delta z} = \lim_{\Delta z \to 0} \dfrac{2z\Delta z + (\Delta z)^2 - \Delta z}{\Delta z} = 2z-1$.

7.3. (a). Since $u = x$, $v = -y^2$, $u_x = 1$, $u_y = 0$, $v_x = 0$, $v_y = -2y$, the function is differentiable only when $1 = -2y$ or $y = -1/2$, and $f' = u_x + iv_x = 1$. (b). Since $u = x^2$, $v = y^2$, $u_x = 2x$, $u_y = 0$, $v_x = 0$, $v_y = 2y$, the function is differentiable only when $2x = 2y$ and $f'(z) = 2x$. (c). Since $u = yx$, $v = y^2$, $u_x = y$, $u_y = x$, $v_x = 0$, $v_y = 2y$, the function is differentiable only when $y = 2y$, $x = 0$ or $x = 0$, $y = 0$, and $f'(z) = 0$. (d). Since $u = x^3$, $v = (1-y)^3$, $u_x = 3x^2$, $u_y = 0$, $v_x = 0$, $v_y = -3(1-y)^2$ the function is differentiable only when $3x^2 = -3(1-y)^2$ or $x = 0$, $y = 1$, and $f'(z) = 0$.

7.4. (a). f is differentiable at every point on the x and y axes, $f'(x+i0) = 3x^2 - 3$, $f'(0+iy) = 3y^2 - 3$, not analytic anywhere. (b). f is differentiable only at the point $z = \frac{3}{16} - \frac{1}{16}i$, $f'(\frac{3}{16} - \frac{1}{16}i) = \frac{1}{4} - \frac{3}{4}i$, not analytic anywhere. (c). f is differentiable everywhere, $f'(z) = (6x+2) + i(6y)$, analytic everywhere (entire). (d). f is differentiable for all $z \neq 0, \pm 2i$,

$f'(z) = -2(z^4 + 5z^2 + 12)/[z^2(z^2+4)^2]$, analytic in $\mathbf{C} - \{0, \pm 2i\}$. (e). f is differentiable everywhere, $f'(z) = -2ze^{-z^2}$, analytic everywhere (entire).

7.5. $a = 1, b = -3, c = -3$, $y = 0$ gives $f(x) = ix^3$, and hence $f(z) = iz^3$ and $dw/dz = 3iz^2$.

7.6. $u_x = 2x$, $u_y = 2y \Rightarrow v_y = 2x$ and $v_x = -2y \Rightarrow v = 2xy + f(y)$ and $v = -2xy + g(x)$, which is impossible.

7.7. LHS $= 2(u_x^2 + u_y^2) + 2(v_x^2 + v_y^2)$.

7.8. (a). $u(x,y) = \begin{cases} (x^3 - 3xy^2)/(x^2 + y^2), & (x,y) \neq (0,0) \\ 0, & (x,y) = (0,0) \end{cases}$ and $v(x,y) = \begin{cases} (y^3 - 3x^2y)/(x^2 + y^2), & (x,y) \neq (0,0) \\ 0, & (x,y) = (0,0). \end{cases}$ $u_x(0,0) = 1$, $u_y(0,0) = 0$, $v_x(0,0) = 0$, $v_y(0,0) = 1$. (b). For $z \neq 0$, $(f(z) - f(0))/(z - 0) = (\bar{z}/z)^2$. Now see Problem 5.6 (a).

7.9. Use rules for differentiation.

7.10. Since $u_r = u_x x_r + u_y y_r$, $u_\theta = u_x x_\theta + u_y y_\theta$, we have
$$u_r = u_x \cos\theta + u_y \sin\theta, \quad u_\theta = -u_x r \sin\theta + u_y r \cos\theta \qquad (7.9)$$
and
$$v_r = v_x \cos\theta + v_y \sin\theta, \quad v_\theta = -v_x r \sin\theta + v_y r \cos\theta,$$
which in view of (6.5) is the same as
$$v_r = -u_y \cos\theta + u_x \sin\theta, \quad v_\theta = u_y r \sin\theta + u_x r \cos\theta. \qquad (7.10)$$
From (7.9) and (7.10), the Cauchy-Riemann conditions (7.7) are immediate. Now, since $f'(z) = u_x + iv_x$ and $u_x = u_r \cos\theta - u_\theta \frac{\sin\theta}{r} = u_r \cos\theta + v_r \sin\theta$, $v_x = v_r \cos\theta - v_\theta \frac{\sin\theta}{r} = v_r \cos\theta - u_r \sin\theta$, it follows that $f'(z) = u_r(\cos\theta - i\sin\theta) + iv_r(\cos\theta - i\sin\theta) = e^{-i\theta}(u_r + iv_r)$.

Since $u = \sqrt{r}\cos\frac{\theta}{2}$, $v = \sqrt{r}\sin\frac{\theta}{2}$, $u_r = \frac{1}{2\sqrt{r}}\cos\frac{\theta}{2}$, $u_\theta = -\frac{1}{2}\sqrt{r}\sin\frac{\theta}{2}$, $v_r = \frac{1}{2\sqrt{r}}\sin\frac{\theta}{2}$, $v_\theta = \frac{1}{2}\sqrt{r}\cos\frac{\theta}{2}$, and hence conditions (7.7) are satisfied. Thus, f is differentiable at all z except $z = 0$. Furthermore, from (7.8) it follows that $f'(z) = e^{-i\theta}\frac{1}{2\sqrt{r}}(\cos\frac{\theta}{2} + i\sin\frac{\theta}{2}) = \frac{1}{2\sqrt{z}}$.

7.11. Since $x = (z + \bar{z})/2$ and $y = (z - \bar{z})/2i$, u and v may be regarded as functions of z and $\bar{z}$. Thus, condition $w_{\bar{z}} = 0$ is the same as $\frac{\partial u}{\partial x}\frac{\partial x}{\partial \bar{z}} + \frac{\partial u}{\partial y}\frac{\partial y}{\partial \bar{z}} + i\left(\frac{\partial v}{\partial x}\frac{\partial x}{\partial \bar{z}} + \frac{\partial v}{\partial y}\frac{\partial y}{\partial \bar{z}}\right) = 0$, which is the same as $\frac{1}{2}\frac{\partial u}{\partial x} - \frac{1}{2i}\frac{\partial u}{\partial y} + \frac{i}{2}\frac{\partial v}{\partial x} - \frac{1}{2}\frac{\partial v}{\partial y} = 0$. Now compare the real and imaginary parts.

7.12. Let $f = u + iv$. From calculus, it suffices to show that Jacobian
$$J(x_0, y_0) = \begin{vmatrix} u_x & v_x \\ u_y & v_y \end{vmatrix} (x_0, y_0) \neq 0.$$
Now use (6.5) and (6.3) to obtain $J(x_0, y_0) = u_x^2(x_0, y_0) + v_x^2(x_0, y_0) = |f'(z_0)|^2 \neq 0$.

7.13. (a). $i/2$, (b). -18, (c). -16.

7.14. $f = u + iv$, $u = c \Rightarrow u_x = u_y = 0 \Rightarrow v_x = v_y = 0 \Rightarrow v$ is also a constant, and hence f is a constant.

7.15. If $f = u + iv$, then $u = 2h^3$ and $v = h$. Now, by the Cauchy-Riemann conditions, we have $6h^2 h_x = h_y$ and $-6h^2 h_y = h_x$. Thus, $-12h^4 h_x = h_x$, or $h_x(12h^4 + 1) = 0$, which implies that $h_x = 0$, and from this $h_y = 0$.

7.16. In (6.1), if we allow $\Delta z \to 0$ along the x-axis, then $f'(z_0)$ is real. However, if we allow $\Delta z \to 0$ along the y-axis, then $f'(z_0)$ is purely imaginary.

7.17. $u_x + v_y = 0$ and $u_x = v_y$ imply that $u_x = v_y = 0$. Thus, $u = \phi(y)$, $v = \psi(x)$. Now, $u_y = -v_x$ implies that $\phi'(y) = -\psi'(x) = c$. Hence, $u = \phi(y) = cy + d_1$, $v = \psi(x) = -cx + d_2$, where d_1 and d_2 are real constants. Thus, $f = u + iv = -icz + d$, where $d = d_1 + id_2$.

7.18. (a). $ie^{-z} + ia$, (b). $\sin x \sinh y + i \cos x \cosh y + a$.

7.19. (a). Since $f = u + iv$ is analytic, $(-i)f = v - iu$ is analytic. Thus, $-u$ is a harmonic conjugate of v.

7.20. The necessary part is Theorem 7.4. To show the sufficiency part, we note that if $u(x,y)$ and $v(x,y)$ are conjugate harmonic functions, then, in particular, they have continuous first derivatives in S, and hence are differentiable in S. Since $u(x,y)$ and $v(x,y)$ also satisfy the Cauchy-Riemann equations in S, it follows that $f(z)$ is analytic in S.

7.21. Verify directly.

7.22. Use $f^{(n)}(t) = \mathrm{Re}\frac{d^n}{dt^n}\left(\frac{1}{t-i}\right)$ and the binomial theorem.

7.23. (a). $f = Re^{i\phi} = R\cos\phi + iR\sin\phi$. By the Cauchy-Riemann equations

$$\frac{\partial R}{\partial x}\cos\phi - R\sin\phi\frac{\partial\phi}{\partial x} = \frac{\partial R}{\partial y}\sin\phi + R\cos\phi\frac{\partial\phi}{\partial y},$$
$$\frac{\partial R}{\partial y}\cos\phi - R\sin\phi\frac{\partial\phi}{\partial y} = -\frac{\partial R}{\partial x}\sin\phi - R\cos\phi\frac{\partial\phi}{\partial x};$$

i.e.,

$$\frac{\partial R}{\partial x}\cos\phi - \frac{\partial R}{\partial y}\sin\phi = R\sin\phi\frac{\partial\phi}{\partial x} + R\cos\phi\frac{\partial\phi}{\partial y}, \tag{7.11}$$

$$\frac{\partial R}{\partial y}\cos\phi + \frac{\partial R}{\partial x}\sin\phi = R\sin\phi\frac{\partial\phi}{\partial y} - R\cos\phi\frac{\partial\phi}{\partial x}. \tag{7.12}$$

Now $(7.11)\cos\phi + (7.12)\sin\phi$ gives the result. (b). Let $z_1 = r_1 e^{i\theta_1}, z_2 = r_2 e^{i\theta_2}$. Then $f(z_1) = f(z_2)$ implies that

$$\left[\left(\frac{1-b}{1-a}\right)r_1 + \frac{b-a}{1-a}\right]e^{i\theta_1} = \left[\left(\frac{1-b}{1-a}\right)r_2 + \frac{b-a}{1-a}\right]e^{i\theta_2};$$

i.e., $\frac{1-b}{1-a}r_1 + \frac{b-a}{1-a} = \frac{1-b}{1-a}r_2 + \frac{b-a}{1-a}$ and $e^{i\theta_1} = e^{i\theta_2}$, and hence $r_1 = r_2$ and $\theta_1 = \theta_2$. Therefore, f is one-to-one. Now, for any $z \in D_b$, let $z = \rho e^{i\theta}$. Then $b < \rho < 1$. Consider $\rho = \frac{1-b}{1-a}r + \frac{b-a}{1-a}$ so that $r = \frac{1-a}{1-b}\rho + \frac{a-b}{1-b}$. Since $\frac{dr}{d\rho} = \frac{1-a}{1-b} > 0$, r is an increasing function of ρ. When $\rho = b, r = a$ and when $\rho = 1, r = 1$, so $a < r < 1$; hence $re^{i\theta} \in D_a$ and $f\left(re^{i\theta}\right) = \rho e^{i\theta} = z$; i.e., f is onto. Now suppose f is analytic then, from part (a), we have

$$\frac{\partial}{\partial x}\left(\frac{1-b}{1-a}r + \frac{b-a}{1-a}\right) = \frac{1-b}{1-a}\frac{\partial r}{\partial x} = \frac{\partial\theta}{\partial y}\left(\frac{1-b}{1-a}r + \frac{b-a}{1-a}\right).$$

From $x = r\cos\theta$, $y = r\sin\theta$, we have $\frac{\partial r}{\partial x} = \cos\theta$, $\frac{\partial\theta}{\partial y} = \frac{\cos\theta}{r}$, and hence $\frac{1-b}{1-a}\cos\theta = \frac{\cos\theta}{r}\left(\frac{1-b}{1-a}r + \frac{b-a}{1-a}\right)$, which implies that $\frac{b-a}{1-a} = 0$, and hence $b = a$. Conversely, if $b = a$, then $f(re^{i\theta}) = re^{i\theta}$ and so f is analytic.

7.24. Since $u_x + u_y dy/dx = 0$ and $v_x + v_y dy/dx = 0$, we have $dy/dx = -u_x/u_y$ and $dy/dx = -v_x/v_y$. Thus, at a point of intersection (x_0, y_0), from the Cauchy-Riemann conditions and the fact that $f'(x_0 + iy_0) \neq 0$ it follows that $(-u_x/u_y)(-v_x/v_y) = -1$.

Lecture 8
Elementary Functions I

We have already seen that the complex exponential function $e^z = e^x(\cos y + i \sin y)$ is entire, and $d(e^z)/dz = e^z$. In this lecture, we shall first provide some further properties of the exponential function, and then define complex trigonometric and hyperbolic functions in terms of e^z.

Let $w = f(z)$ be an analytic function in a domain S. Then, in view of Problem 7.9 and the fact that the exponential function is entire, it follows that the composite function e^w is also analytic in S. Thus, for all $z \in S$, we have

$$\frac{d}{dz}e^w = \frac{dw}{dz}e^w.$$

Hence, in particular, the function e^{z^2-iz+7} is entire, and

$$\frac{d}{dz}e^{z^2-iz+7} = (2z - i)e^{z^2-iz+7}.$$

The polar components of e^z are given by

$$|e^z| = e^x, \quad \arg e^z = y + 2k\pi, \quad k = 0, \pm 1, \pm 2, \cdots.$$

Since e^x is never zero, it follows that e^z is also never zero. However, e^z does assume every other complex value.

In calculus, it is shown that the exponential function is one-to-one on the real axis. However, it is not one-to-one on the complex plane. In fact, we have the following result.

Theorem 8.1. (i). $e^z = 1$ if and only if $z = 2k\pi i$, where k is an integer. (ii). $e^{z_1} = e^{z_2}$ if and only if $z_1 = z_2 + 2k\pi i$, where k is an integer.

Proof. (i). Suppose that $e^z = 1$ with $z = x + iy$. Then, we must have $|e^z| = |e^{x+iy}| = |e^x e^{iy}| = e^x = 1$, and so $x = 0$. This implies that $e^z = e^{iy} = \cos y + i \sin y = 1$. Equating the real and imaginary parts, we have $\cos y = 1$, $\sin y = 0$. These two simultaneous equations are satisfied only when $y = 2k\pi$ for some integer k; i.e., $z = 2k\pi i$. Conversely, if $z = 2k\pi i$, where k is an integer, then $e^z = e^{2k\pi i} = e^0(\cos 2k\pi + i \sin 2k\pi) = 1$.

(ii). We have $e^{z_1} = e^{z_2}$ if and only if $e^{z_1-z_2} = 1$, and hence, from (i) $z_1 - z_2 = 2k\pi i$, where k is an integer. ∎

A function f is said to be *periodic* in a domain D if there exists a constant ω such that $f(z + \omega) = f(z)$ for every z in D. Any constant ω with this property is called a period of f.

Since, for all z, $e^{z+2k\pi i} = e^z$, we find that e^z is periodic with complex period $2\pi i$. Consequently, if we divide the z-plane into the infinite horizontal strips

$$S_n = \{x+iy : \; -\infty < x < \infty, \; (2n-1)\pi < y \le (2n+1)\pi, \; n = 0, \pm1, \pm2, \cdots\}$$

then e^z will behave in the same manner on each strip.

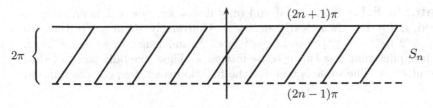

Figure 8.1

From part (ii) of Theorem 8.1, we see that e^z is one-to-one on each strip S_n. Finally, for the function e^z, we note that $\overline{e^z} = e^{\bar{z}}$.

Now, for any given complex number z, we define

$$\sin z = \frac{e^{iz} - e^{-iz}}{2i}, \qquad \cos z = \frac{e^{iz} + e^{-iz}}{2}.$$

Since e^{iz} and e^{-iz} are entire functions, so are $\sin z$ and $\cos z$. In fact,

$$\frac{d}{dz}\sin z = \frac{d}{dz}\left(\frac{e^{iz} - e^{-iz}}{2i}\right) = \frac{1}{2i}\left(ie^{iz} - (-i)e^{-iz}\right) = \cos z.$$

Similarly,

$$\frac{d}{dz}\cos z = -\sin z.$$

Also, $\overline{\sin z} = \sin \bar{z}$, $\overline{\cos z} = \cos \bar{z}$.

The usual trigonometric identities remain valid with complex variables:

$$\sin(z + 2\pi) = \sin z, \qquad \cos(z + 2\pi) = \cos z. \tag{8.1}$$

$$\sin(z+\pi) \;=\; -\sin z, \quad \cos(z+\pi) \;=\; -\cos z, \quad \sin(\pi/2 - z) \;=\; \cos z,$$

$$\sin(-z) \;=\; -\sin z, \quad \cos(-z) \;=\; \cos z, \quad \sin^2 z + \cos^2 z \;=\; 1,$$

$$\sin(z_1 \pm z_2) \;=\; \sin z_1 \cos z_2 \pm \cos z_1 \sin z_2,$$

$$\cos(z_1 \pm z_2) \;=\; \cos z_1 \cos z_2 \mp \sin z_1 \sin z_2,$$

$$\sin 2z \;=\; 2\sin z \cos z, \quad \cos 2z \;=\; \cos^2 z - \sin^2 z,$$

$$2\sin(z_1 + z_2)\sin(z_1 - z_2) \;=\; \cos 2z_2 - \cos 2z_1,$$

$$2\cos(z_1 + z_2)\sin(z_1 - z_2) \;=\; \sin 2z_1 - \sin 2z_2.$$

Equation (8.1) implies that $\sin z$ and $\cos z$ are both periodic with period 2π.

Example 8.1. $\sin z = 0$ if and only if $z = k\pi$, where k is an integer. Indeed, if $z = k\pi$, then clearly $\sin z = 0$. Conversely, if $\sin z = 0$, then we have $(1/2i)(e^{iz} - e^{-iz}) = 0$; i.e., $e^{iz} = e^{-iz}$, and hence $iz = -iz + 2k\pi i$, which implies that $z = k\pi$ for some integer k. Thus, the only zeros of $\sin z$ are real zeros. The same is true for the function $\cos z$; i.e., $\cos z = 0$ if and only if $z = (\pi/2) + k\pi$.

The four other complex trigonometric functions are defined by

$$\tan z \;=\; \frac{\sin z}{\cos z}, \quad \cot z \;=\; \frac{\cos z}{\sin z}, \quad \sec z \;=\; \frac{1}{\cos z}, \quad \operatorname{cosec} z \;=\; \frac{1}{\sin z}.$$

The functions $\cot z$ and $\operatorname{cosec} z$ are analytic for all z except at the points $z = k\pi$, whereas the functions $\tan z$ and $\sec z$ are analytic for all z except at the points $z = (\pi/2) + k\pi$, where k is an integer. Furthermore, the usual rules for differentiation remain valid for these functions:

$$\frac{d}{dz}\tan z \;=\; \sec^2 z, \qquad \frac{d}{dz}\sec z \;=\; \sec z \tan z,$$

$$\frac{d}{dz}\cot z \;=\; -\operatorname{cosec}^2 z, \qquad \frac{d}{dz}\operatorname{cosec} z \;=\; -\operatorname{cosec} z \cot z.$$

For any complex number z, we define

$$\sinh z \;=\; \frac{e^z - e^{-z}}{2}, \qquad \cosh z \;=\; \frac{e^z + e^{-z}}{2}.$$

The functions $\sinh z$ and $\cosh z$ are entire and

$$\frac{d}{dz}\sinh z \;=\; \cosh z, \qquad \frac{d}{dz}\cosh z \;=\; \sinh z.$$

For real x, $\cosh x \geq 1$. Since $d(\sinh x)/dx = \cosh x$, $\sinh x$ is an increasing function.

By comparing the definitions of hyperbolic sine and cosine functions with those of trigonometric functions, we find

$$\cosh(iz) = \cos z, \quad \cos(iz) = \cosh z, \quad \sinh(iz) = i\sin z, \quad \sin(iz) = i\sinh z.$$
(8.2)

Using relations (8.2) and the trigonometric identities, we can show that the following hyperbolic identities are valid in the complex case:

$$\sinh(-z) = -\sinh z, \quad \cosh(-z) = \cosh z, \quad \cosh^2 z - \sinh^2 z = 1,$$
$$\sinh(z_1 \pm z_2) = \sinh z_1 \cosh z_2 \pm \cosh z_1 \sinh z_2,$$
$$\cosh(z_1 \pm z_2) = \cosh z_1 \cosh z_2 \pm \sinh z_1 \sinh z_2,$$
$$\sinh 2z = 2\sinh z \cosh z.$$

From relations (8.2), it follows that $\sinh z$ and $\cosh z$ are both periodic with period $2\pi i$. Furthermore, the zeros of $\sinh z$ are $z = k\pi i$ and the zeros of $\cosh z$ are $z = (k + 1/2)\pi i$, where k is an integer.

The four remaining complex hyperbolic functions are given by

$$\tanh z = \frac{\sinh z}{\cosh z}, \quad \coth z = \frac{\cosh z}{\sinh z}, \quad \operatorname{sech} z = \frac{1}{\cosh z}, \quad \operatorname{cosech} z = \frac{1}{\sinh z}.$$

The functions $\coth z$ and $\operatorname{cosech} z$ are analytic for all z except at the points $z = k\pi i$, whereas the functions $\tanh z$ and $\operatorname{sech} z$ are analytic for all z except at the points $z = (k + 1/2)\pi i$, where k is an integer. Furthermore, for these functions also, the usual rules for differentiation remain valid:

$$\frac{d}{dz}\tanh z = \operatorname{sech}^2 z, \qquad \frac{d}{dz}\operatorname{sech} z = -\operatorname{sech} z \tanh z,$$
$$\frac{d}{dz}\coth z = -\operatorname{cosech}^2 z, \quad \frac{d}{dz}\operatorname{cosech} z = -\operatorname{cosech} z \coth z.$$

Example 8.2. Show that $|\sin z|^2 = \sin^2 x + \sinh^2 y$. Since

$$\sin z = \sin(x + iy) = \sin x \cos(iy) + \cos x \sin(iy)$$
$$= \sin x \cosh y + i\cos x \sinh y,$$

it follows that

$$|\sin z|^2 = \sin^2 x \cosh^2 y + \cos^2 x \sinh^2 y$$
$$= \sin^2 x \cosh^2 y + (1 - \sin^2 x)\sinh^2 y$$
$$= \sin^2 x(\cosh^2 y - \sinh^2 y) + \sinh^2 y = \sin^2 x + \sinh^2 y.$$

Similarly, in view of

$$\cos z = \cos x \cosh y - i\sin x \sinh y,$$

one can show that $|\cos z|^2 = \cos^2 x + \sinh^2 y$.

Example 8.3. As in Example 8.2, we have

$$\begin{aligned} \sinh z &= \sinh x \cos y + i \cosh x \sin y, \\ \cosh z &= \cosh x \cos y + i \sinh x \sin y. \end{aligned}$$

Thus, it follows that

$$\begin{aligned} |\sinh z|^2 &= \sinh^2 x + \sin^2 y, \\ |\cosh z|^2 &= \sinh^2 x + \cos^2 y. \end{aligned}$$

Example 8.4. From the various relations given above, it follows that

$$\begin{aligned} \tan z &= \tan(x + iy) = \frac{\sin(x + iy)}{\cos(x + iy)} \\ &= \frac{\sin x \cosh y + i \cos x \sinh y}{\cos x \cosh y - i \sin x \sinh y} \\ &= \frac{\cos x \sin x + i \cosh y \sinh y}{\cos^2 x \cosh^2 y + \sin^2 x \sinh^2 y} \\ &= \frac{\sin 2x}{\cos 2x + \cosh 2y} + i \frac{\sinh 2y}{\cos 2x + \cosh 2y}. \end{aligned}$$

Lecture 9
Elementary Functions II

In this lecture, we shall introduce the complex logarithmic function, study some of its properties, and then use it to define complex powers and inverse trigonometric functions.

Let $\operatorname{Log} r = \ln r$ denote the natural logarithm of a positive real number r. If $z \neq 0$, then we define $\log z$ to be any of the infinitely many values

$$
\begin{aligned}
\log z &= \operatorname{Log}|z| + i \arg z \\
&= \operatorname{Log}|z| + i \operatorname{Arg} z + 2k\pi i, \quad k = 0, \pm 1, \pm 2, \cdots.
\end{aligned}
$$

Example 9.1. We have

$$
\begin{aligned}
\log 3 &= \operatorname{Log} 3 + i \arg 3 = (1.098 \cdots) + 2k\pi i, \\
\log(-1) &= \operatorname{Log} 1 + i \arg(-1) = (2k+1)\pi i, \\
\log(1+i) &= \operatorname{Log}|1+i| + i \arg(1+i) = \operatorname{Log}\sqrt{2} + i\left(\frac{\pi}{4} + 2k\pi\right),
\end{aligned}
$$

where $k = 0, \pm 1, \pm 2, \cdots$.

Now we shall show the following properties of the logarithmic function:

(i). If $z \neq 0$, then $z = e^{\log z}$. Let $z = re^{i\theta}$. Then, $|z| = r$ and $\arg z = \theta$. Hence, $\log z = \operatorname{Log} r + i\theta$. Thus,

$$
e^{\log z} = e^{(\operatorname{Log} r + i\theta)} = e^{\operatorname{Log} r} e^{i\theta} = re^{i\theta} = z.
$$

(ii). $\log e^z = z + 2k\pi i, \quad k = 0, \pm 1, \pm 2, \cdots$. Let $z = x + iy$. Then $|e^z| = e^x$, $\arg e^z = y + 2k\pi$. Hence,

$$
\begin{aligned}
\log e^z = \operatorname{Log}|e^z| + i \arg e^z &= \operatorname{Log} e^x + i(y + 2k\pi) \\
&= x + iy + 2k\pi i = z + 2k\pi i.
\end{aligned}
$$

(iii). $\log z_1 z_2 = \log z_1 + \log z_2$ \hfill (9.1)

$$
\log\left(\frac{z_1}{z_2}\right) = \log z_1 - \log z_2. \tag{9.2}
$$

Indeed, we have

$$
\begin{aligned}
\log z_1 z_2 &= \operatorname{Log}|z_1 z_2| + i \arg z_1 z_2 \\
&= \operatorname{Log}|z_1| + \operatorname{Log}|z_2| + i \arg z_1 + i \arg z_2 \\
&= \log z_1 + \log z_2.
\end{aligned}
$$

As $\log z$ is a multi-valued function, we must interpret (9.1) and (9.2) to mean that if particular values are assigned to any two of their terms, then one can find a value of the third term so that the equation is satisfied.

Example 9.2.   Let $z_1 = z_2 = -1$. Then, $z_1 z_2 = 1$. Thus, $\log z_1 = (2k_1 + 1)\pi i$, $\log z_2 = (2k_2 + 1)\pi i$, and $\log 1 = 2k_3\pi i$ where k_1, k_2, and k_3 are integers. If we select πi to be the value of $\log z_1$ and $\log z_2$, then equation (9.1) will be satisfied if we use $2\pi i$ for $\log 1$. If we select 0 and πi to be the values of $\log 1$ and $\log z_1$, respectively, then equation (9.1) will be satisfied if we use $-\pi i$ for $\log z_2$.

A *branch* of a multi-valued function is a single-valued function analytic in some domain. The *principal value or branch* of the logarithm, denoted by $\mathrm{Log}\, z$, is the value inherited from the principal value of the argument:

$$\mathrm{Log}\, z \;=\; \mathrm{Log}\,|z| + i\mathrm{Arg}\, z.$$

The value of $\mathrm{Arg}\, z$ jumps by $2\pi i$ as z crosses the negative real axis. Therefore, $\mathrm{Log}\, z$ is not continuous at any point on the nonpositive real axis. However, at all points off the nonpositive real axis, $\mathrm{Log}\, z$ is continuous.

Theorem 9.1.   The function $\mathrm{Log}\, z$ is analytic in the domain D^* consisting of all points of the complex plane except those lying on the nonpositive real axis; i.e., $D^* = \mathbf{C} - (-\infty, 0)$. Furthermore,

$$\frac{d}{dz}\mathrm{Log}\, z \;=\; \frac{1}{z} \quad \text{for} \quad z \quad \text{in} \quad D^*.$$

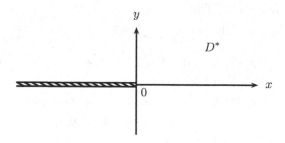

Figure 9.1

Proof.   Set $w = \mathrm{Log}\, z$. Let $z_0 \in D^*$ and $w_0 = \mathrm{Log}\, z_0$. We have to show that the

$$\lim_{z \to z_0} \frac{\mathrm{Log}\, z - \mathrm{Log}\, z_0}{z - z_0}$$

exists and is equal to $1/z_0$. Since $\mathrm{Log}\, z$ is continuous, $w = \mathrm{Log}\, z \to w_0 =$

Log z_0 as $z \to z_0$. Thus,

$$\lim_{z \to z_0} \frac{\text{Log}\, z - \text{Log}\, z_0}{z - z_0} = \lim_{w \to w_0} \frac{w - w_0}{e^w - e^{w_0}} = \lim_{w \to w_0} \frac{1}{\dfrac{e^w - e^{w_0}}{w - w_0}} \quad (9.3)$$

$$= \frac{1}{e^{w_0}} = \frac{1}{e^{\text{Log}\, z_0}} = \frac{1}{z_0}.$$

Note that (9.3) is meaningful since, for $z \neq z_0$, w will not coincide with w_0. This follows from the fact that $z = e^{\text{Log}\, z} = e^w$. Thus, $w = \text{Log}\, z$ is differentiable at every point in D^*, and hence is analytic there. ∎

A line used to create a domain of analyticity is called a *branch line* or *branch cut*. Any point that must lie on a branch cut-no matter what branch is used-is called a *branch point* of a multi-valued function. For example, the nonpositive real axis shown in Figure 9.1 is a *branch cut* for Log z, and the point $z = 0$ is a *branch point*.

If α is a complex constant and $z \neq 0$, then we define z^α by $z^\alpha = e^{\alpha \log z}$. Powers of z are, in general, multi-valued.

Example 9.3. Find all values of 1^i. Since $\log 1 = \text{Log}\, 1 + 2k\pi i = 2k\pi i$, we have $1^i = e^{i \log 1} = e^{-2k\pi}$, where $k = 0, \pm 1, \pm 2, \cdots$.

Example 9.4. Find all values of $(-2)^i$. Since $\log(-2) = \text{Log}\, 2 + (\pi + 2k\pi)i$, we have $(-2)^i = e^{i \log(-2)} = e^{i\text{Log}\, 2} e^{-(\pi + 2k\pi)}$, where $k = 0, \pm 1, \pm 2, \cdots$. Thus, $(-2)^i$ has infinitely many values.

Example 9.5. Find all values of i^{-2i}. Since $\log i = \text{Log}\, 1 + \left(2k + \frac{1}{2}\right)\pi i = \left(2k + \frac{1}{2}\right)\pi i$, we have $i^{-2i} = e^{-2i \log i} = e^{-2i(2k+1/2)\pi i} = e^{(4k+1)\pi}$, where $k = 0, \pm 1, \pm 2, \cdots$.

Since $\log z = \text{Log}\, |z| + i\text{Arg}\, z + 2k\pi i$, we can write

$$z^\alpha = e^{\alpha(\text{Log}\,|z| + i\text{Arg}\, z + 2k\pi i)} = e^{\alpha(\text{Log}\,|z| + i\text{Arg}\, z)} e^{\alpha 2k\pi i}, \quad (9.4)$$

where $k = 0, \pm 1, \pm 2, \cdots$. The value of z^α obtained by taking $k = k_1$ and $k = k_2 (\neq k_1)$ in equation (9.4) will be the same when $e^{\alpha 2k_1 \pi i} = e^{\alpha 2k_2 \pi i}$. This occurs if and only if $\alpha 2k_1 \pi i = \alpha 2k_2 \pi i + 2m\pi i$ (m is an integer); i.e., $\alpha = m/(k_1 - k_2)$. Hence, formula (9.4) yields some identical values of z^α only when α is a real rational number. Consequently, if α is not a real rational number, we obtain infinitely many different values of z^α, one for each choice of the integer k in (9.4).

On the other hand, if $\alpha = m/n$, where m and $n > 0$ are integers having no common factor, then there are exactly n distinct values (branches) of $z^{m/n}$, namely

$$z^{m/n} = e^{(m/n)\text{Log}\,|z|} e^{(m(\text{Arg}\, z + 2k\pi)i)/n}, \quad k = 0, 1, \cdots, n - 1.$$

In summary, we find:

1. z^α is single-valued when α is a real integer.
2. z^α takes finitely many values when α is a real rational number.
3. z^α takes infinitely many values in all other cases.

If we use the principal value of $\log z$, we obtain the *principal branch* of z^α, namely $e^{\alpha \mathrm{Log}\, z}$.

Example 9.6. The principal value of $(-i)^i$ is $e^{i\mathrm{Log}\,(-i)} = e^{i(-\pi i/2)} = e^{\pi/2}$.

Since e^z is entire and $\mathrm{Log}\, z$ is analytic in the domain $D^* = \mathbf{C}\backslash(-\infty, 0]$, the chain rule implies that the principal branch of z^α is also analytic in D^*. Furthermore, for z in D^*, we have

$$\frac{d}{dz}\left(e^{\alpha \mathrm{Log}\, z}\right) \;=\; e^{\alpha \mathrm{Log}\, z}\frac{d}{dz}(\alpha \mathrm{Log}\, z) \;=\; e^{\alpha \mathrm{Log}\, z}\left(\frac{\alpha}{z}\right).$$

Now we shall use logarithms to describe the inverse of the trigonometric and hyperbolic functions. For this, we recall that if f is a one-to-one complex function with domain S and range S', then the *inverse function* of f, denoted as f^{-1}, is the function with domain S' and range S defined by $f^{-1}(w) = z$ if $f(z) = w$. It is clear that if the function f is bijective, then f^{-1} maps S' onto S. Furthermore, both the compositions $f \circ f^{-1}$ and $f^{-1} \circ f$ are the identity function. For example, the inverse of the function $f(z) = az + b$, $a \neq 0$, is $f^{-1}(z) = (z - b)/a$.

The inverse sine function $w = \sin^{-1} z$ is defined by the equation $z = \sin w$. We shall show that $\sin^{-1} z$ is a multi-valued function given by

$$\sin^{-1} z \;=\; -i \log[iz + (1 - z^2)^{1/2}]. \tag{9.5}$$

From the equation

$$z \;=\; \sin w \;=\; \frac{e^{iw} - e^{-iw}}{2i},$$

we have $2iz = e^{iw} - e^{-iw}$. Multiplying both sides by e^{-iw}, we deduce that

$$e^{2iw} - 2ize^{iw} - 1 \;=\; 0,$$

which is quadratic in e^{iw}. Solving for e^{iw}, we find

$$e^{iw} \;=\; iz + (1 - z^2)^{1/2}, \tag{9.6}$$

where $(1 - z^2)^{1/2}$ is a double-valued function of z. Taking logarithms of each side of (9.6) and recalling that $w = \sin^{-1} z$, we arrive at the representation (9.5).

Example 9.7. From (9.5), we have

$$\sin^{-1}(-i) = -i\log(1 \pm \sqrt{2}).$$

However, since

$$\log(1 + \sqrt{2}) = \text{Log}\,(1 + \sqrt{2}) + 2k\pi i, \quad k = 0, \pm 1, \pm 2, \cdots, \qquad (9.7)$$

$$\log(1 - \sqrt{2}) = \text{Log}\,(\sqrt{2} - 1) + (2k + 1)\pi i, \quad k = 0, \pm 1, \pm 2, \cdots, \qquad (9.8)$$

and

$$\text{Log}(\sqrt{2} - 1) = \text{Log}\frac{1}{1 + \sqrt{2}} = -\text{Log}\,(1 + \sqrt{2}),$$

the lists (9.7) and (9.8) can be combined in one list as

$$(-1)^k \text{Log}\,(1 + \sqrt{2}) + k\pi i, \quad k = 0, \pm 1, \pm 2, \cdots.$$

Thus, it follows that

$$\sin^{-1}(-i) = k\pi + i(-1)^{k+1}\text{Log}(1 + \sqrt{2}), \quad k = 0, \pm 1, \pm 2, \cdots.$$

Similar to the expression for $\sin^{-1} z$ in (9.5), it is easy to show that

$$\cos^{-1} z = -i\log[z + i(1 - z^2)^{1/2}]$$

and

$$\tan^{-1} z = \frac{i}{2}\log\frac{i + z}{i - z}. \qquad (9.9)$$

Clearly, the functions $\cos^{-1} z$ and $\tan^{-1} z$ are also multi-valued.

The derivatives of these three functions are readily obtained from the representations above and appear as

$$\frac{d}{dz}\sin^{-1} z = \frac{1}{(1 - z^2)^{1/2}}, \quad \frac{d}{dz}\cos^{-1} z = -\frac{1}{(1 - z^2)^{1/2}}, \quad \frac{d}{dz}\tan^{-1} z = \frac{1}{1 + z^2}.$$

Finally, we note that the inverse hyperbolic functions can be treated in a corresponding manner. It turns out that

$$\sinh^{-1} z = \log[z + (z^2 + 1)^{1/2}],$$
$$\cosh^{-1} z = \log[z + (z^2 - 1)^{1/2}],$$
$$\tanh^{-1} z = \frac{1}{2}\log\frac{1 + z}{1 - z},$$

and

$$\frac{d}{dz}\sinh^{-1} z = \frac{1}{(z^2 + 1)^{1/2}},$$
$$\frac{d}{dz}\cosh^{-1} z = \frac{1}{(z^2 - 1)^{1/2}},$$
$$\frac{d}{dz}\tanh^{-1} z = \frac{1}{1 - z^2}.$$

Problems

9.1. Find all values of z such that

(a). $e^z = -2$,   (b). $e^z = 1 + \sqrt{3}i$,   (c). $\exp(2z - 1) = 1$,   (d). $\sin z = 2$.

9.2. If z_1, $z_2 \neq (\pi/2) + k\pi$, show that $\tan z_1 = \tan z_2$ if and only if $z_1 = z_2 + k\pi$, where k is an integer.

9.3. Use (7.7) and (7.8) to prove Theorem 9.1.

9.4. Evaluate the following:

(a). $\text{Log}(-ei)$,   (b). $\text{Log}(1 - i)$,   (c). $\log(-1 + \sqrt{3}i)$.

9.5. Show that

(a). $\text{Log}(1 + i)^2 = 2\text{Log}(1 + i)$ but   (b). $\text{Log}(-1 + i)^2 \neq 2\text{Log}(-1 + i)$.

9.6. Find the limit $\lim_{y \to 0^+} [\text{Log}(a + iy) - \text{Log}(a - iy)]$ when $a > 0$, and when $a < 0$.

9.7. Evaluate the following and find their principal values:

(a). $(1 + i)^i$,   (b). $(-1)^\pi$,   (c). $(1 - i)^{4i}$,   (d). $(-1 + i\sqrt{3})^{3/2}$.

9.8. Establish (9.9).

9.9. Evaluate the following:

(a). $\sin^{-1} \sqrt{5}$,   (b). $\sinh^{-1} i$.

9.10. Prove the following inequalities:

$$\frac{e^{-2y}}{1 + e^{-2y}} \quad < \quad |\tan(x + iy) - i| \quad < \quad \frac{e^{-2y}}{1 - e^{-2y}}, \quad y > 0$$

$$\frac{e^{-2y}}{1 + e^{-2y}} \quad < \quad |\cot(x + iy) + i| \quad < \quad \frac{e^{-2y}}{1 - e^{-2y}}, \quad y > 0$$

$$\frac{e^{2y}}{1 + e^{2y}} \quad < \quad |\tan(x + iy) + i| \quad < \quad \frac{e^{2y}}{1 - e^{2y}}, \quad y < 0$$

$$\frac{e^{2y}}{1 + e^{2y}} \quad < \quad |\cot(x + iy) - i| \quad < \quad \frac{e^{2y}}{1 - e^{2y}}, \quad y > 0$$

Answers or Hints

9.1. (a). $e^z = -2$ iff $e^z = e^{\ln(2)}e^{i\pi}$ iff $z = \ln 2 + i\pi + i(2k\pi)$, $k \in \mathbf{Z}$, (b). $e^z = 1 + \sqrt{3}i = 2e^{i\pi/3}$ iff $z = \ln 2 + i\frac{\pi}{3} + i(2k\pi)$, $k \in \mathbf{Z}$, (c). $e^{2z-1} = e^{i0}$ iff

$2z - 1 = i(2k\pi)$ iff $z = \frac{1}{2} + i(k\pi)$, $k \in \mathbf{Z}$, (d). $\sin z = 2$ iff $e^{iz} - e^{-iz} = 4i$ iff $e^{2iz} - 4ie^{iz} - 1 = 0$ iff $e^{iz} = (4i \pm \sqrt{-16+4})/2 = (4i \pm \sqrt{12}i)/2 = (2 \pm \sqrt{3})i = (2 + \sqrt{3})e^{i\pi/2}$, $(2 - \sqrt{3})e^{i\pi/2}$ so $z = \ln(2 + \sqrt{3}) + i\left(\frac{\pi}{2} + 2k\pi\right)$, or $\ln(2 - \sqrt{3}) + i\left(\frac{\pi}{2} + 2k\pi\right)$, $k \in \mathbf{Z}$.

9.2. $\tan z_1 - \tan z_2 = 0$ if and only if $\sin(z_1 - z_2) = 0$.

9.3. If $\text{Log}\, z = \text{Log}\, |z| + i\text{Arg}\, z = \text{Log}\, r + i\Theta$, then $u = \text{Log}\, r$, $v = \Theta$. Thus, $u_r = 1/r$, $u_\Theta = 0$, $v_r = 0$, $v_\Theta = 1$.

9.4. (a). $\text{Log}(-ei) = \ln e + i(-\pi/2) = 1 - i\pi/2$, (b). $\text{Log}(1 - i) = \ln \sqrt{2} - i\pi/4$, (c). $\log(-1 + \sqrt{3}i) = \ln 2 + i\frac{2\pi}{3} + 2k\pi i$, $k \in \mathbf{Z}$.

9.5. (a). $\text{Log}(1 + i)^2 = \text{Log}(2i) = \ln 2 + i\frac{\pi}{2} = 2\left(\ln\sqrt{2} + i\frac{\pi}{4}\right) = 2\text{Log}(1 + i)$, (b). $\text{Log}(-1 + i)^2 = \text{Log}(-2i) = \ln 2 - i\pi/2$ and $2\text{Log}(-1 + i) = 2\left(\ln\sqrt{2} + i\frac{3\pi}{4}\right) = \ln 2 + i\frac{3\pi}{2}$.

9.6. 0 when $a > 0$, and $2\pi i$ when $a < 0$.

9.7. (a). $(1 + i)^i = e^{i\log(1+i)} = e^{i(\ln\sqrt{2} + i(\pi/4 + 2k\pi))} = e^{-(\pi/4 + 2k\pi)}e^{i\ln\sqrt{2}}$, $k \in \mathbf{Z}$, (b). $(-1)^\pi = e^{\pi\log(-1)} = e^{\pi(\ln 1 + i(\pi + 2k\pi))} = e^{i\pi^2(1+2k)}$, $k \in \mathbf{Z}$, (c). $e^\pi e^{i(2\ln 2)}$, (d). $e^{3/2(\ln 2 + i2\pi/3)} = 2\sqrt{2}e^{i\pi} = -2\sqrt{2}$.

9.8. $1 + i\tan w = 2e^{iw}/(e^{iw} + e^{-iw})$, $1 - i\tan w = 2e^{-iw}/(e^{iw} + e^{-iw})$.

9.9. (a). $(4k + 1)\pi/2 \pm i\text{Log}(\sqrt{5} + 2)$, (b). $(4k + 1)\pi i/2$.

9.10. Follow the same method as in Example 8.2.

Lecture 10
Mappings by Functions I

In this lecture, we shall present a graphical representation of some elementary functions. For this, we will need two complex planes representing, respectively, the domain and the image of the function.

Consider z- and w-planes with the points as usual denoted as $z = x + iy$ and $w = u + iv$. We shall visualize the function $w = f(z)$ as a mapping (transformation) from a subset of the z-plane (domain of f) to the w-plane (range of f).

The mapping

$$w = Az \tag{10.1}$$

is known as *dilation*. Here, A is a nonzero complex constant and $z \neq 0$. We write A and z in exponential form; i.e., $A = ae^{i\alpha}$, $z = re^{i\theta}$. Then,

$$w = (ar)e^{i(\alpha+\theta)}. \tag{10.2}$$

From (10.2), it follows that the transformation (10.1) expands or contracts the radius vector representing z by the factor $a = |A|$ and rotates it through an angle $\alpha = \arg A$ about the origin. The image of a given region is therefore geometrically similar to that region. Thus, in particular, a dilation maps a straight line onto a straight line and a circle onto a circle.

The mapping

$$w = z + B \tag{10.3}$$

is known as *translation*; here, B is any complex constant. It is a translation, as can be seen by means of the vector representation of B; i.e., if $w = u + iv$, $z = x + iy$, and $B = b_1 + ib_2$, then the image of any point (x, y) in the z-plane is the point $(u, v) = (x + b_1, y + b_2)$ in the w-plane. Since each point in any given region of the z-plane is mapped into the w-plane in this manner, the image region is geometrically congruent to the original one. Thus, in particular, a translation also maps a straight line onto a straight line and a circle onto a circle.

The general *linear mapping*

$$w = Az + B, \quad A \neq 0, \tag{10.4}$$

is an expansion or contraction and a rotation, followed by a translation.

Example 10.1. The mapping $w = (1+i)z+2$ transforms the rectangular region in Figure 10.1 into the rectangular region shown in the w-plane. This is clear by writing it as a composition of the transformations

$$Z - (1+i)z \quad \text{and} \quad w = Z + 2.$$

Since $1+i = \sqrt{2}\exp(i\pi/4)$, the first of these transformations is an expansion by the factor $\sqrt{2}$ and a rotation through the angle $\pi/4$. The second is a translation two units to the right.

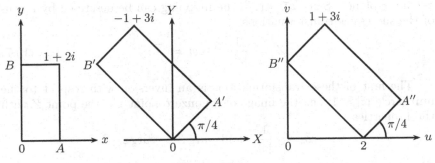

Figure 10.1

The mapping

$$w = z^n, \quad n \in \mathbb{N},\tag{10.5}$$

in polar coordinates can be written as

$$\rho e^{i\phi} = r^n e^{in\theta}.$$

Thus, it maps the annular region $r \geq 0$, $0 \leq \theta \leq \pi/n$, of the z-plane onto the upper half $\rho \geq 0$, $0 \leq \phi \leq \pi$, of the w-plane. Clearly, this mapping is one-to-one.

Example 10.2. Let S be the sector $S = \{z : |z| \leq 2, \ 0 \leq \arg z \leq \pi/6\}$. Find the image of S under the mapping $w = f(z) = z^3$. Clearly, we have

$$f(S) = \{w : |w| \leq 8, \ 0 \leq \arg w \leq \pi/2\}.$$

Example 10.3. Let S be the vertical strip $S = \{z = x+iy : 2 \leq x \leq 3\}$. Find the image of S under the mapping $w = f(z) = z^2$. Since $w = x^2 - y^2 + 2ixy$, a point (x, y) of the z-plane maps into $(u, v) = (x^2 - y^2, 2xy)$ in the w-plane. Now, eliminating y from the equations $u = x^2 - y^2$ and $v = 2xy$, we get

$$u = x^2 - \frac{v^2}{4x^2}.$$

Thus, a vertical line in the z-plane; i.e., $x = x_0$ fixed, maps into a leftward-facing parabola with the vertex at $(x_0^2, 0)$ and v-intercepts at $(0, \pm 2x_0^2)$.

Hence, as $2 \le x \le 3$, the corresponding parabolas in the w-plane describe a parabolic region

$$f(S) = \left\{ w = u + iv : 4 - \frac{v^2}{16} \le u \le 9 - \frac{v^2}{36} \right\}.$$

The mapping

$$w = \frac{1}{z} \tag{10.6}$$

establishes a one-to-one correspondence between the nonzero points of the z- and w-planes. Since $z\overline{z} = |z|^2$, the mapping can be described by means of the successive transformations

$$Z = \frac{1}{|z|^2}z, \quad w = \overline{Z}. \tag{10.7}$$

The first of these transformations is an inversion with respect to the unit circle $|z| = 1$; i.e., the image of a nonzero point z is the point Z with the properties

$$|Z| = \frac{1}{|z|} \quad \text{and} \quad \arg Z = \arg z.$$

Thus, the points exterior to the circle $|z| = 1$ are mapped onto the nonzero points interior to it, and conversely. Any point on the circle is mapped onto itself. The second transformation in (10.7) is simply a reflection in the real axis.

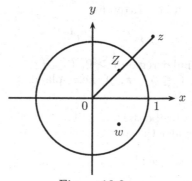

Figure 10.2

Since $\lim_{z \to 0} 1/z = \infty$ and $\lim_{z \to \infty} 1/z = 0$, we can define a one-to-one transformation $w = T(z)$ from the extended z-plane onto the extended w-plane by writing

$$T(z) = \begin{cases} \infty, & z = 0, \\ 0, & z = \infty, \\ \dfrac{1}{z}, & \text{otherwise.} \end{cases}$$

Clearly, T is then continuous throughout the extended z-plane.

When a point $w = u + iv$ is the image of a nonzero point $z = x + iy$ under the transformation $w = \dfrac{1}{z} = \dfrac{\bar{z}}{z^2}$, then

$$u = \frac{x}{x^2 + y^2}, \quad v = \frac{-y}{x^2 + y^2}. \tag{10.8}$$

Also, since $z = \dfrac{1}{w} = \dfrac{\bar{w}}{|w|^2}$,

$$x = \frac{u}{u^2 + v^2}, \quad y = \frac{-v}{u^2 + v^2}. \tag{10.9}$$

Let A, B, C, and D be real numbers such that

$$B^2 + C^2 > 4AD. \tag{10.10}$$

The equation

$$A(x^2 + y^2) + Bx + Cy + D = 0 \tag{10.11}$$

represents an arbitrary circle or line according to whether $A \neq 0$ or $A = 0$. Since (10.11) is the same as

$$\left(x + \frac{B}{2A}\right)^2 + \left(y + \frac{C}{2A}\right)^2 = \left(\frac{\sqrt{B^2 + C^2 - 4AD}}{2A}\right)^2,$$

condition (10.10) is clear when $A \neq 0$. Also, when $A = 0$, condition (10.10) reduces to $B^2 + C^2 > 0$, which means B and C are not both zero.

Now, if x and y satisfy equation (10.11), we can use (10.9) to obtain the equation

$$D(u^2 + v^2) + Bu - Cv + A = 0, \tag{10.12}$$

which also represents a circle or line. Conversely, if u and v satisfy equation (10.12), it follows from relations (10.8) that x and y satisfy equation (10.11).

From (10.11) and (10.12), it is clear that:

(1). A circle ($A \neq 0$) not passing through the origin ($D \neq 0$) in the z-plane is transformed into a circle not passing through the origin in the w-plane.

(2). A circle ($A \neq 0$) through the origin ($D = 0$) in the z-plane is transformed into a line that does not pass through the origin in the w-plane.

(3). A line ($A = 0$) not passing through the origin ($D \neq 0$) in the z-plane is transformed into a circle through the origin in the w-plane.

(4). A line ($A = 0$) through the origin $D = 0$ in the z-plane is transformed into a line through the origin in the w-plane.

Example 10.4. In view of (10.11) and (10.12), a vertical line $x = c_1$ $(c_1 \neq 0)$ is transformed by $w = 1/z$ into the circle $-c_1(u^2 + v^2) + u = 0$, or

$$\left(u - \frac{1}{2c_1}\right)^2 + v^2 = \left(\frac{1}{2c_1}\right)^2,$$

which is centered on the u-axis and tangent to the v-axis.

Example 10.5. The transformation $w = 1/z$ maps a horizontal line $y = c_2$ $(c_2 \neq 0)$ onto the circle

$$u^2 + \left(v + \frac{1}{2c_2}\right)^2 = \left(\frac{1}{2c_2}\right)^2,$$

which is centered on the v-axis and tangent to the u-axis.

Example 10.6. When $w = 1/z$, the half-plane $x \geq c_1$ $(c_1 > 0)$ is mapped onto the disk

$$\left(u - \frac{1}{2c_1}\right)^2 + v^2 \leq \left(\frac{1}{2c_1}\right)^2.$$

Lecture 11
Mappings by Functions II

In this lecture, we shall study graphical representations of the Möbius transformation, the trigonometric mapping $\sin z$, and the function $z^{1/2}$.

The transformation

$$w = \frac{az + b}{cz + d}, \quad ad - bc \neq 0, \tag{11.1}$$

where a, b, c, and d are complex numbers, is called a *linear fractional transformation* or *Möbius transformation*. Clearly, (11.1) can be written as

$$Azw + Bz + Cw + D = 0, \quad AD - BC \neq 0, \tag{11.2}$$

and, conversely, any equation of type (11.2) can be put in the form (11.1). Since (11.2) is linear in z and linear in w, or bilinear in z and w, another name for a linear fractional transformation is *bilinear transformation*.

When $c = 0$, the condition $ad - bc \neq 0$ reduces to $ad \neq 0$ and (11.1) becomes a nonconstant linear function. When $c \neq 0$, (11.1) can be written as

$$w = \frac{a}{c} + \frac{bc - ad}{c} \frac{1}{cz + d}, \quad ad - bc \neq 0. \tag{11.3}$$

Once again, the condition $bc - ad \neq 0$ ensures that we do not have a constant function.

Equation (11.3) reveals that, when $c \neq 0$, a linear transformation is a composition of the mappings

$$Z = cz + d, \quad W = \frac{1}{Z}, \quad w = \frac{a}{c} + \frac{bc - ad}{c} W, \quad ad - bc \neq 0.$$

Thus, a linear fractional transformation always transforms circles and lines into circles and lines because these special linear fractional transformations do this.

Solving (11.1) for z, we find

$$z = \frac{-dw + b}{cw - a}, \quad ad - bc \neq 0. \tag{11.4}$$

Thus, when a given point w is the image of some point z under transformation (11.1), the point z is retrieved by means of equation (11.4). If $c = 0$, so that a and d are both nonzero, each point in the w-plane is evidently the image of one and only one point in the z-plane. The same is true if $c \neq 0$, except when $w = a/c$, since the denominator in equation (11.4) vanishes if w has that value. We can, however, enlarge the domain of definition of (11.1) in order to define a linear fractional transformation T on the extended z-plane such that the point $w = a/c$ is the image of $z = \infty$ when $c \neq 0$. We first write

$$T(z) = \frac{az + b}{cz + d}, \quad ad - bc \neq 0. \tag{11.5}$$

We then write $T(\infty) = \infty$ if $c = 0$ and $T(\infty) = a/c$ and $T(-d/c) = \infty$ if $c \neq 0$.

It can be shown that T is continuous on the extended z-plane. It also agrees with the way in which we enlarged the domain of definition of the transformation $w = 1/z$.

When its domain of definition is enlarged in this way, the linear transformation (11.5) is a one-to-one mapping of the extended z-plane onto the extended w-plane. Hence, associated with the transformation T there is an inverse transformation T^{-1} that is defined on the extended w-plane as follows: $T^{-1}(w) = z$ if and only if $T(z) = w$. From (11.4), we have

$$T^{-1}(w) = \frac{-dw + b}{cw - a}, \quad ad - bc \neq 0. \tag{11.6}$$

Clearly, T^{-1} is itself a linear fractional transformation, where $T^{-1}(\infty) = \infty$ if $c = 0$ and $T^{-1}(a/c) = \infty$ and $T^{-1}(\infty) = -d/c$ if $c \neq 0$.

If T and T' are two linear fractional transformations given by

$$T(z) = \frac{a_1 z + b_1}{c_1 z + d_1} \quad \text{and} \quad T'(z) = \frac{a_2 z + b_2}{c_2 z + d_2},$$

where $a_1 d_1 - b_1 c_1 \neq 0$ and $a_2 d_2 - b_2 c_2 \neq 0$, then their composition,

$$T'[T(z)] = \frac{(a_1 a_2 + c_1 b_2)z + (a_2 b_1 + d_1 b_2)}{(a_1 c_2 + c_1 d_2)z + (c_2 b_1 + d_1 d_2)}, \quad (a_1 d_1 - b_1 c_1)(a_2 d_2 - b_2 c_2) \neq 0,$$

is also a linear fractional transformation. Note that, in particular, $T^{-1}[T(z)] = z$ for each point z in the extended plane.

There is always a linear fractional transformation that maps three given distinct points z_1, z_2 and z_3 onto three specified distinct points w_1, w_2, and w_3, respectively. In fact, it can be written as

$$\frac{(w - w_1)(w_2 - w_3)}{(w - w_3)(w_2 - w_1)} = \frac{(z - z_1)(z_2 - z_3)}{(z - z_3)(z_2 - z_1)}. \tag{11.7}$$

To verify this, we write (11.7) as

$$(z - z_3)(z_2 - z_1)(w - w_1)(w_2 - w_3) = (z - z_1)(z_2 - z_3)(w - w_3)(w_2 - w_1).$$
(11.8)

If $z = z_1$, the right-hand side of (11.8) is zero and it follows that $w = w_1$. Similarly, if $z = z_3$, the left-hand side of (11.8) is zero and we get $w = w_3$. If $z = z_2$, we have the equation

$$(w - w_1)(w_2 - w_3) = (w - w_3)(w_2 - w_1),$$

whose unique solution is $w = w_2$. One can see that the mapping defined by equation (11.7) is actually a linear fractional transformation by expanding the products in (11.8) and writing the result in the form

$$Azw + Bz + Cw + D = 0. \tag{11.9}$$

The condition $AD - BC \neq 0$, which is needed with (11.9), is clearly satisfied because (11.7) does not define a constant function. We also note that (11.7) defines the only linear fractional transformation mapping the points z_1, z_2, and z_3 onto w_1, w_2, and w_3, respectively.

Example 11.1. Find the bilinear transformation that maps the points $z_1 = -1$, $z_2 = 0$, and $z_3 = 1$ onto the points $w_1 = -i$, $w_2 = 1$, and $w_3 = i$. Using equation (11.7), we have

$$\frac{(w + i)(1 - i)}{(w - i)(1 + i)} = \frac{(z + 1)(0 - 1)}{(z - 1)(0 + 1)},$$

which on solving for w in terms of z gives the transformation

$$w = \frac{i - z}{i + z}.$$

If (11.7) is modified properly, it can also be used when the point at infinity is one of the prescribed points in either the (extended) z- or w-plane. Suppose, for example, that $z_1 = \infty$. Since any linear fractional transformation is continuous on the extended plane, we need only replace z_1 on the right-hand side of (11.7) by $1/z_1$, clear fractions, and let z_1 tend to zero

$$\lim_{z_1 \to 0} \frac{(z - 1/z_1)(z_2 - z_3)\, z_1}{(z - z_3)(z_2 - 1/z_1)\, z_1} = \lim_{z_1 \to 0} \frac{(z_1 z - 1)(z_2 - z_3)}{(z - z_3)(z_1 z_2 - 1)} = \frac{z_2 - z_3}{z - z_3}.$$

Thus, the desired modification of (11.7) is

$$\frac{(w - w_1)(w_2 - w_3)}{(w - w_3)(w_2 - w_1)} = \frac{z_2 - z_3}{z - z_3}. \tag{11.10}$$

Note that this modification is obtained by simply deleting the factors involving z_1 in (11.7). Furthermore, the same formal approach applies when any of the other prescribed points is ∞.

Example 11.2. Find the bilinear transformation that maps the points $z_1 = 1$, $z_2 = 0$, and $z_3 = -1$ onto the points $w_1 = i$, $w_2 = \infty$, and $w_3 = 1$. In this case, we use the modification

$$\frac{w - w_1}{w - w_3} = \frac{(z - z_1)(z_2 - z_3)}{(z - z_3)(z_2 - z_1)}$$

to obtain

$$\frac{w - i}{w - 1} = \frac{(z - 1)(0 + 1)}{(z + 1)(0 - 1)},$$

which gives

$$w = \frac{(i + 1)z + (i - 1)}{2z}.$$

Now let S be the semi-infinite strip $S = \{z = x + iy : -\pi/2 \leq x \leq \pi/2,\ y \geq 0\}$. We shall find the image of S under the mapping $w = f(z) = \sin z$. For this, since

$$w = u + iv = \sin x \cosh y + i \cos x \sinh y,$$

we have

$$u = \sin x \cosh y \quad \text{and} \quad v = \cos x \sinh y. \tag{11.11}$$

Thus, if $y = 0$, then $v = 0$ and $u = \sin x$, and hence $w = \sin z$ maps the interval $-\pi/2 \leq x \leq \pi/2$ into the interval $-1 \leq u \leq 1$. If $x = \pi/2$, then $u = \cosh y$ and $v = 0$, and hence $w = \sin z$ maps the positive-vertical line $x = \pi/2$ onto the part $u \geq 1$ of the real axis of the w-plane. Similarly, if $x = -\pi/2$, then $w = \sin z$ maps the positive-vertical line $x = -\pi/2$ onto the part $u \leq -1$ of the real axis of the w-plane. Hence, under the mapping $w = \sin z$, the boundary of S is mapped to the entire u-axis. If $y = y_0 > 0$ and $-\pi/2 \leq x \leq \pi/2$, equations (11.11) imply that $v \geq 0$ and

$$\frac{u^2}{\cosh^2 y_0} + \frac{v^2}{\sinh^2 y_0} = 1, \tag{11.12}$$

and hence $w = \sin z$ maps the interval $-\pi/2 \leq x \leq \pi/2$ into the upper semi-ellipse. The u-intercepts of this ellipse are at $\pm \cosh y_0$, and the v-intercept is at $\sinh y_0$. Since $\lim_{y_0 \to 0} \sinh y_0 = 0$, $\lim_{y_0 \to 0} \cosh y_0 = 1$, $\lim_{y_0 \to \infty} \sinh y_0 = \infty$, and $\lim_{y_0 \to \infty} \cosh y_0 = \infty$, as y_0 varies in the interval $0 < y_0 < \infty$, the upper semi-ellipses fill the upper half w-plane. Thus, the image of S under the mapping $w = \sin z$ is the upper half w-plane including the u-axis.

From the considerations above it is clear that the image of the infinite strip $\{z = x + iy : -\pi/2 \leq x \leq \pi/2, \ -\infty < y < \infty\}$ under the mapping $w = \sin z$ is the entire w-plane.

Finally, we shall consider the mapping $w = f(z) = z^{1/2}$. For this, we recall that the mapping $g(z) = z^2$ is not one-to-one; however, if we restrict the domain of g to $S = \{z : -\pi/2 < \arg z \leq \pi/2\}$, then it is one-to-one. To see this, let $z_1, z_2 \in S$. If $z_1^2 = z_2^2$; i.e., $z_1^2 - z_2^2 = (z_1 - z_2)(z_1 + z_2) = 0$, then either $z_1 = z_2$ or $z_1 = -z_2$. Since for $z = 0$ the $\arg z$ is not defined, both z_1 and z_2 cannot be zero. Furthermore, since the points z and $-z$ are symmetric about the origin, if $z_2 \in S$, then $-z_2 \notin S$. Hence, $z_1 \neq -z_2$, and we must have $z_1 = z_2$. Thus, the mapping $g(z) = z^2$ is one-to-one on S to $\mathbf{C} - \{0\}$, and therefore the inverse function $g^{-1}(z) = f(z) = z^{1/2}$ exists. The domain of g^{-1} is thus $\mathbf{C} - \{0\}$ and the range is the domain of g; i.e., S. In conclusion, if $z = re^{i\theta}$, then the principal branch f_1 of the function $w = f(z) = z^{1/2}$ is

$$f_1(z) = \sqrt{r}e^{i\theta/2}, \quad r > 0, \quad -\pi < \theta \leq \pi,$$

and it takes the square root of the modulus of a point and halves the principal argument. In particular, under this mapping the image of the sector $\{z : |z| \leq 4, \ \pi/3 \leq \arg z \leq \pi/2\}$ is the sector $\{z : |z| \leq 2, \ \pi/6 \leq \arg z \leq \pi/4\}$.

From the arguments above it is clear that the principal branch f_1 of the function $w = f(z) = z^{1/n}$ is

$$f_1(z) = r^{1/n}e^{i\theta/n}, \quad r > 0, \quad -\pi < \theta \leq \pi,$$

and it takes the nth root of the modulus of a point and divides the principal argument by n.

Problems

11.1. Find images of the following sets under the mapping $w = e^z$:

(a). the vertical line segment $x = a$, $-\pi < y \leq \pi$,

(b). the horizontal line $-\infty < x < \infty$, $y = b$,

(c). the rectangular area $-1 \leq x \leq 1$, $0 \leq y \leq \pi$,

(d). the region $-\infty < x < \infty$, $-\pi < y \leq \pi$.

11.2. Find images of the following sets under the mapping $w = 1/z$:

(a). $\{z : 0 < |z| < 1, \ 0 \leq \arg z \leq \pi/2\}$,

(b). $\{z : 3 \leq |z|, \ 0 \leq \arg z \leq \pi\}$.

11.3. Find images of the following sets under the mapping $w = \operatorname{Log} z$:

(a). $\{z : |z| > 0\}$,   (b). $\{z : |z| = r\}$,

(c). $\{z : \arg z = \theta\}$,   (d). $\{z : 2 \le |z| \le 5\}$.

11.4. Show that as z moves on the real axis from -1 to $+1$ the point $w = \dfrac{1 - iz}{z - i}$ moves on part of the unit circle with center $(0, 0)$ and radius 1.

11.5. Show that when $\dfrac{z - i}{z - 1}$ is purely imaginary, the locus of z is a circle with center at $(1/2, 1/2)$ and radius $1/\sqrt{2}$, but when it is purely real the locus is a straight line.

11.6. A point $\alpha \in \mathbf{C}$ is called a *fixed point* of the mapping f provided $f(\alpha) = \alpha$.

(a). Show that except the unit mapping $w = z$, (11.1) can have at most two fixed points.

(b). If (11.1) has two distinct fixed points, α and β, then

$$\frac{w - \alpha}{w - \beta} = \frac{z - \alpha}{z - \beta} \frac{a - c\alpha}{a - c\beta}.$$

(c). If (11.1) has only one fixed point, α, then

$$\frac{1}{w - \alpha} = \frac{1}{z - \alpha} + \frac{c}{a - c\alpha}.$$

(d). Find the fixed points of $\dfrac{z - 1}{z + 1}$ and $\dfrac{5z + 3}{2z - 1}$.

11.7. Find the entire linear transformation with fixed point $1 + 2i$ that maps the point i into the point $-i$.

11.8. Show that a necessary and sufficient condition for two Möbius transformations

$$T_1(z) = \frac{a_1 z + b_1}{c_1 z + d_1} \quad \text{and} \quad T_2(z) = \frac{a_2 z + b_2}{c_2 z + d_2}$$

to be identical is that $a_2 = \lambda a_1$, $b_2 = \lambda b_1$, $c_2 = \lambda c_1$, $d_2 = \lambda d_1$, $\lambda \ne 0$.

11.9. For any four complex numbers z_1, z_2, z_3, z_4 in the extended plane, the *cross ratio* is denoted and defined as

$$(z_1, z_2, z_3, z_4) = \frac{z_4 - z_1}{z_4 - z_3} : \frac{z_2 - z_1}{z_2 - z_3} = \frac{(z_4 - z_1)(z_2 - z_3)}{(z_4 - z_3)(z_2 - z_1)}.$$

Show that

(a). the cross ratio is invariant under any linear fractional transformation T; i.e.,

$$(T(z_1), T(z_2), T(z_3), T(z_4)) = (z_1, z_2, z_3, z_4),$$

(b). the complex numbers z_1, z_2, z_3, z_4 lie on a line or a circle in the complex plane if and only if their cross ratio is a real number.

11.10. Show that the image of the vertical line $x = x_0$, where $-\pi/2 < x_0 < \pi/2$, under the mapping $w = \sin z$ is the right-half of the hyperbola

$$\frac{u^2}{\sin^2 x_0} - \frac{v^2}{\cos^2 x_0} = 1$$

if $x_0 > 0$ and the left-half if $x_0 < 0$.

11.11. Find the image of the rectangle $S = \{z = x + iy : -\pi/2 \le x \le \pi/2, \ 0 < c \le y \le d\}$ under the mapping $w = \sin z$.

11.12. Find the image of the semi-infinite strip $S = \{z = x + iy : 0 \le x \le \pi/2, \ y \ge 0\}$ under the mapping $w = \cos z$.

Answers or Hints

11.1. (a). circle $|w| = e^a$, (b). ray $\arg w = b$, (c). semi-annular area between semi-circles of radii e^{-1} and e with center at 0, (d). $\mathbf{C} - \{0\}$.

11.2. (a). $\{w : 1 < |w|, \ -\pi/2 \le \arg w \le 0\}$, (b). $\{w : 0 < |w| < 1/3, \ -\pi \le \arg w \le 0\}$.

11.3. (a). $\{w : -\infty < u < \infty, \ -\pi < v \le \pi\}$, (b). $\{w : u = \operatorname{Log} r, \ -\pi < v \le \pi\}$, (c). $\{w : -\infty < u < \infty, \ v = \theta\}$, (d). $\{w : \operatorname{Log} 2 \le u \le \operatorname{Log} 5, \ -\pi < v \le \pi\}$.

11.4. $w = \frac{2x}{x^2+(y-1)^2} + i\frac{1-x^2-y^2}{x^2+(y-1)^2} = u + iv$. On the real axis, $y = 0$, $-1 \le x \le 1$, $u = \frac{2x}{x^2+1}$, $v = \frac{1-x^2}{1+x^2}$, and hence $u^2 + v^2 = \frac{4x^2+(1+x^4-2x^2)}{(1+x^2)^2} = 1$. If $x = -1$, then $u = -1$, $v = 0$; if $x = 0$, then $u = 0$, $v = 1$; if $x = 1$ then $u = 1$, $v = 0$. Hence, w moves on the upper half of the unit circle.

11.5. $w = \frac{x(x-1)+y(y-1)+i(1-x-y)}{(x-1)^2+y^2}$, and hence w is imaginary provided $x^2 - x + y^2 - y = 0$; i.e., $(x - 1/2)^2 + (y - 1/2)^2 = 1/2$ and w is real if $x + y = 1$.

11.6. (a). Find all possible solutions of $z = (az + b)/(cz + d)$. (b). Verify directly. (c). Verify directly. (d). $\pm i$ and $(3 \pm \sqrt{15})/2$.

11.7. $w = (2 + i)z + 1 - 3i$.

11.8. The sufficiency is obvious. If $T_1(z) = T_2(z)$, then in particular $T_1(0) = T_2(0)$, $T_1(1) = T_2(1)$, $T_1(\infty) = T_2(\infty)$, which give

$$\frac{b_1}{d_1} = \frac{b_2}{d_2} = \mu, \quad \frac{a_1+b_1}{c_1+d_1} = \frac{a_2+b_2}{c_2+d_2}, \quad \frac{a_1}{c_1} = \frac{a_2}{c_2} = \nu.$$

Substituting $b_1 = \mu d_1$, $b_2 = \mu d_2$, $a_1 = \nu c_1$, and $a_2 = \nu c_2$ in the second relation, we find $(c_1 d_2 - c_2 d_1)(\nu - \mu) = 0$. But $\nu \neq \mu$, since otherwise $a_1/c_1 = b_1/d_1$; i.e., $a_1 d_1 - b_1 c_1 = 0$. Thus, $c_1/d_1 = c_2/d_2$.

11.9. (a). Define T' by $(T(z_1), T(z_2), T(z_3), w) = (z_1, z_2, z_3, z)$, that maps z_j to $T(z_j)$, $j = 1, 2, 3$. But $T(z)$ itself also maps z_j to $T(z_j)$. By uniqueness, $w = T'(z) = T(z)$. Therefore, this equality holds for $w = T(z)$ also. In particular, it holds for $z = z_4$ and $w = T(z_4)$. (b). Suppose that z_1, z_2, z_3, z_4 lie on a line or circle γ. We can find a linear fractional transformation $w = f(z)$ that maps γ onto the real axis. Then, each $w_j = f(z_j)$ will be real. Consequently, the cross ratio (w_1, w_2, w_3, w_4) will be real. But then, from (11.7), (z_1, z_2, z_3, z_4) must also be real. Conversely, suppose that (z_1, z_2, z_3, z_4) is real. We need to show that the points z_1, z_2, z_3, z_4 lie on some line or circle. Clearly, any three points lie on some line or circle (a line if the points are collinear, a circle otherwise). Hence, there is a unique line or circle γ_1 passing through the points z_1, z_2, z_3. Therefore, it suffices to show that z_4 also lies on γ_1. Let $w = g(z)$ be a linear fractional transformation that maps γ_1 to the real axis, and let $w_j = g(z_j)$. Since z_1, z_2, z_3 lie on γ_1, w_1, w_2, w_3 will be on the real axis. Now, since $(w_1, w_2, w_3, w_4) = (z_1, z_2, z_3, z_4)$ is real, we can solve it for w_4 in terms of (z_1, z_2, z_3, z_4) and w_1, w_2, w_3, and hence w_4 must also be real. Finally, since $w = g(z)$ is one-to-one, the only values in the z-plane that can map to the real axis in the w-plane are on γ_1. Hence, $z_4 = g^{-1}(w_4)$ is on γ_1.

11.10. Use (11.11).

11.11. The region between upper semi-ellipses given by (11.12) with $y_0 = c$ and $y_0 = d$.

11.12. Since $u = \cos x \cosh y$, $v = -\sin x \sinh y$, the image region is $\{w = u + iv : u \geq 0, v \leq 0\}$.

Lecture 12
Curves, Contours, and Simply Connected Domains

In this lecture, we define a few terms that will be used repeatedly in complex integration. We shall also state Jordan's Curve Theorem, which seems to be quite obvious; however, its proof is rather complicated.

Let $x(t)$ and $y(t)$ be continuous real-valued functions defined on $[a, b]$. A *curve* or *path* γ in the complex plane is the range of the continuous function $z : [a, b] \to \mathbf{C}$ given by $z(t) = x(t) + iy(t)$, $t \in [a, b]$. The curve γ begins with its initial point $z(a) = x(a) + iy(a)$ and goes all the way to its terminal point $z(b) = x(b) + iy(b)$. If we write the function $z(t)$ in its parametric form; i.e., $z(t) = (x(t), y(t))$, then the curve γ is the set of points $\{z(t) = (x(t), y(t)) : t \in [a, b]\}$. This set is called the *track* of γ, and is denoted as $\{\gamma\}$.

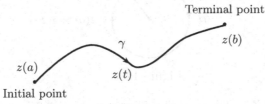

Figure 12.1

The curve γ is said to be *simple* if for all different $t_1, t_2 \in [a, b]$, $z(t_1) \neq z(t_2)$; i.e., γ does not cross itself.

Simple   Not simple

Figure 12.2

The curve γ is said to be *closed* if $z(a) = z(b)$. The interior of a closed curve γ is denoted as $I(\gamma)$. The curve γ is called a *simple closed curve* (or *Jordan curve*) if it is closed and $a < t_1 < t_2 < b$ implies that $z(t_1) \neq z(t_2)$; i.e., γ does not cross itself except at the end points.

Example 12.1. Let z_1, $z_2 \in \mathbf{C}$ be different points. The line segment γ_1, denoted as $[z_1, z_2]$ and given by

$$z(t) = z_1 + t(z_2 - z_1), \quad 0 \le t \le 1,$$

is a simple curve.

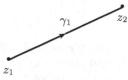

Figure 12.3

Example 12.2. The unit circle γ_2 given by

$$z(t) = e^{it} = \cos t + i \sin t, \quad 0 \le t \le 2\pi$$

is a simple closed curve.

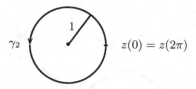

Figure 12.4

Example 12.3. Consider the functions $z_1(t) = e^{it}$, $t \in [0, 2\pi]$ and $z_2(t) = e^{2\pi it}$, $t \in [0, 1]$. Both curves trace the unit circle. Thus, different functions may represent the same curve.

A curve γ given by the range of $z : [a, b] \to \mathbf{C}$ is called *smooth* if

(i). $z'(t) = x'(t) + iy'(t)$ exists and is continuous on $[a, b]$,

(ii). $z'(t) \neq 0$ for all $t \in (a, b)$.

Example 12.4. Clearly, the curves γ_1 and γ_2 are smooth; however, the curve γ_3 given by

$$z(t) = \begin{cases} t + 2ti, & 0 \le t \le 1 \\ t + 2i, & 1 \le t \le 2 \end{cases}$$

is simple but not smooth.

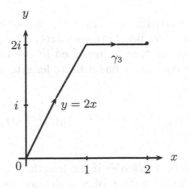

Figure 12.5

A curve γ given by the range of $z : [a, b] \to \mathbf{C}$ is said to be *piecewise continuous* if

(i). $z(t)$ exists and is continuous for all but finitely many points in (a, b),

(ii). at any point $c \in (a, b)$ where z fails to be continuous, both the left limit $\lim_{t \to c-} z(t)$ and the right limit $\lim_{t \to c+} z(t)$ exist and are finite, and

(iii). at the end points the right limit $\lim_{t \to a+} z(t)$ and the left limit $\lim_{t \to b-} z(t)$ exist and are finite.

The curve γ is called *piecewise smooth* if z and z' both are piecewise continuous.

A *contour* γ is a sequence of smooth curves $\{\gamma_1, \cdots, \gamma_n\}$ such that the terminal point of γ_k coincides with the initial point of γ_{k+1} for $1 \leq k \leq n-1$. In this case, we write

$$\gamma = \gamma_1 + \gamma_2 + \cdots + \gamma_n.$$

It is clear that a contour is a continuous piecewise smooth curve.

Figure 12.6

The *opposite contour* of γ is

$$-\gamma = (-\gamma_n) + (-\gamma_{n-1}) + \cdots + (-\gamma_1).$$

Example 12.5. Let γ_1 be the curve $z_1(t) = t + 2ti$, $0 \leq t \leq 1$, and γ_2 be the curve $z_2 = t + 2i$, $1 \leq t \leq 2$. Then, $\gamma = \gamma_1 + \gamma_2$ is a contour (see Figure 12.5).

Now consider any partition $a = t_0 < t_1 < \cdots < t_{n-1} < t_n = b$ of the interval $[a, b]$. Then, the line segments $[z(t_{j-1}), z(t_j)]$, $j = 1, 2, \cdots, n$ form a polygonal contour σ that is inscribed in the curve γ. Clearly, such an inscribed polygon a contour σ has a finite length, which we denote and define as

$$ \ell_\sigma = \sum_{j=1}^{n} \left[(x(t_j) - x(t_{j-1}))^2 + (y(t_j) - y(t_{j-1}))^2 \right]^{1/2}. $$

A curve γ is said to be *rectifiable* if the lengths ℓ_σ of all inscribed polygons σ are bounded. The length ℓ of γ is defined as $\ell = \sup_\sigma \ell_\sigma$. If γ is a piecewise smooth curve, then the Mean Value Theorem ensures that the curve γ is rectifiable.

A simple closed contour γ is called *positively oriented* (anticlockwise) if the interior domain lies to the left of an observer tracing out the points in order; otherwise it is *negatively oriented* (clockwise). With this convention, it is clear that the positive direction of traversing a curve surrounding the point at infinity is the clockwise direction. A simple open contour γ is said to be positively oriented if we traverse it from its initial point to its terminal point.

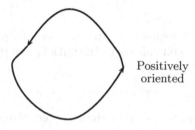

Positively oriented

Figure 12.7

Theorem 12.1 (Jordan Curve Theorem). The points on any simple closed curve or simple closed contour γ are boundary points of two distinct domains, one of which is the interior of γ and is bounded. The other, which is the exterior of γ, is unbounded.

Example 12.6. Consider the disjoint open disks $S_1 = \{z : |z - z_0| < r\}$ and $S_2 = \{z : |z - z_0| > r\}$. Clearly, the circle $\gamma = \{z : |z - z_0| = r\}$ is a closed contour, and the points on γ are the boundary points of S_1 and S_2. The interior of γ is S_1, which is bounded, and the exterior of γ is S_2, which is unbounded.

A *simply connected domain* S is a domain having the following property: If γ is any simple closed contour lying in S, then the domain interior to γ

lies wholly in S. An immediate consequence of Theorem 12.1 is that the interior of a simple closed curve is simply connected. A domain that is not simply connected is called *multiply connected.*

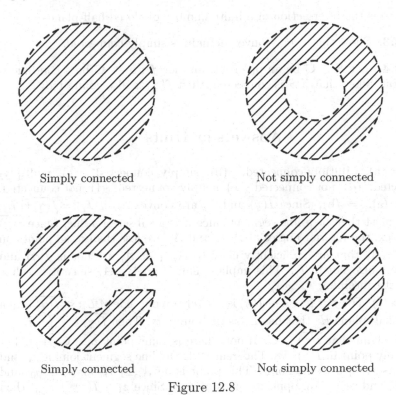

Simply connected	Not simply connected
Simply connected	Not simply connected

Figure 12.8

Problems

12.1. For each of the following domains, determine whether it is simply connected, multiply connected, or neither (not connected):

(a). $A = \{z : 0 < |2z - 1| < 3\}$,

(b). $B = \{z : \operatorname{Arg} z \neq 0\}$,

(c). $C = \{z : |z - i| > 1, \; |z - 3| > 3 \; \text{and} \; |z| < 10\}$,

(d). $D = \{z : \operatorname{Im} z \neq 0\}$,

(e). $E = \{z : |z| < 1 \; \text{and} \; \operatorname{Re} z < 0\}$,

(f). $F = \mathbf{C} - \{x + iy : 0 \leq x \leq 3\}$.

12.2. Let $\gamma = A_0 A_1 \cdots A_{n-1} A_n$ be a closed polygonal Jordan curve with interior $I(\gamma)$. Show that the following are equivalent:

(a). The polygon $\overline{I(\gamma)}$ is convex.

(b). For $j = 1, \cdots, n$, the line L_j containing $A_{j-1}A_j$ does not intersect $I(\gamma)$.

(c). $\overline{I(\gamma)}$ is the intersection of a finite number of closed half-planes.

12.3. Show that every convex domain is simply connected.

12.4. Let $S \subset \mathbf{C}$ be an open set, and let $f : S \to \mathbf{C}$ be a continuous function. Show that if S is connected, then $f(S)$ is also connected.

Answers or Hints

12.1. (a). Multiply-connected, (b). simply connected, (c). multiply-connected, (d). not connected, (e). simply connected, (f). not connected.

12.2. (a). $\Rightarrow$ (b). Since $\overline{I(\gamma)}$ and L_j are convex, so is $I_j := \overline{I(\gamma)} \cap L_j$, which must then be a line segment since it is a subset of a line. Since $\overline{I(\gamma)}$ is convex, the interior angles at A_{j-1} and A_j are less than π, so points on $L_j \backslash A_{j-1}A_j$ that are sufficiently close to A_{j-1} or A_j are not in $\overline{I(\gamma)}$ and therefore also not in I_j. This implies that $I_j = A_{j-1}A_j$ since it is a line segment containing $A_{j-1}A_j$.

(b). $\Rightarrow$ (c). For $j = 1, \ldots, n$, $I(\gamma)$ is entirely on one side of L_j, so $\overline{I(\gamma)}$ is in a closed half-plane H_j that has L_j as its boundary. Then $\overline{I(\gamma)} \subset \bigcap_{j=1}^{n} H_j =:$ $\mathbf{C}$. We claim that $\overline{I(\gamma)} = \mathbf{C}$. If not, there is some point $z_0 \in \mathbf{C} \backslash \overline{I(\gamma)}$. Let z_1 be any point in $I(\gamma)$. By Theorem 12.1, the line segment joining z_0 and z_1 intersects γ at some point. This point is on $A_{j_0-1}A_{j_0}$ for some j_0, and then z_0 and z_1 are on opposite sides of L_{j_0}. Since $z_1 \in I(\gamma) \subset H_{j_0}$, then $z_0 \notin H_{j_0}$, which is a contradiction since $z_0 \in \mathbf{C}$.

(c). $\Rightarrow$ (a). This follows since every closed half-plane is convex and an intersection of convex sets is convex.

12.3. Let S be a convex domain and γ a simple closed curve lying in S. We want to show that the domain $I(\gamma)$ interior to γ lies in S. If not, there is some point $z_0 \in I(\gamma) \backslash S$. Let L be any line through z_0. Since $\overline{I(\gamma)}$ is bounded, we can take two points z_1 and z_2 on $L \backslash \overline{I(\gamma)}$ such that z_0 is on the line segment joining z_1 and z_2. By Theorem 12.1, the line segment joining z_0 and z_1 (resp. z_0 and z_2) intersects γ at some point z_1' (resp. z_2'). Then z_1' and z_2' are in S and z_0 is on the line segment joining them, so $z_0 \in S$ since S is convex, a contradiction.

12.4. Let $w_1, w_2 \in f(S)$. Then, there exist $z_1, z_2 \in S$ such that $w_1 = f(z_1)$, $w_2 = f(z_2)$. Since S is connected, there exists a curve γ in S connecting z_1 and z_2. Suppose $z = z(t)$, $a \le t \le b$ represents γ. Then, $f(z(t))$, $a \le t \le b$ defines a curve in $f(S)$ connecting w_1 and w_2.

Lecture 13
Complex Integration

In this lecture, we shall introduce integration of complex-valued functions along a directed contour. For this, we shall begin with the integration of complex-valued functions of a real variable. Our approach is based on Riemann integration from calculus. We shall also prove an inequality that plays a fundamental role in our later lectures.

Recall that a complex-valued function w of a real variable $t \in [a, b] \subset \mathbb{R}$ is defined as $w(t) = u(t) + iv(t)$; i.e., $w : [a, b] \subset \mathbb{R} \to \mathbf{C}$. The derivative of $w(t)$ at a point t is defined as

$$\frac{d}{dt}w(t) = w'(t) = u'(t) + iv'(t),$$

provided each of the derivatives u' and v' exists at t. Similarly, the integral of $w(t)$ over $[a, b]$ is defined as

$$\int_a^b w(t)dt = \int_a^b u(t)dt + i\int_a^b v(t)dt$$

provided the individual integrals on the right exist. For this, it is sufficient to assume that the functions u and v are piecewise continuous in $[a, b]$. The following properties of differentiation and integration hold:

(i). $\dfrac{d}{dt}[w_1(t) + w_2(t)] = w_1'(t) + w_2'(t).$

(ii). $\dfrac{d}{dt}[z_0 w_1(t)] = z_0 \dfrac{d}{dt}w_1(t), \quad z_0 \in \mathbf{C}.$

(iii). $\displaystyle\int_a^b [w_1(t) + w_2(t)]dt = \int_a^b w_1(t)dt + \int_a^b w_2(t)dt.$

(iv). $\displaystyle\int_a^b z_0 w_1(t)dt = z_0 \int_a^b w_1(t)dt, \quad z_0 \in \mathbf{C}.$

(v). $\text{Re}\displaystyle\int_a^b w_1(t)dt = \int_a^b \text{Re}[w_1(t)]dt, \quad \text{Im}\int_a^b w_1(t)dt = \int_a^b \text{Im}[w_1(t)]dt.$

Here we shall show only (iv). For this, let $z_0 = x_0 + iy_0$ and $w_1(t) = u_1(t) + iv_1(t)$. Then, we have

$$z_0 w_1(t) = [x_0 u_1(t) - y_0 v_1(t)] + i[x_0 v_1(t) + y_0 u_1(t)]$$

and

$$\int_a^b z_0 w_1(t) dt = \int_a^b [x_0 u_1(t) - y_0 v_1(t)] dt + i \int_a^b [x_0 v_1(t) + y_0 u_1(t)] dt$$

$$= \left[x_0 \int_a^b u_1(t) dt - y_0 \int_a^b v_1(t) dt \right]$$

$$\quad + i \left[x_0 \int_a^b v_1(t) dt + y_0 \int_a^b u_1(t) dt \right]$$

$$= (x_0 + i y_0) \left[\int_a^b u_1(t) dt + i \int_a^b v_1(t) dt \right]$$

$$= z_0 \int_a^b w_1(t) dt.$$

Example 13.1. Suppose $w(t)$ is continuous on $[a, b]$ and $w'(t)$ exists on (a, b). For such functions, the Mean Value Theorem for derivatives no longer applies; i.e., it is not necessarily true that there is a number $c \in (a, b)$ such that $w'(c) = (w(b) - w(a))/(b - a)$. To show this, we consider the function $w(t) = e^{it}$, $0 \le t \le 2\pi$. Clearly, $|w'(t)| = |ie^{it}| = 1$, and hence $w'(t)$ is never zero, while $w(2\pi) - w(0) = 0$.

We recall that if $f : [a, b] \to \mathbb{R}$, then $|\int_a^b f(x) dx| \le \int_a^b |f(x)| dx$. We shall now show that the same inequality holds for $w : [a, b] \to \mathbb{C}$; i.e.,

$$\left| \int_a^b w(t) dt \right| \le \int_a^b |w(t)| dt, \quad a \le b < \infty.$$

For this, let $|\int_a^b w(t) dt| = r$, so that $\int_a^b w(t) dt = re^{i\theta}$ in polar form. Now we have

$$r = e^{-i\theta} \int_a^b w(t) dt = \int_a^b e^{-i\theta} w(t) dt$$

$$= \int_a^b \operatorname{Re}\left[e^{-i\theta} w(t) \right] dt + i \int_a^b \operatorname{Im}\left[e^{-i\theta} w(t) \right] dt$$

$$\qquad\qquad (= 0 \text{ since LHS is real})$$

$$= \int_a^b \operatorname{Re}\left[e^{-i\theta} w(t) \right] dt$$

$$\le \left| \int_a^b \operatorname{Re}\left[e^{-i\theta} w(t) \right] dt \right| \le \int_a^b \left| \operatorname{Re}\left[e^{-i\theta} w(t) \right] \right| dt$$

$$\le \int_a^b \left| e^{-i\theta} w(t) \right| dt = \int_a^b |w(t)| dt \quad (\text{since } |e^{-i\theta}| = 1).$$

We note that the Fundamental Theorem of Calculus also holds: If $W(t) = U(t) + iV(t)$, $w(t) = u(t) + iv(t)$ and $W'(t) = w(t)$, $t \in [a, b]$, then

$$\int_a^b w(t)dt = W(b) - W(a).$$

Thus, in particular, we have

$$\int_a^b e^{z_0 t}dt = \frac{1}{z_0}e^{z_0 t}\Big|_a^b = \frac{1}{z_0}\left(e^{z_0 b} - z^{z_0 a}\right), \quad z_0 \neq 0.$$

The *length* of a smooth curve γ given by the range of $z : [a, b] \to \mathbf{C}$ is defined by

$$L(\gamma) = \int_a^b |z'(t)|dt.$$

Example 13.2. For $z(t) = z_1 + t(z_2 - z_1)$, $0 \leq t \leq 1$, we have

$$z'(t) = z_2 - z_1, \quad \text{and hence} \quad L(\gamma) = \int_0^1 |z_2 - z_1|dt = |z_2 - z_1|.$$

Example 13.3. For $z(t) = z_0 + re^{it}$, $0 \leq t \leq 2\pi$, we have

$$z'(t) = ire^{it}, \quad \text{and hence} \quad L(\gamma) = \int_0^{2\pi} |ire^{it}|dt = 2\pi r.$$

Now let S be an open set, and let γ, given by the range of $z : [a, b] \to \mathbf{C}$, be a smooth curve in S. If $f : S \to \mathbf{C}$ is continuous, then the *integral of f along γ* is defined by

$$\int_\gamma f(z)dz = \int_a^b f(z(t))z'(t)dt.$$

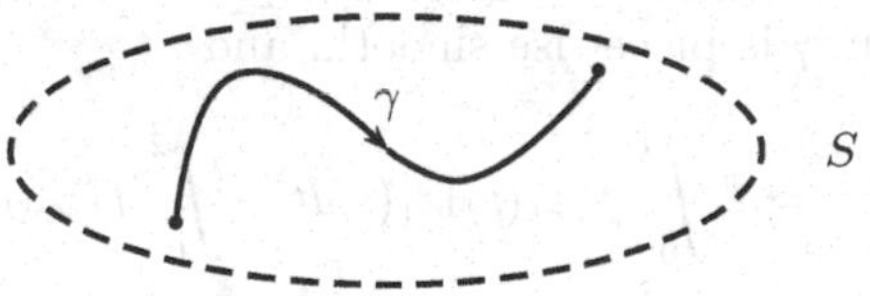

Figure 13.1

The following properties of the integration above are immediate:

$$\int_\gamma [f(z) \pm g(z)]dz = \int_\gamma f(z)dz \pm \int_\gamma g(z)dz,$$

$$\int_\gamma z_0 f(z)dz = z_0 \int_\gamma f(z)dz, \quad z_0 \in \mathbf{C},$$

$$\int_{-\gamma} f(z)dz = -\int_\gamma f(z)dz.$$

Example 13.4. Let $f(z) = z - 1$ and let γ be the curve given by $z(t) = t + it^2$, $0 \le t \le 1$. Clearly, γ is smooth and

$$
\begin{aligned}
\int_\gamma f(z)dz &= \int_0^1 f(z(t))z'(t)dt = \int_0^1 (t + it^2 - 1)(1 + 2it)dt \\
&= \int_0^1 (t - 1 - 2t^3)dt + i\int_0^1 (2t(t-1) + t^2)dt,
\end{aligned}
$$

which can now be easily evaluated.

Example 13.5. Let $f(z) = z + (1/z)$ for $z \ne 0$ and γ be the upper semi-circle at the origin of radius 1; i.e., $z(t) = e^{\pi it}$, $0 \le t \le 1$. Clearly, γ is smooth and

$$
\int_\gamma f(z)dz = \int_0^1 \pi i \left(e^{\pi it} + e^{-\pi it}\right) e^{\pi it} dt = \pi i.
$$

The *length* of a contour $\gamma = \gamma_1 + \cdots + \gamma_n$ is defined by

$$
L(\gamma) = \sum_{j=1}^n (\text{length of } \gamma_j) = \sum_{j=1}^n L(\gamma_j).
$$

If $f : S \to \mathbf{C}$ is continuous on $\{\gamma\}$, then the *contour integral of f along* γ is defined by

$$
\int_\gamma f(z)dz = \sum_{j=1}^n \int_{\gamma_j} f(z)dz.
$$

Example 13.6. Let $f(z) = z - 1$ and $\gamma = \gamma_1 + \gamma_2$, where γ_1 is given by $z_1(t) = t$, $0 \le t \le 1$ and γ_2 is given by $z_2(t) = 1 + i(t - 1)$, $1 \le t \le 2$. Clearly, the contour γ is piecewise smooth, and

$$
\begin{aligned}
\int_\gamma f(z)dz &= \int_0^1 f(z_1(t))z_1'(t)dt + \int_1^2 f(z_2(t))z_2'(t)dt \\
&= \int_0^1 (t - 1)dt + \int_1^2 i(t - 1)idt,
\end{aligned}
$$

which can be evaluated.

Theorem 13.1 (ML-Inequality). Suppose that f is continuous on an open set containing a contour γ, and $|f(z)| \le M$ for all $z \in \{\gamma\}$. Then, the following inequality holds

$$
\left|\int_\gamma f(z)dz\right| \le ML,
$$

where L is the length of γ.

Proof. First assume that γ given by the range of $z : [a, b] \to \mathbf{C}$ is a smooth curve. Then, we have

$$\left| \int_{\gamma} f(z)dz \right| = \left| \int_{a}^{b} f(z(t))z'(t)dt \right| \leq \int_{a}^{b} |f(z(t))||z'(t)|dt$$

$$\leq M \int_{a}^{b} |z'(t)|dt = ML.$$

If $\gamma = \gamma_1 + \gamma_2 + \cdots + \gamma_n$, where $\gamma_1, \gamma_2, \cdots, \gamma_n$ are smooth, then we find

$$\left| \int_{\gamma} f(z)dz \right| = \left| \sum_{j=1}^{n} \int_{\gamma_j} f(z)dz \right| \leq \sum_{j=1}^{n} \left| \int_{\gamma_j} f(z)dz \right|$$

$$\leq \sum_{j=1}^{n} ML(\gamma_j) = ML(\gamma). \qquad \blacksquare$$

Example 13.7. Let γ be given by $z(t) = 2e^{it}$, $0 \leq t \leq 2\pi$. Show that

$$\left| \int_{\gamma} \frac{e^z}{z^2 + 1} dz \right| \leq \frac{4\pi e^2}{3}.$$

Since $|e^z| = e^x \leq e^2$ and $|z^2 + 1| \geq ||z^2| - 1| = |4 - 1| = 3$, it follows that

$$\left| \int_{\gamma} \frac{e^z}{z^2 + 1} dz \right| \leq \frac{e^2}{3} \times 2 \times 2\pi = \frac{4\pi e^2}{3}.$$

Problems

13.1. Evaluate the following integrals

(a). $\int_{0}^{1} (1 + it^2)dt$, (b). $\int_{-\pi}^{\pi/4} te^{-it^2} dt$,

(c). $\int_{0}^{\pi} (\sin 2t + i \cos 2t)dt$, (d). $\int_{1}^{2} \text{Log}(1 + it)dt$.

13.2. Find the length of the arch of the cycloid given by $z(t) = a(t - \sin t) + a i(1 - \cos t)$, $0 \leq t \leq 2\pi$ where a is a positive real number.

13.3. Let γ be the curve given by $z(t) = t + it^2$, $0 \leq t \leq 2\pi$. Evaluate $\int_{\gamma} |z|^2 dz$.

13.4. Let $f(z) = y - x - 3ix^2$ and γ be given by the line segment $z = 0$ to $z = 1 + i$. Evaluate $\int_\gamma f(z)dz$.

13.5. Let γ be given by the semicircle $z = 2e^{i\theta}$, $0 \le \theta \le \pi$. Evaluate $\int_\gamma \dfrac{z-2}{z}dz$ and $\int_{-\gamma} |z^{1/2}| \exp(i\operatorname{Arg} z)dz$.

13.6. Show that if m and n are integers, then

$$\int_0^{2\pi} e^{im\theta} e^{-in\theta} d\theta = \begin{cases} 0 & \text{when } m \ne n \\ 2\pi & \text{when } m = n. \end{cases}$$

Hence, evaluate $\int_\gamma z^m \bar{z}^n dz$, where γ is the circle given by $z = \cos t + i \sin t$, $0 \le t \le 2\pi$.

13.7. Let γ be the positively oriented ellipse $\dfrac{x^2}{a^2} + \dfrac{y^2}{b^2} = 1$ with $a^2 - b^2 = 1$. Show that

$$\int_\gamma \frac{dz}{\sqrt{1-z^2}} = \pm 2\pi,$$

where a continuous branch of the integrand is chosen.

13.8. Let $z_1(t)$, $a \le t \le b$ and $z_2(t)$, $c \le t \le d$ be equivalent parameterizations of the same curve γ; i.e., there is an increasing continuously differentiable function $\phi : [c, d] \to [a, b]$ such that $\phi(c) = a$ and $\phi(d) = b$ and $z_2(t) = z_1 \circ \phi(t)$ for all $t \in [c, d]$. Show that if f is continuous on an open set containing γ, then

$$\int_a^b f(z_1(t))z_1'(t)dt = \int_c^d f(z_2(t))z_2'(t)dt.$$

13.9. Let γ be the arc of the circle $|z| = 2$ from $z = 2$ to $z = 2i$ that lies in the first quadrant. Without evaluating the integral, show that

$$\left| \int_\gamma \frac{dz}{z^2 - 1} \right| \le \frac{\pi}{3}.$$

13.10. Let γ be the circle $|z| = 2$. Show that

$$\left| \int_\gamma \frac{1}{z^2 - 1}dz \right| \le \frac{4\pi}{3}.$$

13.11. Show that if γ is the boundary of the triangle with vertices at $z = 0$, $z = 3i$, and $z = -4$ oriented in the counterclockwise direction, then

$$\left| \int_\gamma (e^z - \bar{z})dz \right| \le 48.$$

13.12. Let γ_R be the circle $|z| = R$ described in the counterclockwise direction, where $R > 0$. Suppose $\operatorname{Log} z$ is the principal branch of the logarithm function. Show that

$$\left| \int_{\gamma_R} \frac{\operatorname{Log} z}{z^2} dz \right| \leq 2\pi \left(\frac{\pi + \operatorname{Log} R}{R} \right).$$

13.13. Let γ_R be the circle $|z| = R$ described in the counterclockwise direction, where $R > 2$. Suppose $\operatorname{Log} z$ is the principal branch of the logarithm function. Show that

$$\left| \int_{\gamma_R} \frac{\operatorname{Log} z^2}{z^2 + z + 1} dz \right| \leq 2\pi R \left(\frac{\pi + 2\operatorname{Log} R}{R^2 - R - 1} \right).$$

13.14. Let S be an open connected set and γ a closed curve in S. Suppose $f(z)$ is analytic on S and the derivative $f'(z)$ is continuous on S. Show that

$$I = \int_\gamma \overline{f(z)} f'(z) dz$$

is purely imaginary.

13.15. Let $f(z)$ be a continuous function in the region $|z| \geq 1$, and suppose the limit $\lim_{z \to \infty} z f(z) = A$ exists. Let α be a fixed real number such that $0 < \alpha \leq 2\pi$. Denote by γ_R the circular arc given by parametric equation $z = Re^{i\theta}$, $0 \leq \theta \leq \alpha$ with $R \geq 1$. Find $\lim\limits_{R \to \infty} \int_{\gamma_R} f(z) dz$.

Answers or Hints

13.1. (a). $1 + i\frac{1}{3}$, (b). $\frac{i}{2}\left(e^{-i\pi^2/16} - e^{-i\pi^2}\right)$, (c). 0, (d). $\int_1^2 \frac{1}{2}\ln(1 + t^2)dt + i\int_1^2 \tan^{-1} t\, dt$.

13.2. $8a$.

13.3. $\int_\gamma |z|^2 dz = \int_0^{2\pi} |t + it^2|^2(1 + 2ti)dt = \int_0^{2\pi}(t^2 + t^4)(1 + 2ti)dt$.

13.4. $\int_\gamma (y - x - 3ix^2)dz = \int_0^1 (t - t - 3it^2)(1 + i)dt = 1 - i$ ($\gamma(t) = (1 + i)t$, $t \in [0, 1]$).

13.5. $\int_\gamma \frac{z-2}{z} dz = \int_0^\pi \frac{2e^{i\theta}-2}{2e^{i\theta}} 2ie^{i\theta} d\theta = -4 - 2i\pi, \int_{-\gamma} |z|^{1/2} \exp\left(i\operatorname{Arg} z\right) dz = -\int_0^\pi |2e^{i\theta}|^{1/2} e^{i\theta} 2ie^{i\theta} d\theta = -\sqrt{2}e^{2i\theta}\big|_0^\pi = 0$.

13.6. $\int_0^{2\pi} e^{i(m-n)\theta} d\theta = 2\pi$ if $m = n$ and $= \left\{ e^{i(m-n)\theta}/i(m-n) \right\}\big|_0^{2\pi} = 0$ if $m \neq n$. Now $\int_\gamma z^m \bar{z}^n dz = \int_0^{2\pi} e^{imt} e^{-int} ie^{it} dt = 2\pi i$ if $m - n + 1 = 0$ and 0 if $m - n + 1 \neq 0$.

13.7. In parametric form, the equation of ellipse is $x = a\cos t,\ y = b\sin t,\ t \in [0, 2\pi]$. Thus, $z(t) = a\cos t + ib\sin t$. Since $a^2 - b^2 = 1$, we find $\sqrt{1 - z^2} = \pm(a\sin t - ib\cos t)$ and $z'(t) = -a\sin t + ib\cos t$.

13.8. $\int_c^d f(z_2(t))z_2'(t)dt = \int_c^d f(z_1(\phi(t)))z_1'(\phi(t))\phi'(t)dt = \int_a^b f(z_1(s))z_1'(s)ds.$

13.9. $\left|\int_\gamma \frac{dz}{z^2-1}\right| \leq ML(\gamma),\ |z| = 2,\ |z^2 - 1| \geq |z|^2 - 1 \geq 3$, so $\left|\frac{1}{z^2-1}\right| \leq \frac{1}{3}$ if $z \in \gamma$, $L(\gamma) = \frac{4\pi}{4} = \pi$.

13.10. $|z^2 - 1| \geq |z^2| - 1 = 3$ for $|z| = 2$. Thus, $\left|\int_\gamma \frac{dz}{z^2-1}\right| \leq \frac{4\pi}{3}.$

13.11. $\left|\int_\gamma (e^z - \bar{z})dz\right| \leq \left|\int_\gamma e^z dz\right| + \left|\int_\gamma \bar{z}dz\right| \leq 0 + 4(5 + 4 + 3) = 48.$

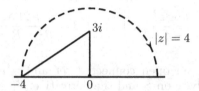

13.12. $\left|\frac{\text{Log}\,z}{z^2}\right| = \left|\frac{\text{Log}\,R + i\text{Arg}\,z}{R^2}\right| \leq \frac{|\text{Log}\,R| + \pi}{R^2} = \frac{\text{Log}\,R + \pi}{R^2}$, so $\left|\int_{\gamma_R} \frac{\text{Log}\,z}{z^2}dz\right| \leq 2\pi R\left(\frac{\text{Log}\,R + \pi}{R^2}\right) = 2\pi\left(\frac{\text{Log}\,R + \pi}{R}\right).$

13.13. Similar to Problem 13.12.

13.14. Let the parametric equation of γ be $z = z(t),\ a \leq t \leq b$. Since γ is a closed curve, $z(a) = z(b)$. Let $f(z) = u(z) + iv(z)$, where u, v are real, so $f'(z) = u_x + iv_x = v_y - iu_y$. Thus, $I = \int_a^b (u(z(t)) - iv(z(t)))(u_x + iv_x)(x_t + iy_t)dt$. Hence, $\text{Re}\,I = \int_a^b (uu_x x_t - uv_x y_t + vu_x y_t + vv_x x_t)dt = \int_a^b [(uu_x x_t + uu_y y_t) + (vv_y y_t + vv_x x_t)]dt = (1/2)\int_a^b [\frac{d}{dt}(u^2 + v^2)]dt = 0.$

13.15. Clearly, $\int_{\gamma_R} f(z)dz = \int_0^\alpha f(Re^{i\theta})Rie^{i\theta}d\theta = i\int_0^\alpha Re^{i\theta}f(Re^{i\theta})d\theta$ and $iA\alpha = i\int_0^\alpha Ad\theta$. Hence, $|\int_{\gamma_R} f(z)dz - iA\alpha| \leq \int_0^\alpha |Re^{i\theta}f(Re^{i\theta}) - A|d\theta$. Now since $\lim_{z\to\infty} zf(z) = A$, for any $\epsilon > 0$ there exists a $\delta > 0$ such that $|Re^{i\theta}f(Re^{i\theta}) - A| < \epsilon/\alpha$ for all θ. Therefore, it follows that $|\int_{\gamma_R} f(z)dz - iA\alpha| < \epsilon$, and hence $\lim_{R\to\infty} \int_{\gamma_R} f(z)dz = iA\alpha.$

Lecture 14
Independence of Path

The main result of this lecture is to provide conditions on the function f so that its contour integral is independent of the path joining the initial and terminal points. This result, in particular, helps in computing the contour integrals rather easily.

Let f be a continuous function in a domain S. A function F such that $F'(z) = f(z)$ for all $z \in S$ is called an *antiderivative* of f on S. Since f is continuous, F is analytic, and hence continuous. Furthermore, any two antiderivatives of f differ by a constant.

Example 14.1. Clearly,

$f(z)$	z^n	e^z	$\cos z$	$\sin z$
$F(z)$	$\dfrac{z^{n+1}}{n+1} + c$	$e^z + c$	$\sin z + c$	$-\cos z + c,$

where c is an arbitrary constant.

The main result of this lecture is the following.

Theorem 14.1. Suppose that the function f is continuous and has an antiderivative F in a domain S. If $\alpha,\ \beta \in S$ and γ is a contour in S joining α and β, then

$$\int_\gamma f(z)dz = F(\beta) - F(\alpha);$$

i.e., the integral only depends on the end points and not on the choice of γ. In particular, if γ is a closed contour in S, then

$$\int_\gamma f(z)dz = 0.$$

Proof. Suppose that $\gamma = \gamma_1 + \gamma_2 + \cdots + \gamma_n$, where, for each $1 \le j \le n$, γ_j is a smooth curve given by $z_j : [a_{j-1}, a_j] \to \mathbf{C}$ such that $z_1(a_0) = \alpha$, $z_n(a_n) = \beta$.

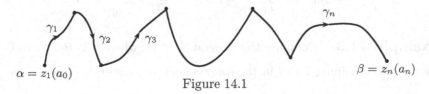

$$\alpha = z_1(a_0) \qquad\qquad\qquad\qquad \beta = z_n(a_n)$$

Figure 14.1

Now, since

$$\frac{d}{dt}F(z_j(t)) = F'(z_j(t))z_j'(t) = f(z_j(t))z_j'(t),$$

it follows that

$$\int_{\gamma_j} f(z)dz = \int_{a_{j-1}}^{a_j} f(z_j(t))z_j'(t)dt = F(z_j(t))\Big|_{a_{j-1}}^{a_j}$$
$$= F(z_j(a_j)) - F(z_j(a_{j-1})).$$

Therefore, we have

$$\int_{\gamma} f(z)dz = \sum_{j=1}^{n}\int_{\gamma_j} f(z)dz = \sum_{j=1}^{n}\{F(z_j(a_j)) - F(z_j(a_{j-1}))\}$$
$$= \{F(z_1(a_1)) - F(z_1(a_0))\} + \cdots + \{F(z_n(a_n)) - F(z_n(a_{n-1}))\}$$
$$= F(\beta) - F(\alpha).$$

If γ is closed, then $\alpha = \beta$, so we have

$$\int_{\gamma} f(z)dz = F(\beta) - F(\alpha) = 0. \qquad \blacksquare$$

Example 14.2. Compute the integral $\int_{\gamma}(z^2 - 2)dz$, where γ is the contour in Figure 14.2.

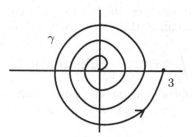

Figure 14.2

Since $z^2 - 2$ has the antiderivative $(z^3/3) - 2z$, we find

$$\int_{\gamma}(z^2 - 2)dz = \int_0^3 (z^2 - 2)dz = \frac{z^3}{3} - 2z\Big|_0^3 = 3.$$

Example 14.3. Compute the integral $\int_{\gamma} e^z dz$, where γ is the part of the unit circle joining 1 to i in the counterclockwise direction. Since e^z is the derivative of e^z, by the theorem above, we have

$$\int_\gamma e^z dz = e^i - e.$$

Example 14.4. Compute the integral $\int_\gamma \sin z\, dz$, where γ is the contour in Figure 14.3.

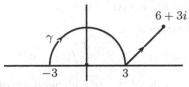

Figure 14.3

Since $\sin z$ has the antiderivative $F(z) = -\cos z$, by Theorem 14.1, we have

$$\int_\gamma \sin z\, dz = -\cos z \Big|_{-3}^{6+3i} = -\cos(6+3i) + \cos(-3).$$

Example 14.5. Compute $\int_\gamma dz/z$, where γ is the contour in Figure 14.4.

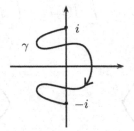

Figure 14.4

At each point of the contour γ, the function $1/z$ is the derivative of the principal branch of $\log z$. Hence, we have

$$\int_\gamma \frac{1}{z} dz = \operatorname{Log} z \Big|_i^{-i} = \operatorname{Log}(-i) - \operatorname{Log}(i) = -\frac{\pi}{2}i - \frac{\pi}{2}i = -\pi i.$$

Example 14.6. Compute the integral $\int_{\gamma_r}(z - z_0)^n dz$, where n is an integer not equal to -1 and γ_r is the circle $|z - z_0| = r$ traversed once in the counterclockwise direction.

Let $S = \mathbf{C} - \{z_0\}$. Then, S is a domain and the function $(z - z_0)^n$ is continuous throughout S. Moreover, $(z - z_0)^n$ is the derivative of the function $(z - z_0)^{n+1}/(n+1)$. Since γ_r is a closed contour that lies in S, we deduce that

$$\int_{\gamma_r} (z - z_0)^n dz = 0, \quad n \neq -1.$$

We shall now prove the following theorem.

Theorem 14.2. Let f be a continuous function in a domain S. Then, the following statements are equivalent:

(i). f has an antiderivative in S.

(ii). For any closed contour γ in S, $\displaystyle\int_{\gamma} f(z) = 0$.

(iii). The contour integrals of f are independent of paths in S; i.e., if α, $\beta \in S$ and γ_1 and γ_2 are contours in S joining α and β, then

$$\int_{\gamma_1} f(z)dz = \int_{\gamma_2} f(z)dz.$$

Proof. Theorem 14.1 shows that (i) $\Rightarrow$ (iii) $\Rightarrow$ (ii), so we need to prove that (ii) $\Rightarrow$ (iii) $\Rightarrow$ (i).

(ii) $\Rightarrow$ (iii). $\gamma_1 + (-\gamma_2)$ is a closed contour in S (see Figure 14.5). Hence,

$$\int_{\gamma_1+(-\gamma_2)} f(z)dz = \int_{\gamma_1} f(z)dz - \int_{\gamma_2} f(z)dz = 0.$$

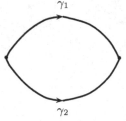

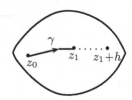

Figure 14.5

Figure 14.6

(iii) $\Rightarrow$ (i). Fix $z_0 \in S$. For any $z_1 \in S$, define $F(z_1) = \displaystyle\int_{\gamma} f(z)dz$, where γ is a contour in S joining z_0 to z_1. By (iii), the function F is well-defined. Let $h \in \mathbf{C}$ be such that $|h|$ is sufficiently small, so that $[z_1, z_1 + h] \subset S$ (see Figure 14.6). Then, we have

$$\frac{F(z_1 + h) - F(z_1)}{h} = \frac{1}{h}\left[\int_{\gamma+[z_1,z_1+h]} - \int_{\gamma}\right] f(z)dz = \frac{1}{h}\int_{[z_1,z_1+h]} f(z)dz$$

and

$$f(z_1) = f(z_1)\frac{1}{h}\int_{[z_1,z_1+h]} 1dz = \frac{1}{h}\int_{[z_1,z_1+h]} f(z_1)dz.$$

Therefore, it follows that

$$\frac{F(z_1 + h) - F(z_1)}{h} - f(z_1) = \frac{1}{h}\int_{[z_1,z_1+h]} (f(z) - f(z_1))dz. \qquad (14.1)$$

We shall show that the right hand side of (14.1) tends to 0 as $h \to 0$. For this, let $\epsilon > 0$ be given. Since f is continuous at z_1, there exists a $\delta > 0$ such that

$$|z - z_1| < \delta \Rightarrow |f(z) - f(z_1)| < \epsilon. \qquad (14.2)$$

Since the length of $[z_1, z_1 + h]$ is equal to $|h|$, if $|h| < \delta$ then for all $z \in (z_1, z_1 + h]$, $|z - z_1| \leq |h| < \delta$. Thus, if $|h| < \delta$, then

$$\left| \frac{F(z_1 + h) - F(z_1)}{h} - f(z_1) \right| = \left| \frac{1}{h}\int_{[z_1,z_1+h]} (f(z) - f(z_1))dz \right|$$

$$\leq \frac{1}{|h|}\epsilon|h| \quad \text{by (14.2) and the } ML\text{-inequality}$$

$$= \epsilon.$$

Thus, $F'(z_1) = f(z_1)$; i.e., F is an antiderivative of f in S. ∎

Example 14.7. Let $\gamma = \gamma_1 + \gamma_2$, where γ_1 and γ_2 respectively are given by $z_1(t) = ti$ and $z_2(t) = t + i$, $t \in [0,1]$. Furthermore, let $f(z) = (y - x) + 3ix^2$. It follows that

$$\int_\gamma f(z)dz = \int_0^1 f(z_1(t))z_1'(t)dt + \int_0^1 f(z_2(t))z_2'(t)dt$$

$$= \int_0^1 tidt + \int_0^1 (1 - t + 3t^2i)dt = \frac{1}{2} + \frac{3}{2}i.$$

Now let γ_3 be given by $z_3(t) = i + t(1 + i)$, $t \in [0,1]$. Then, for the same function f, we have

$$\int_{\gamma_3} f(z)dz = 2i.$$

Hence, although γ and γ_3 have the same end points,

$$\int_\gamma f(z)dz \neq \int_{\gamma_3} f(z)dz.$$

Thus, from Theorem 14.2, this function f cannot have an antiderivative.

Lecture 15
Cauchy-Goursat Theorem

In this lecture, we shall prove that the integral of an analytic function over a simple closed contour is zero. This result is one of the most fundamental theorems in complex analysis.

We begin with the following theorem from calculus.

Theorem 15.1 (Green's Theorem). Let C be a piecewise smooth, simple closed curve that bounds a domain D in the complex plane. Let P and Q be two real-valued functions defined on an open set U that contains D, and suppose that P and Q have continuous first order partial derivatives. Then,

$$\int_C Pdy - Qdx = \int\int_D \left(\frac{\partial P}{\partial x} + \frac{\partial Q}{\partial y}\right) dxdy,$$

where C is taken along the positive direction.

The left-hand side of this equality is the line integral, whereas the right-hand side is the double integral.

Now consider a function $f(z) = u(x, y) + iv(x, y)$ that is analytic in a simply connected domain S. Suppose that γ is a simple, closed, positively oriented contour lying in S and given by the function $z(t) = x(t) + iy(t)$, $t \in [a, b]$. Then, we have

$$\int_\gamma f(z)dz = \int_a^b f(z(t))\frac{dz(t)}{dt}dt$$

$$= \int_a^b [u(x(t), y(t)) + iv(x(t), y(t))]\left(\frac{dx}{dt} + i\frac{dy}{dt}\right) dt$$

$$= \int_a^b \left[u(x(t), y(t))\frac{dx}{dt} - v(x(t), y(t))\frac{dy}{dt}\right] dt$$

$$+ i\int_a^b \left[v(x(t), y(t))\frac{dx}{dt} + u(x(t), y(t))\frac{dy}{dt}\right] dt$$

$$= \int_\gamma (udx - vdy) + i\int_\gamma (vdx + udy).$$

Thus, if the partial derivatives of u and v are continuous, then, by Green's

Theorem, we have

$$\int_\gamma f(z)dz = \int\int_{S'}\left(-\frac{\partial v}{\partial x} - \frac{\partial u}{\partial y}\right)dxdy + i\int\int_{S'}\left(\frac{\partial u}{\partial x} - \frac{\partial v}{\partial y}\right)dxdy,$$

where S' is the domain interior to γ. Since $f(z)$ is analytic in S, the first partial derivatives of u and v satisfy the Cauchy-Riemann equations. Hence, it follows that

$$\int_\gamma f(z)dz = 0.$$

Thus, we have shown that if f is analytic in a simply connected domain and its derivative $f'(z)$ is continuous (recall that analyticity ensures the existence of $f'(z)$; however, it does not guarantee the continuity of $f'(z)$), then its integral around any simple closed contour in the domain is zero. Goursat was the first to prove that the *condition of continuity on f' can be omitted.*

Theorem 15.2 (Cauchy-Goursat Theorem). If f is analytic in a simply connected domain S and γ is any simple, closed, rectifiable contour in S, then $\int_\gamma f(z)dz = 0$.

Proof. The proof is divided into the following three steps.

Step 1. If γ is the boundary $\partial\Delta$ of a triangle Δ, then $\int_{\partial\Delta} f(z)dz = 0$: We construct four smaller triangles Δ_j, $j = 1, 2, 3, 4$ by joining the midpoints of the sides of Δ by straight lines. Then, from Figure 15.1, it is clear that

$$\int_{\partial\Delta} f(z)dz = \sum_{j=1}^{4}\int_{\partial\Delta_j} f(z)dz,$$

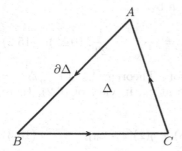
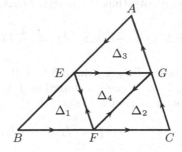

Figure 15.1

which in view of the triangle inequality leads to

$$\left|\int_{\partial\Delta} f(z)dz\right| \le \sum_{j=1}^{4}\left|\int_{\partial\Delta_j} f(z)dz\right|.$$

Let Δ^1 be the triangle among Δ_j, $j = 1, 2, 3, 4$ such that

$$\left| \int_{\partial \Delta^1} f(z)dz \right| = \max_{1 \leq j \leq 4} \left| \int_{\partial \Delta_j} f(z)dz \right|,$$

so that

$$\left| \int_{\partial \Delta} f(z)dz \right| \leq 4 \left| \int_{\partial \Delta^1} f(z)dz \right|,$$

which is the same as

$$\left| \int_{\partial \Delta^1} f(z)dz \right| \geq \frac{1}{4} \left| \int_{\partial \Delta} f(z)dz \right|.$$

Following the process above we divide the triangle Δ^1 into four smaller triangles Δ_j^1, $j = 1, 2, 3, 4$, and obtain the triangle Δ^2 such that

$$\left| \int_{\partial \Delta^2} f(z)dz \right| \geq \frac{1}{4} \left| \int_{\partial \Delta^1} f(z)dz \right| \geq \frac{1}{4^2} \left| \int_{\partial \Delta} f(z)dz \right|.$$

Continuing in this way, we obtain a sequence of triangles $\Delta = \Delta^0 \supset \Delta^1 \supset \Delta^2 \supset \cdots$ such that

$$\left| \int_{\partial \Delta^n} f(z)dz \right| \geq \frac{1}{4} \left| \int_{\partial \Delta^{n-1}} f(z)dz \right|$$

for all $n \geq 1$. This inequality from induction gives

$$\left| \int_{\partial \Delta^n} f(z)dz \right| \geq \frac{1}{4^n} \left| \int_{\partial \Delta} f(z)dz \right|. \tag{15.1}$$

We also note that the length $L(\partial \Delta_n)$ is given by

$$L(\partial \Delta^n) = \frac{1}{2} L(\partial \Delta^{n-1}) = \frac{1}{2^2} L(\partial \Delta^{n-2}) = \cdots = \frac{1}{2^n} L(\partial \Delta^0), \tag{15.2}$$

and hence $\lim_{n \to \infty} \operatorname{diam} \Delta^n = 0$. Thus, from Theorem 4.1, $\bigcap_{n=0}^{\infty} \Delta^n = \{z_0\}$. Now, since the function f is analytic at z_0, in view of (6.2), there exists a function $\eta(z)$ such that

$$f(z) = f(z_0) + f'(z_0)(z - z_0) + \eta(z)(z - z_0), \tag{15.3}$$

where $\lim_{z \to z_0} \eta(z) = 0$. Since the functions 1 and $(z - z_0)$ are analytic and their derivatives are continuous, $\int_{\partial \Delta^n} 1\, dz = \int_{\partial \Delta^n} (z - z_0)dz = 0$, and hence, from (15.3), we find

$$\int_{\partial \Delta^n} f(z)dz = \int_{\partial \Delta^n} \eta(z)(z - z_0)dz. \tag{15.4}$$

Next, since $\lim_{z \to z_0} \eta(z) = 0$, for a given $\epsilon > 0$ we can find a $\delta > 0$ such that

$$|z - z_0| < \delta \quad \text{implies that} \quad |\eta(z)| < \frac{2}{L^2(\partial\Delta^0)}\epsilon. \qquad (15.5)$$

We choose an integer n such that $\partial\Delta^n$ lies in the neighborhood $|z - z_0| < \delta$, as in Figure 15.2.

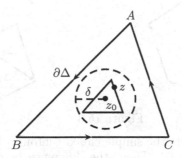

Figure 15.2

Finally, for all $z \in \partial\Delta^n$, it is clear that

$$|z - z_0| < \frac{1}{2}L(\partial\Delta^n),$$

which from (15.2) is the same as

$$|z - z_0| < \frac{1}{2^{n+1}}L(\partial\Delta^0). \qquad (15.6)$$

Now, successively from (15.1), (15.4)-(15.6), Theorem 13.1, and (15.2), we have

$$\left| \int_{\partial\Delta} f(z)dz \right| \leq 4^n \left| \int_{\partial\Delta^n} f(z)dz \right|$$

$$\leq 4^n \left| \int_{\partial\Delta^n} \eta(z)(z - z_0)dz \right|$$

$$\leq 4^n \frac{2}{L^2(\partial\Delta^0)}\epsilon \frac{1}{2^{n+1}}L(\partial\Delta^0)L(\partial\Delta^n)$$

$$\leq \epsilon 2^n \frac{1}{L(\partial\Delta^0)} \frac{1}{2^n}L(\partial\Delta^0) = \epsilon.$$

Since ϵ is arbitrary, it follows that $\int_{\partial\Delta} f(z)dz = 0$.

Step 2. If γ is the boundary ∂C of a polygonal contour C, then $\int_{\partial C} f(z)dz = 0$: We can add interior edges of C so that the interior is subdivided into a finite number of triangles (see Figure 15.3). From Step 1, the integral around each triangle is zero, and since the sum of all these integrals is equal to the integral around C; i.e.,

$$\int_{\partial c} f(z)dz = \sum_{j=1}^{n} \int_{\partial \Delta_j} f(z)dz,$$

Step 2 follows.

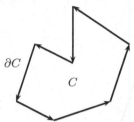
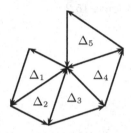

Figure 15.3

Step 3. We approximate the simple closed contour γ by a polygonal contour C_n (see Figure 15.4). Then, the difference between $\int_\gamma f(z)dz$ and $\int_{C_n} f(z)dz$ can be made arbitrarily small as $n \to \infty$.

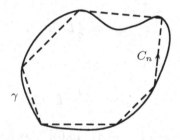

Figure 15.4

Remark 15.1 (i). If γ is a closed contour but not simple, then the integration over γ can always be decomposed into integrations over simple closed curves.

(ii). Since the interior of a simple closed contour is a simple connected domain, Theorem 15.2 can be stated in a more practical form: If γ is a simple closed contour and f is analytic at each point on and inside γ, then $\int_\gamma f(z)dz = 0$.

(iii). Theorem 14.2 can be stated as follows: In a simple connected domain, any analytic function has an antiderivative, its contour integrals are independent of the path, and its integrals over a closed contour vanish.

Example 15.1. Evaluate $\int_\gamma \dfrac{e^z}{z^2 - 16}dz$ where γ is the circle $|z| = 2$ traversed once counterclockwise. Since $e^z/(z^2 - 16)$ is analytic everywhere except at $z = \pm 4$, where the denominator vanishes, and since these points lie exterior to the contour, the integral is zero by Theorem 15.2.

Example 15.2. The function $f(z) = e^{z^2}$ is analytic in **C**. The function $F(z) = \int_{[0,z]} e^{w^2} dw$, where $[0, z]$ is the line segment joining 0 and z is an antiderivative of f. However, we do not know how to express this as an algebraic function of elementary functions.

Lecture 16
Deformation Theorem

In this lecture, we shall show that the integral of a given function along some given path can be replaced by the integral of the same function along a more amenable path.

Let γ_1 be a simple closed contour that can be *continuously deformed* into another simple closed contour γ_2 without passing through a point where f is not analytic. Then the value of the integral $\int_{\gamma_1} f(z)dz$ is the same as $\int_{\gamma_2} f(z)dz$. (Here, in Figure 16.1, we have taken γ_2 as a circle for simplicity and applications in mind.)

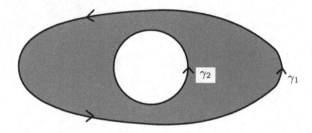

Figure 16.1

We state this result precisely in the following theorem.

Theorem 16.1. Let γ_1 and γ_2 be positively oriented, simple, closed contours with γ_2 interior to γ_1. If f is analytic on the closed region containing $\{\gamma_1\}$ and $\{\gamma_2\}$ and the points between them, then

$$\int_{\gamma_1} f(z)dz = \int_{\gamma_2} f(z)dz.$$

Proof. We draw two line segments from γ_1 to γ_2. Then, the end points P_1 and P_2 of the segments divide γ_1 into two contours, γ_{1a} and γ_{1b}, and the end points P_3 and P_4 divide γ_2 into γ_{2a} and γ_{2b}. Let P_1P_3 and P_2P_4 denote the line segments from P_1 to P_3 and P_2 to P_4, respectively. Now, by Remark 15.1 (ii), we have

$$\int_{\gamma_{1a}} f(z)dz = \left(\int_{P_1P_3} + \int_{\gamma_{2a}} + \int_{P_4P_2}\right) f(z)dz$$

and

$$\int_{\gamma_{1b}} f(z)dz = \left(\int_{P_2 P_4} + \int_{\gamma_{2b}} + \int_{P_3 P_1}\right) f(z)dz.$$

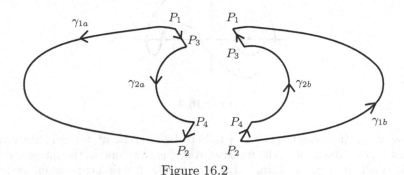

γ_{1a} P_1 P_1

P_3

P_3

γ_{2a} γ_{2b}

P_4 P_4 γ_{1b}

P_2 P_2

Figure 16.2

Adding these equations, we find

$$\int_{\gamma_1} f(z)dz = \left(\int_{\gamma_{1a}} + \int_{\gamma_{1b}}\right) f(z)dz$$

$$= \left(\int_{\gamma_{2a}} + \int_{\gamma_{2b}}\right) f(z)dz = \int_{\gamma_2} f(z)dz. \quad \blacksquare$$

Example 16.1. Determine the possible values of $\int_{\gamma} \dfrac{1}{(z-a)}dz$, where γ is any positively oriented, simple, closed contour not passing through $z = a$. Observe that the function $1/(z-a)$ is analytic everywhere except at the point $z = a$. If a lies exterior to γ, then the integral is zero by Theorem 15.2. If a lies inside γ, we choose a small circle γ_r centered at a and lying within γ. Then, from Theorem 16.1, it follows that

$$\int_{\gamma} \frac{dz}{z-a} = \int_{\gamma_r} \frac{dz}{z-a}.$$

Now, since, on γ_r, $z = a + re^{i\theta}$ (r is fixed), $dz = rie^{i\theta}d\theta$, and hence

$$\int_{\gamma_r} \frac{dz}{z-a} = \int_0^{2\pi} \frac{1}{r}e^{-i\theta}rie^{i\theta}d\theta = \int_0^{2\pi} id\theta = 2\pi i.$$

Hence, we have

$$\int_{\gamma} \frac{dz}{z-a} = \begin{cases} 2\pi i & \text{if } a \text{ is inside } \gamma \\ 0 & \text{if } a \text{ is outside } \gamma. \end{cases}$$

Example 16.2. Evaluate $\int_\gamma \frac{1}{z^2-1} dz$, where the contour γ is given in Figure 16.3.

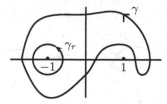

Figure 16.3

Clearly, the integrand $1/(z^2-1)$ fails to be analytic at $z = \pm 1$. Since the point 1 lies outside of γ, the integral over γ is the same as the integral over that small circle γ_r enclosing -1. Using partial fraction expansion, we find

$$\int_\gamma \frac{dz}{z^2-1} = \int_{\gamma_r} \frac{dz}{z^2-1} = \int_\gamma \left[\frac{1}{2(z-1)} - \frac{1}{2(z+1)}\right] dz$$

$$= 0 - \frac{1}{2} 2\pi i = -\pi i.$$

Now we state the following result that extends Theorem 16.1.

Theorem 16.2. Let $\gamma, \gamma_1, \cdots, \gamma_n$ be simple, closed, positively oriented contours such that each γ_j, $j = 1, \cdots, n$ lies interior to γ, and the interior of γ_j has no points in common with the interior of γ_k if $j \neq k$ (see Figure 16.4, where for simplicity we have taken each γ_j to be a circle). If f is analytic on the closed region containing $\{\gamma\}$, $\{\gamma_1\}, \cdots, \{\gamma_n\}$ and the points between them, then

$$\int_\gamma f(z)dz = \sum_{j=1}^{n} \int_{\gamma_j} f(z)dz. \tag{16.1}$$

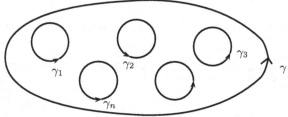

Figure 16.4

Proof. The proof of Theorem 16.2 is exactly the same as that of Theorem 16.1, except here we need to divide γ into several parts. ∎

Example 16.3. Find $\displaystyle\int_\gamma \frac{(5z-2)}{z^2-z}dz$, where γ is given in Figure 16.5.

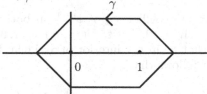

Figure 16.5

The integrand $f(z) = (5z-2)/(z^2-z)$ is analytic everywhere except for the zeros of the denominator, $z = 0$ and $z = 1$. Let γ_1 and γ_2 be two small circles enclosing these points. Then, using Theorem 16.2, we get

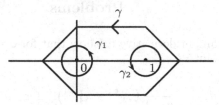

Figure 16.6

$$\int_\gamma f(z)dz = \int_{\gamma_1} f(z)dz + \int_{\gamma_2} f(z)dz.$$

Thus, we have

$$\int_\gamma \frac{5z-2}{z(z-1)}dz = \int_{\gamma_1}\left(\frac{2}{z}+\frac{3}{z-1}\right)dz + \int_{\gamma_2}\left(\frac{2}{z}+\frac{3}{z-1}\right)dz$$
$$= 2\times 2\pi i + 0 + 0 + 3\times 2\pi i = 10\pi i.$$

Example 16.4. Let γ be a simple closed contour that contains the distinct points $z_1, z_2, \cdots, z_n$ in its interior. We shall show that

$$\int_\gamma \frac{dz}{(z-z_1)(z-z_2)\cdots(z-z_n)} = 0.$$

For this, we write the partial fractional decomposition

$$\frac{1}{(z-z_1)(z-z_2)\cdots(z-z_n)} = \frac{A_1}{z-z_1} + \frac{A_2}{z-z_2} + \cdots + \frac{A_n}{z-z_n},$$

which is the same as

$$1 = \sum_{j=1}^{n} A_j(z - z_1) \cdots (z - z_{j-1})(z - z_{j+1}) \cdots (z - z_n).$$

Thus, comparing the coefficients of z^{n-1} on both the sides, we find that $A_1 + A_2 + \cdots + A_n = 0$. Next, let γ_j, $j = 1, 2, \cdots, n$ be small circles with center z_j, that lie in the interior of γ. Then, from Theorem 16.2 and Example 16.1, it follows that

$$\int_{\gamma} \frac{dz}{(z - z_1)(z - z_2) \cdots (z - z_n)} = \sum_{j=1}^{n} \int_{\gamma_j} \frac{A_j}{z - z_j} = 2\pi i \sum_{j=1}^{n} A_j = 0.$$

Problems

16.1. Use an antiderivative to show that for every γ extending from a point z_1 to a point z_2

$$\int_{\gamma} z^n dz = \frac{1}{n+1} \left(z_2^{n+1} - z_1^{n+1}\right) \quad \text{for} \quad n = 0, 1, 2, \cdots.$$

16.2. Let γ be the semicircle from $-3i$ to $3i$ in anticlockwise direction. Show that $\int_{\gamma} \frac{dz}{z} = \pi i$.

16.3. Evaluate each of the following integrals where the path is an arbitrary contour between the limits of integrations

(a). $\displaystyle\int_{i}^{i/2} e^{\pi z} dz$, (b). $\displaystyle\int_{0}^{\pi+2i} \cos\left(\frac{z}{2}\right) dz$, (c). $\displaystyle\int_{1}^{3} (z - 3)^3 dz$.

16.4. Show that if z^i is the principal value, then

$$\int_{\gamma} z^i dz = \frac{1 + e^{-\pi}}{2}(1 - i),$$

where γ is the upper semicircle from $z = 1$ to $z = -1$.

16.5. Let f' and g' be analytic for all z, and let γ be any contour joining the points z_1 and z_2. Show that

$$\int_{\gamma} f(z)g'(z)dz = f(z_2)g(z_2) - f(z_1)g(z_1) - \int_{\gamma} f'(z)g(z)dz.$$

16.6. Let $f = u + iv$ be continuous and possess continuous partial derivatives with respect to x and y in a neighborhood of $z = 0$. Show that f is differentiable at $z = 0$ if and only if

$$\lim_{r \to 0} \frac{1}{\pi r^2} \int_{|z|=r} f(z)dz = 0. \tag{16.2}$$

16.7. Use Green's Theorem to show that

$$\text{Area of } D = -\frac{i}{2} \int_C \bar{z}dz.$$

16.8. For each of the following functions f, describe the domain of analyticity and apply the Cauchy-Goursat Theorem to show that $\int_\gamma f(z)dz = 0$, where γ is the circle $|z| = 1$:

(a). $f(z) = \dfrac{1}{z^2 + 2z + 2}$, (b). $f(z) = ze^{-z}$.

What about (c). $f(z) = (2z - i)^{-2}$?

16.9. Let γ denote the boundary of the domain between the circle $|z| = 4$ and the square whose sides lie along the lines $x = \pm 1$, $y = \pm 1$. Assuming that γ is oriented so that the points of the domain lie to the left of γ, state why $\int_\gamma f(z)dz = 0$ when

(a). $f(z) = \dfrac{z+2}{\sin z/2}$, (b). $f(z) = \dfrac{z}{1 - e^z}$.

16.10. Let γ denote the boundary of the domain between the circles $|z| = 1$ and $|z| = 2$. Assuming that γ is oriented so that the points of the domain lie to the left of γ, show that $\int_\gamma f(z)dz = 0$ for the following functions:

(a). $f(z) = \dfrac{e^z}{z^2 + 9}$, (b). $f(z) = \cot z$.

16.11. Let γ be the unit circle $|z| = 1$ traversed twice in the clockwise direction. Evaluate

(a). $\int_\gamma \text{Log}\,(z + 2)dz$, (b). $\int_\gamma \dfrac{dz}{3z^2 + 1}$.

16.12. Let r, R be constants such that $0 < r < R$. Denote by γ_r the circle $z = re^{i\theta}$, $0 \leq \theta \leq 2\pi$. Show that

$$\frac{1}{2\pi i} \int_{\gamma_r} \frac{R + z}{(R - z)z}dz = 1.$$

Hence, deduce that

$$\frac{1}{2\pi}\int_0^{2\pi} \frac{R^2 - r^2}{R^2 + r^2 - 2rR\cos\theta}d\theta = 1.$$

16.13. Suppose that f is of the form

$$f(z) = \sum_{j=1}^{n} \frac{A_j}{z^j} + g(z),$$

where g is analytic inside and on the simple, closed, positively oriented contour γ containing 0 in its interior. Show that $\int_\gamma f(z)dz = 2\pi i A_1$.

16.14. Establish the following Wallis formulas

(a). $\displaystyle\int_0^{2\pi} \cos^{2k} t\, dt = 2\pi\frac{(2k)!}{2^{2k}(k!)^2}$,   (b). $\displaystyle\int_0^{2\pi} \cos^{2k+1} t\, dt = 0$,

(c). $\displaystyle\int_0^{2\pi} \sin^{2k} t\, dt = 2\pi\frac{(2k)!}{2^{2k}(k!)^2}$,   (d). $\displaystyle\int_0^{2\pi} \sin^{2k+1} t\, dt = 0$.

Answers or Hints

16.1. $\int_\gamma z^n dz = \frac{1}{n+1}(z_2^{n+1} - z_1^{n+1})$.

16.2. $\int_\gamma dz/z = \mathrm{Log}\, z\Big|_{-3i}^{3i}$ (since $\frac{d}{dz}\mathrm{Log}\, z = \frac{1}{z}$ on γ) $= \ln|3| + i\frac{\pi}{2} -$
$\left(\ln|3| - \frac{i\pi}{2}\right) = i\pi$.

16.3. (a). $(1+i)/\pi$, (b). $e + e^{-1}$, (c). -4.

16.4. Principal value $z^i = e^{i\,\mathrm{Log}\, z}$. Hence, $\int_\gamma z^i dz = -\int_\pi^0 e^{i(i\theta)}ie^{i\theta}d\theta =$
$-\frac{(1-i)}{2}(e^{-\pi} + 1)$.

16.5. Use $[f(z)g(z)]' = f'(z)g(z) + f(z)g'(z)$.

16.6. As in Lecture 15, we have

$$\int_{|z|=r} f(z)dz = \int\int_{|z|\leq r}(-v_x - u_y)dxdy + i\int\int_{|z|\leq r}(u_x - v_y)dxdy,$$

and hence, from the Mean Value Theorem of integral calculus, we find

$$\frac{1}{\pi r^2}\int_{|z|=r} f(z)dz = -(v_x(\overline{x},\overline{y}) + u_y(\overline{x},\overline{y})) + i(u_x(\hat{x},\hat{y}) - v_y(\hat{x},\hat{y})),$$

where $(\overline{x},\overline{y})$ and $(\hat{x},\hat{y})$ are suitable points. Thus, (16.2) holds if and only if the Cauchy-Riemann conditions (6.5) hold at $z = 0$. Now, since the partial derivatives of u and v are continuous, it is equivalent to the differentiability of f at $z = 0$.

16.7. In Green's Theorem, we let $P = x$ and $Q = y$, to obtain $2\int\int_D dx dy = \int_C x dy - y dx$, and hence $2D = \int_C x dy - y dx$. Suppose the parametric equation of C is $x = x(t)$, $y = y(t)$, so that $z(t) = x(t) + iy(t)$ and $x(a) - x(b)$, $y(a) = y(b)$ (C is closed). Then $2D = \int_a^b (xy' - yx')dt$. Now $\int_C \overline{z} dz = \int_a^b (x - iy)(x' + iy')dt = \int_a^b [(xx' + yy') + i(xy' - yx')]dt = \frac{1}{2}[x^2(t) + y^2(t)]\big|_a^b + i\int_a^b (xy' - yx')dt = 0 + i\int_a^b (xy' - yx')dt = 2iD$, and hence $D = \frac{1}{2i}\int_C \overline{z} dz = -\frac{i}{2}\int_C \overline{z} dz$.

16.8. (a). $f(z) = \frac{1}{(z-(-1+i))(z-(-1-i))}$ and thus analytic if $z \neq -1 + i$ or $-1 - i$; furthermore, $\int_\gamma \frac{dz}{z^2+2z+2} = 0$, (b). ze^{-z} is entire. (c). Analytic if $z \neq i/2$, cannot apply Cauchy-Goursat Theorem; however, since f has an antiderivative on γ, $\int_\gamma \frac{dz}{(2z-i)^2} = 0$.

16.9. (a). $f'(z) = [(\sin z/2)(1) - (z+2)(1/2)\cos z/2]/\sin^2 z/2$ if $\sin z/2 \neq 0$; $\sin z/2 = 0$ if and only if $e^{-iz/2}(e^{iz} - 1) = 0$; i.e., if and only if $e^{iz} = 1$, or $z = 2k\pi$, $k \in \mathbf{Z}$; none of this point is in the domain. (b). $f'(z) = [(1-e^z) - z(-e^z)]/(1-e^z)^2$ if $e^z \neq 1$; $e^z = 1$ if and only if $z = 2k\pi i$, $k \in \mathbf{Z}$; none of this point is in the domain. Finally, note that $\gamma = \gamma_1 + \gamma_2$ and $\int_{\gamma_1} f dz = \int_{\gamma_2} f dz = 0$.

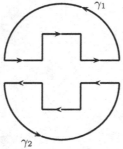

16.10. (a). $f(z) = e^z/(z^2 + 9)$ is analytic if $z^2 + 9 \neq 0$; i.e., $z = \pm 3i$. But $\pm 3i$ are not in the domain, and hence, by the Cauchy-Goursat Theorem, $\int_{\gamma_1} f(z)dz = \int_{\gamma_2} f(z)dz = 0$. Hence, $\int_\gamma f(z)dz = 0$.

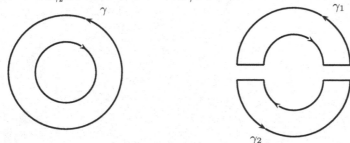

(b). $\cot z$ is analytic if $\sin z \neq 0$; i.e., $z \neq k\pi$, $k \in \mathbf{Z}$. But $k\pi$ are not in the domain for all $k \in \mathbf{Z}$. Hence, as in (a), $\int_\gamma \cot z dz = 0$.

16.11. Let γ be the circle $|z| = 1$ taken in the anticlockwise direction.

(a). $\int_\gamma \text{Log}\,(z+2)dz = 0$ by the Cauchy-Goursat Theorem, (b). $\int_\gamma \frac{dz}{3z^2+1} =$
$\frac{1}{3}\int_\gamma \left[\frac{\sqrt{3}}{2i(z-i/\sqrt{3})} - \frac{\sqrt{3}}{2i(z+i/\sqrt{3})}\right]dz = \frac{\sqrt{3}}{6i}(2\pi i) - \frac{\sqrt{3}}{6i}(-2\pi i) = 0.$ Thus,
$-\int_\gamma \text{Log}(z+2)dz = \int_\gamma \text{Log}(z+2)dz + \int_\gamma \text{Log}(z+2)dz = 0$ and $-\int_\gamma \frac{dz}{3z^2+1} =$
$\int_\gamma \frac{dz}{3z^2+1} + \int_\gamma \frac{dz}{3z^2+1} = 0.$

16.12. Use partial fractions. Compare the real parts.

16.13. Use Theorem 16.1 and Examples 14.6 and 16.1.

16.14. Let γ be the unit circle $|z| = 1$. Then, $\frac{1}{2\pi i}\int_\gamma \left(z + \frac{1}{z}\right)^n \frac{dz}{z} = \frac{2^n}{2\pi} \times$
$\int_0^{2\pi} \cos^n t\,dt.$ Now expand $(z + 1/z)^n$ using the binomial formula, and use Problem 16.13.

Lecture 17
Cauchy's Integral Formula

In this lecture, we shall present Cauchy's integral formula that expresses the value of an analytic function at any point of a domain in terms of the values on the boundary of this domain, and has numerous important applications. We shall also prove a result that paves the way for the Cauchy's integral formula for derivatives given in the next lecture.

Theorem 17.1 (Cauchy's Integral Formula). Let γ be a simple, closed, positively oriented contour. If f is analytic in some simply connected domain S containing γ and z_0 is any point inside γ, then

$$f(z_0) = \frac{1}{2\pi i} \int_\gamma \frac{f(z)}{z - z_0} dz. \qquad (17.1)$$

Proof. The function $f(z)/(z - z_0)$ is analytic everywhere in S except at the point z_0. Hence, in view of Theorem 16.1, the integral over γ is the same as the integral over some small positively oriented circle $\gamma_r : |z - z_0| = r$; i.e.,

$$\int_\gamma \frac{f(z)}{z - z_0} dz = \int_{\gamma_r} \frac{f(z)}{z - z_0} dz.$$

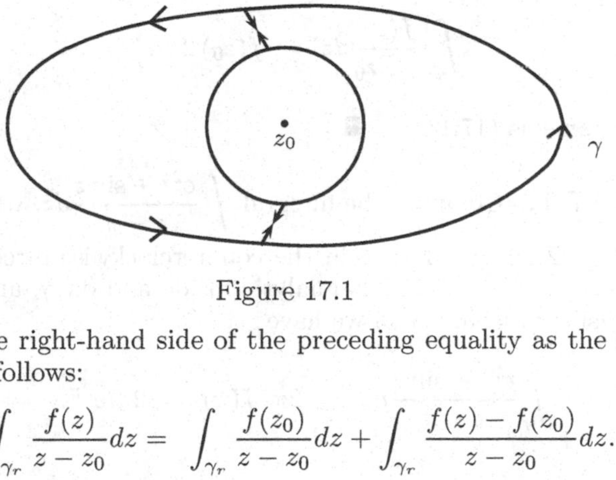

Figure 17.1

We write the right-hand side of the preceding equality as the sum of two integrals as follows:

$$\int_{\gamma_r} \frac{f(z)}{z - z_0} dz = \int_{\gamma_r} \frac{f(z_0)}{z - z_0} dz + \int_{\gamma_r} \frac{f(z) - f(z_0)}{z - z_0} dz.$$

However, since from Example 16.1

$$\int_{\gamma_r} \frac{f(z_0)}{z - z_0} dz = f(z_0) \int_{\gamma_r} \frac{dz}{z - z_0} = f(z_0)\, 2\pi i,$$

it follows that

$$\int_\gamma \frac{f(z)}{z - z_0} dz = f(z_0)\, 2\pi i + \int_{\gamma_r} \frac{f(z) - f(z_0)}{z - z_0} dz.$$

The first two terms in the equation above are independent of r, and hence the value of the last term does not change if we allow $r \to 0$; i.e.,

$$\int_\gamma \frac{f(z)}{z - z_0} dz = f(z_0)\, 2\pi i + \lim_{r \to 0^+} \int_{\gamma_r} \frac{f(z) - f(z_0)}{z - z_0} dz. \tag{17.2}$$

Let $M_r = \max\{|f(z) - f(z_0)| : z \in \gamma_r\}$. Since f is continuous, such a finite number M_r exists, and clearly, $M_r \to 0$ as $r \to 0$. Now, for z on γ_r, we have

$$\left| \frac{f(z) - f(z_0)}{z - z_0} \right| = \frac{|f(z) - f(z_0)|}{r} \leq \frac{M_r}{r}.$$

Hence, from Theorem 13.1, we find

$$\left| \int_{\gamma_r} \frac{f(z) - f(z_0)}{z - z_0} dz \right| \leq \frac{M_r}{r} L(\gamma_r) = \frac{M_r}{r} 2\pi r = 2\pi M_r,$$

which implies that

$$\lim_{r \to 0^+} \int_{\gamma_r} \frac{f(z) - f(z_0)}{z - z_0} dz = 0.$$

Therefore, equation (17.2) reduces to

$$\int_\gamma \frac{f(z)}{z - z_0} dz = f(z_0)\, 2\pi i,$$

which is the same as (17.1). ∎

Example 17.1. Compute the integral $\displaystyle\int_\gamma \frac{e^{2z} + \sin z}{z - \pi} dz$, where γ is the circle $|z - 2| = 2$ traversed once in the counterclockwise direction. Since the function $f(z) = e^{2z} + \sin z$ is analytic inside and on γ, and the point $z_0 = \pi$ lies inside γ, from (17.1) we have

$$\int_\gamma \frac{e^{2z} + \sin z}{z - \pi} dz = 2\pi i\, f(\pi) = 2\pi i e^{2\pi}.$$

Example 17.2. Compute the integral $\displaystyle\int_\gamma \frac{\cos z + \sin z}{z^2 - 9} dz$ along the contour given in Figure 17.2. Clearly, the integrand fails to be analytic at the points $z = \pm 3$. However, only $z = 3$ lies inside γ. If we write $(\cos z +$

$\sin z)/(z^2 - 9) = [(\cos z + \sin z)/(z + 3)]/(z - 3)$, then we can apply (17.1) to the function $f(z) = (\cos z + \sin z)/(z + 3)$. Hence, it follows that

$$\int_\gamma \frac{\cos z + \sin z}{z^2 - 9} dz = 2\pi i f(3) = \frac{1}{3}\pi i(\cos 3 + \sin 3).$$

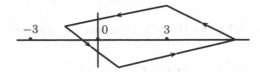

Figure 17.2

Example 17.3. Compute $\displaystyle\int_\gamma \frac{e^z}{z(z-2)} dz$, where γ is the following figure-eight contour.

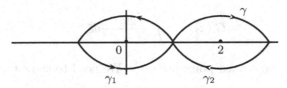

Figure 17.3

Let γ_1 and γ_2 be the positively oriented left lobe and the negatively oriented right lobe, respectively. Then, we have

$$\int_\gamma \frac{e^z}{z(z-2)} dz = \int_{\gamma_1} \frac{e^z/(z-2)}{z} dz + \int_{\gamma_2} \frac{e^z/z}{z-2} dz$$

$$= 2\pi i \frac{e^z}{z-2}\Big|_{z=0} + (-2\pi i) \frac{e^z}{z}\Big|_{z=2} = -\pi i - e^2\pi i.$$

Example 17.4. Compute $\displaystyle\int_\gamma \frac{e^z}{z(z-2)} dz$, where γ is the following contour.

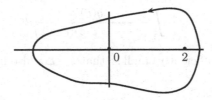

Figure 17.4

Clearly, we have

$$\int_\gamma \frac{e^z}{z(z-2)}dz = \int_\gamma \frac{e^z}{2(z-2)}dz - \int_\gamma \frac{e^z}{2z}dz$$

$$= 2\pi i \left.\frac{e^z}{2}\right|_{z=2} - 2\pi i \left.\frac{e^z}{2}\right|_{z=0} = \pi i e^2 - \pi i.$$

Now we shall prove the following general result, which we shall need in the next lecture.

Theorem 17.2. Let the function g be continuous on the contour γ. For each z not on γ, we define

$$G(z) = \int_\gamma \frac{g(\xi)}{\xi - z}d\xi.$$

Then, the function G is analytic at each point not on γ, and its derivative is given by

$$G'(z) = \int_\gamma \frac{g(\xi)}{(\xi - z)^2}d\xi.$$

Proof. Let z be a fixed point not on γ. We need to show that

$$\lim_{\Delta z \to 0} \frac{G(z + \Delta z) - G(z)}{\Delta z} = \int_\gamma \frac{g(\xi)}{(\xi - z)^2}d\xi$$

or, equivalently, the difference

$$\Lambda = \frac{G(z + \Delta z) - G(z)}{\Delta z} - \int_\gamma \frac{g(\xi)}{(\xi - z)^2}d\xi$$

approaches zero as $\Delta z \to 0$.

From the definition of $G(z)$, we have

$$\frac{G(z + \Delta z) - G(z)}{\Delta z} = \frac{1}{\Delta z}\int_\gamma \left[\frac{1}{(\xi - (z + \Delta z))} - \frac{1}{(\xi - z)}\right]g(\xi)d\xi$$

$$= \int_\gamma \frac{g(\xi)}{(\xi - z - \Delta z)(\xi - z)}d\xi,$$

where Δz is chosen sufficiently small so that $z + \Delta z$ also lies off of γ. Then, we get

$$\Lambda = \int_\gamma \frac{g(\xi)d\xi}{(\xi - z - \Delta z)(\xi - z)} - \int_\gamma \frac{g(\xi)d\xi}{(\xi - z)^2} = \Delta z \int_\gamma \frac{g(\xi)d\xi}{(\xi - z - \Delta z)(\xi - z)^2}.$$

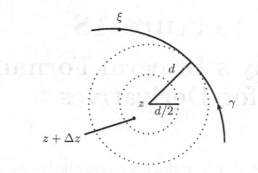

Figure 17.5

Let $M = \max_{\xi \in \gamma} |g(\xi)|$ and let d equal the shortest distance from z to γ, so that $|\xi - z| \geq d > 0$ for all ξ on γ. Since we are letting Δz approach zero, we may assume that $|\Delta z| < d/2$. Then, by the triangle inequality, we have

$$|\xi - z - \Delta z| \geq |\xi - z| - |\Delta z| \geq d - (d/2) = d/2$$

for all ξ on γ. Hence, it follows that

$$\left| \frac{g(\xi)}{(\xi - z - \Delta z)(\xi - z)^2} \right| \leq \frac{M}{\frac{d}{2}d^2} = \frac{2M}{d^3}$$

for all ξ on γ. Now, from Theorem 13.1, we find

$$|\Lambda| = \left| \Delta z \int_{\gamma} \frac{g(\xi)d\xi}{(\xi - z - \Delta z)(\xi - z)^2} \right| \leq \frac{|\Delta z|2ML(\gamma)}{d^3},$$

where $L(\gamma)$ denotes the length of γ. Thus, Λ must approach zero as $\Delta z \to 0$. This completes the proof. ∎

Lecture 18
Cauchy's Integral Formula for Derivatives

In this lecture, we shall show that, for an analytic function in a given domain, all the derivatives exist and are analytic. This result leads to Cauchy's integral formula for derivatives. Next, we shall prove Morera's Theorem, which is a converse of the Cauchy-Goursat Theorem. We shall also establish Cauchy's inequality for the derivatives, which plays an important role in proving Liouville's Theorem.

The arguments employed to prove Theorem 17.2 can be repeated. In fact, starting with the function

$$H(z) = \int_\gamma \frac{g(\xi)}{(\xi - z)^2} d\xi \quad (z \text{ not on } \gamma), \tag{18.1}$$

it can be shown that H is analytic at each point not on γ and that

$$H'(z) = 2 \int_\gamma \frac{g(\xi)}{(\xi - z)^3} d\xi \quad (z \text{ not on } \gamma),$$

which is obtained formally from (18.1) by differentiating with respect to z under the integral sign.

Now we shall apply these results to analytic functions. Suppose that f is analytic at some point z_0. Then, f is differentiable in some neighborhood U of z_0. Choose a positively oriented circle $\gamma_r : |\xi - z_0| = r$ in U.

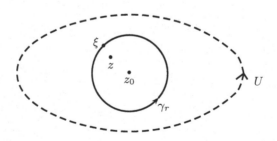

Figure 18.1

116

By Cauchy's Integral Theorem, we have

$$f(z) = \frac{1}{2\pi i} \int_{\gamma_r} \frac{f(\xi)}{\xi - z} d\xi \quad (z \text{ inside } \gamma_r), \tag{18.2}$$

and hence it follows from Theorem 17.2 that

$$f'(z) = \frac{1}{2\pi i} \int_{\gamma_r} \frac{f(\xi)}{(\xi - z)^2} d\xi \quad (z \text{ inside } \gamma_r). \tag{18.3}$$

Clearly, the right-hand side of (18.3) is a function of the form (18.1), and hence it has a derivative at each point inside γ_r. Since the domain interior to γ_r is a neighborhood of z_0, f' is analytic at z_0.

We summarize these considerations in the following theorem.

Theorem 18.1 (Differentiation of Analytic Functions). If f is analytic in a domain S, then all its derivatives $f', f'', \cdots, f^{(n)}, \cdots$ exist and are analytic in S.

Remark 18.1. The analogue of Theorem 18.1 for real functions does not hold, for example, the function $f(x) = x^{3/2}$, $x \in \mathbb{R}$ is differentiable for all real x, but $f'(x) = (3/2)x^{1/2}$ has no derivative at $x = 0$.

Now, repeated differentiation of (18.3) with respect to z under the integral sign leads to the following result.

Theorem 18.2 (Cauchy's Integral Formula for Derivatives). If f is analytic inside and on a simple, closed, positively oriented contour γ, and if z is any point inside γ, then

$$f^{(n)}(z) = \frac{n!}{2\pi i} \int_{\gamma} \frac{f(\xi)}{(\xi - z)^{n+1}} d\xi, \quad n = 1, 2, \cdots. \tag{18.4}$$

From an applications point of view, it is better to write (18.2) and (18.4) in the equivalent form

$$\int_{\gamma} \frac{f(z)}{(z - z_0)^n} dz = \frac{2\pi i}{(n-1)!} f^{(n-1)}(z_0), \quad n = 1, 2, \cdots; \tag{18.5}$$

here, z_0 is inside γ.

Example 18.1. Compute $\int_{\gamma} \frac{\sin 3z}{z^4} dz$, where γ is the circle $|z| = 1$

traversed once counterclockwise. Since $f(z) = \sin 3z$ is analytic inside and on γ, from (18.5) with $z_0 = 0$ and $n = 4$, we have

$$\int_{\gamma} \frac{\sin 3z}{z^4} dz = \frac{2\pi i}{3!} f'''(0) = -9\pi i.$$

Example 18.2. Compute $\int_\gamma \dfrac{3z+1}{z(z-2)^2}dz$ along the contour γ given in Figure 17.3. As in Example 17.3, let γ_1 and γ_2 be the positively oriented left lobe and the negatively oriented right lobe, respectively. Then, we have

$$\int_\gamma \frac{3z+1}{z(z-2)^2}dz = \int_{\gamma_1}\frac{(3z+1)/(z-2)^2}{z}dz + \int_{\gamma_2}\frac{(3z+1)/z}{(z-2)^2}dz.$$

Applying (18.5) to the right-hand side, we get

$$\int_\gamma \frac{3z+1}{z(z-2)^2}dz = 2\pi i\left.\frac{3z+1}{(z-2)^2}\right|_{z=0} - \frac{2\pi i}{1!}\frac{d}{dz}\left.\left(\frac{3z+1}{z}\right)\right|_{z=2}$$

$$= \frac{1}{2}\pi i + \frac{1}{2}\pi i = \pi i.$$

Example 18.3. Compute $\int_\gamma \dfrac{\cosh z}{z(z+1)^2}dz$, where the contour γ is given in Figure 18.2.

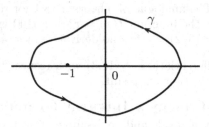

Figure 18.2

Using partial fraction expansion, we have

$$\frac{1}{z(z+1)^2} = \frac{1}{z} - \frac{1}{z+1} - \frac{1}{(z+1)^2},$$

and hence, from (18.5), it follows that

$$\int_\gamma \frac{\cosh z}{z(z+1)^2}dz = \int_\gamma \frac{\cosh z}{z}dz - \int_\gamma \frac{\cosh z}{z+1}dz - \int_\gamma \frac{\cosh z}{(z+1)^2}dz$$

$$= 2\pi i\left.\cosh z\right|_{z=0} - 2\pi i\left.\cosh z\right|_{z=-1} - 2\pi i\left.\frac{d}{dz}\cosh z\right|_{z=-1}$$

$$= 2\pi i - 2\pi i\cosh 1 + 2\pi i\sinh 1.$$

Now recall that if $f(z) = u(x,y) + iv(x,y)$ is analytic, then

$$f'(z) = \frac{\partial u}{\partial x} + i\frac{\partial v}{\partial x} = \frac{\partial v}{\partial y} - i\frac{\partial u}{\partial y}.$$

We now know from Theorem 18.1 that f' is analytic, and hence continuous. Therefore, the equations above imply that all first order partial derivatives of u and v are continuous. Similarly, since f'' exists and is given by

$$
\begin{aligned}
f''(z) &= \frac{\partial^2 u}{\partial x^2} + i\frac{\partial^2 v}{\partial x^2} = \frac{\partial^2 v}{\partial y \partial x} - i\frac{\partial^2 u}{\partial y \partial x} \\
&= \frac{\partial^2 v}{\partial x \partial y} - i\frac{\partial^2 u}{\partial x \partial y} = -\frac{\partial^2 u}{\partial y^2} - i\frac{\partial^2 v}{\partial y^2}
\end{aligned}
$$

the continuity of f'' implies that all second-order partial derivatives of u and v are continuous at the point where f is analytic. Continuing with this process, we obtain the following interesting theorem.

Theorem 18.3. If $f = u + iv$ is analytic in a domain S, then all partial derivatives of u and v exist and are continuous in S.

The following result is a converse of Theorem 15.2.

Theorem 18.4 (Morera's Theorem). If f is continuous in a domain S and $\int_\gamma f(z)dz = 0$ for every closed contour γ in S, then f is analytic.

Proof. By Theorem 14.2, f has an antiderivative F in S; i.e., $F'(z) = f(z)$, $z \in S$. This means that $F(z)$ is analytic. But then, from Theorem 18.1, $F'(z)$ is analytic; i.e., $f(z)$ is analytic. ∎

Example 18.4. Consider the function $f(z) = \int_0^1 e^{-z^2 t} dt$. Let γ be any simple closed contour in the complex plane. Changing the order of integration, we have

$$
\int_\gamma f(z)dz = \int_0^1 \left(\int_\gamma e^{-z^2 t} dz \right) dt = 0.
$$

Hence, in view of Theorem 18.4, the function $f(z) = (1 - e^{-z^2})/z^2$, $z \neq 0$ is analytic.

Now we shall prove the following theorem.

Theorem 18.5 (Cauchy's Inequality). Let f be an analytic function inside and on a circle γ_R of radius R centered at z_0. If $|f(z)| \leq M$ for all z on γ_R, then the following inequality holds

$$
\left| f^{(n)}(z_0) \right| \leq \frac{n!\,M}{R^n}, \quad n = 1, 2, \cdots.
$$

Proof. From (18.4), with $z = z_0$, $\xi = z$, and $\gamma = \gamma_R$, assumed to be positively oriented, we have

$$f^{(n)}(z_0) = \frac{n!}{2\pi i} \int_{\gamma_R} \frac{f(z)}{(z - z_0)^{n+1}} dz. \tag{18.6}$$

For z on γ_R, the integrand is bounded by M/R^{n+1} and the length of γ_R is $2\pi R$. Thus, from Theorem 13.1, it follows that

$$\left| f^{(n)}(z_0) \right| \leq \frac{n!}{2\pi} \frac{M}{R^{n+1}} 2\pi R = \frac{n!M}{R^n}. \quad \blacksquare$$

Finally, we shall prove the following result.

Theorem 18.6 (Liouville's Theorem). The only bounded entire functions are the constant functions.

Proof. Suppose f is analytic and bounded by some number M over the whole complex plane $\mathbf{C}$. By Theorem 18.5, for the case $n = 1$, we have $|f'(z_0)| \leq M/R$ for any $z_0 \in \mathbf{C}$ and for any $R > 0$. Letting $R \to \infty$, we find $f'(z_0) = 0$ for any z_0. Hence, f' vanishes everywhere; i.e., f must be a constant. $\quad \blacksquare$

Problems

18.1. Show that the Cauchy integral formula implies the Cauchy-Goursat Theorem.

18.2. Suppose that f is analytic in a simply connected domain S, and let $z_0 \in S$ be a fixed point. Define the integral along any contour connecting the points z_0 and $z \in S$ as $F(z) = \int_{z_0}^{z} f(\xi)d\xi$. Show that $F'(z) = f(z)$; i.e., the integral is also an analytic function of its upper limit.

18.3. Suppose that f is analytic in a domain S, and let $z_0 \in S$ be a fixed point. Define

$$g(z) = \begin{cases} \dfrac{f(z) - f(z_0)}{z - z_0} & \text{if } z \neq z_0 \\ f'(z_0) & \text{if } z = z_0. \end{cases}$$

Show that g is analytic in S. Hence, deduce that the function

$$g(z) = \begin{cases} \dfrac{\sin z}{z} & \text{if } z \neq 0 \\ 1 & \text{if } z = 0 \end{cases} \tag{18.7}$$

is entire.

18.4. Let f be analytic within and on a positively oriented closed contour γ, and the point z_0 is not on γ. Show that

$$\int_\gamma \frac{f(z)}{(z-z_0)^2}\,dz = \int_\gamma \frac{f'(z)}{z-z_0}\,dz.$$

18.5. Let $u = u(x,y)$ be a harmonic function in a domain S. Show that all partial derivatives $u_x, u_y, u_{xx}, u_{xy}, u_{yy}, \cdots$ exist and are harmonic.

18.6. Let γ be a simple closed contour described in the positive sense, and write $g(z) = \int_\gamma \frac{\xi^3 + 7\xi}{(\xi - z)^3}\,d\xi$. Show that $g(z) = 6\pi i z$ when z is inside γ and that $g(z) = 0$ when z is outside γ.

18.7. Let $f(z) = (c^z + e^{-z})/2$. Evaluate $\int_\gamma \frac{f(z)}{z^4}\,dz$, where γ is any simple closed curve enclosing 0.

18.8. (a). Let γ be the contour $z = e^{i\theta}$, $-\pi \le \theta \le \pi$ traversed in the positive direction. Show that, for any real constant a, $\int_\gamma \frac{e^{az}}{z}\,dz = 2\pi i$.

(b). Deduce that $\int_0^\pi e^{a\cos\theta}\cos(a\sin\theta)\,d\theta = \pi$.

18.9. Let γ denote the boundary of the rectangle whose vertices are $-2 - 2i, 2 - 2i, 2 + i$ and $-2 + i$ in the positive direction. Evaluate each of the following integrals:

(a). $\int_\gamma \frac{e^{-z}}{z - \frac{\pi i}{4}}\,dz$, (b). $\int_\gamma \frac{\cos z}{z^4}\,dz$, (c). $\int_\gamma \frac{z}{(2z+1)^2}\,dz$,

(d). $\int_\gamma \frac{e^{-z}}{z^2 + 2}\,dz$, (e). $\int_\gamma \frac{dz}{z(z+1)}$, (f). $\int_\gamma \left[e^z \sin z + \frac{1}{(z^2+3)^2} \right]\,dz$.

18.10. Let f be analytic inside and on the unit circle γ. Show that, for $0 < |z| < 1$,

$$2\pi i f(z) = \int_\gamma \frac{f(\xi)}{\xi - z}\,d\xi - \int_\gamma \frac{f(\xi)}{\xi - 1/\bar{z}}\,d\xi.$$

Hence, deduce the *Poisson integral formula*

$$f(re^{i\theta}) = \frac{1}{2\pi}\int_0^{2\pi} \frac{(1 - r^2)}{1 - 2r\cos(\theta - t) + r^2}f(e^{it})\,dt, \quad 0 < r < 1.$$

18.11. If $f(z)$ is analytic and $|f(z)| \le 1/(1 - |z|)$ in $|z| < 1$, show that $|f'(0)| \le 4$.

18.12. Let the function f be entire and $f(z) \to \infty$ as $z \to \infty$. Show that f must have at least one zero.

18.13. Let $f(z)$ be an entire function such that $|f'(z)| \leq |z|$. Show that $f(z) = a + bz^2$ with some constants $a, b \in \mathbf{C}$ such that $|b| \leq 1$.

18.14. Suppose $f(z)$ is an entire function with $f(z) = f(z+1) = f(z+i)$ for all $z \in \mathbf{C}$. Show that $f(z)$ is a constant.

18.15. Suppose $f(z)$ and $g(z)$ are entire functions, $g(z) \neq 0$ and $|f(z)| \leq |g(z)|$, $z \in \mathbf{C}$. Show that there is a constant c such that $f(z) = cg(z)$.

18.16. Show that an entire function whose real part is nonpositive is constant.

18.17. A subset S of $\mathbf{C}$ is said to be *dense* in $\mathbf{C}$ if its closure $\overline{S} = \mathbf{C}$. Show that the image of a nonconstant entire function is dense in $\mathbf{C}$.

18.18. Suppose $f(z, t)$ is continuous, and continuously differentiable with respect to t, and suppose γ is a smooth curve in the domain of definition of f. Then, the function $F(t) = \displaystyle\int_{\gamma} f(z, t) dz$ is continuously differentiable with respect to t, and its derivative is

$$\frac{dF(t)}{dt} = \int_{\gamma} \frac{\partial f(z, t)}{\partial t} dz.$$

18.19. Let $f(z, \lambda)$ be a continuous function of z on the bounded closed region S for each value of λ in $|\lambda - \lambda_0| < \rho$, and $f(z, \lambda) \to F(z)$ uniformly as $\lambda \to \lambda_0$. Show that $F(z)$ is continuous on S. Moreover, if γ is any closed contour lying in S, then

$$\lim_{\lambda \to \lambda_0} \int_{\gamma} f(z, \lambda) dz = \int_{\gamma} F(z) dz.$$

18.20. Let $f(z, \lambda)$ be an analytic function of z on the domain S for each value of λ in $|\lambda - \lambda_0| < \rho$, and $f(z, \lambda) \to F(z)$ uniformly as $\lambda \to \lambda_0$ in every closed region G of S. Show that $F(z)$ is analytic on S. Moreover, as $\lambda \to \lambda_0$, $\partial f(z, \lambda)/\partial z \to F'(z)$ uniformly on G.

Answers or Hints

18.1. Let f be a complex function that is analytic throughout the region enclosed by a simple closed contour γ. Let z_0 be a point inside γ. Then $\int_{\gamma} f(z) dz = \int_{\gamma} \frac{f(z)(z-z_0)}{z-z_0} dz = 2\pi i (f(z_0)(z_0 - z_0)) = 0$.

18.2. If $z + \Delta z \in S$, then $\frac{F(z+\Delta z)-F(z)}{\Delta z} - f(z) = \frac{1}{\Delta z}\int_z^{z+\Delta z}[f(\xi)-f(z)]d\xi$.
Since f is continuous at z, $|\xi - z| < \delta$, or in particular $|\Delta z| < \delta$ implies
$|f(\xi) - f(z)| < \epsilon$. Thus, $\left|\frac{F(z+\Delta z)-F(z)}{\Delta z} - f(z)\right| < \frac{\epsilon}{|\Delta z|}\int_z^{z+\Delta z}|d\xi| = \epsilon$.

18.3. Let $B(z_0, R)$ be an open disk contained in S. For $z \neq z_0$ in $B(z_0, R)$
consider the segment $\xi(t) = z_0(1 - t) + tz$, $t \in [0,1]$. Since f' is analytic in
$B(z_0, R)$ we can write $g(z)$ as $g(z) = \int_0^1 f'((z - z_0)t + z_0)dt$. This is also
defined at $z = z_0$, in fact, $g(z_0) = \int_0^1 f'(z_0)dt = f'(z_0)$. Now use the fact
that f'' is analytic in $B(z_0, R)$.

18.4. Use (17.1) for $f'(z)$ and (18.6) for $n = 1$.

18.5. For each point $z_0 = (x_0, y_0)$ in S there exists a disk $B(z_0, R) \subset S$.
In this disk, a conjugate harmonic function v exists, so that the function
$f = u + iv$ is analytic. Now use Theorem 18.1.

18.6. If z is inside γ, we have $\int_\gamma \frac{\xi^3+7\xi}{(\xi-z)^3}d\xi = \frac{2\pi i}{2!}f''(z) = 6\pi iz$ ($f(\xi) = \xi^3 + 7\xi$). If z is outside γ, then $\frac{\xi^3+7\xi}{(\xi-z)^3}$ is analytic throughout the region
enclosed by γ (including γ) and therefore $\int_\gamma \frac{\xi^3+7\xi}{(\xi-z)^3}d\xi = 0$.

18.7. 0.

18.8. (a). $2\pi i(e^{a\cdot 0})$, (b). $\int_\gamma \frac{e^{az}}{z}dz = \int_{-\pi}^{\pi}\frac{e^{a\cos\theta}e^{ia\sin\theta}}{e^{i\theta}}ie^{i\theta}d\theta$
$= \int_{-\pi}^{\pi}ie^{a\cos\theta}\cos(a\sin\theta)d\theta - \int_{-\pi}^{\pi}e^{a\cos\theta}\sin(a\sin\theta)d\theta = 2\pi i$.

18.9. (a). $2\pi i(e^{-\pi i/4})$, (b). 0, (c). $\pi i/2$, (d). $-\pi e^{\sqrt{2}i}/\sqrt{2}$, (e). 0,
(f). $-\pi\sqrt{3}/18$.

18.10. $1/\bar{z}$ is outside γ. Take $z = re^{i\theta}$ and $\xi = e^{it}$.

18.11. For any $r < 1$, let γ_r denote the circle $|z| = r$. By (18.6), we have
$|f'(0)| \leq \frac{1}{2\pi}\int_{\gamma_r}\frac{1}{(1-|z|)|z|^2}|dz| \leq \frac{1}{2\pi}\frac{1}{r^2(1-r)}2\pi r = \frac{1}{r(1-r)}$. Now let $r = 1/2$.

18.12. If f has no zero in $\mathbf{C}$, then $g(z) = 1/f(z)$ is entire and $g(z) \to 0$
as $z \to \infty$. Now show that g is bounded, and then apply Theorem 18.6 to
conclude that g is a constant.

18.13. Note that $f''(z_0) = \frac{1}{2\pi i}\int_{\gamma_R}[f'(z)/(z - z_0)^2]dz$, where γ_R is the
circle $|z - z_0| = R$ taken in the positive direction. Hence, $|f''(z_0)| \leq M_R/R$
where $M_R \geq |f'(z)|$ on $|z - z_0| = R$. Now, since $|f'(z)| \leq |z|$, it follows
that $|f''(z_0)| \leq (R + |z_0|)/R$ for all R. Thus, in particular, for all $z_0 \in \mathbf{C}$, $|f''(z_0)| \leq 2$. Therefore, in view of Theorem 18.6, $f''(z)$ must be a
constant, and so $f(z) = a + \alpha z + bz^2$. However, $|f'(z)| \leq |z|$ implies that
$f'(0) = 0$, and hence we should have $\alpha = 0$; i.e., $f(z) = a + bz^2$. Finally,
since $|f''(z)| = |2b| \leq 2$, $|b| \leq 1$.

18.14. Observe that $f(z)$ is bounded, and apply Theorem 18.6.

18.15. Observe that $f(z)/g(z)$ is entire and $|f(z)/g(z)| \leq 1$. Now use Theorem 18.6.

18.16. If $f = u + iv$, then $u \leq 0$ implies $|e^f| = e^u \leq 1$. Thus, in view of
Theorem 18.6, $e^f = c$. Hence, $e^f f' = 0$, which gives $f' = 0$.

18.17. Let $f : \mathbf{C} \to \mathbf{C}$ be an entire function. If $f(\mathbf{C})$ is not dense in $\mathbf{C}$,
then $\mathbf{C}\backslash\overline{f(\mathbf{C})}$ is open and $\neq \emptyset$. Thus, there exist $w \in \mathbf{C}\backslash\overline{f(\mathbf{C})}$ and $r > 0$
such that $B(w, r) \subset \mathbf{C}\backslash\overline{f(\mathbf{C})}$. Hence, $|f(z) - w| \geq r > 0$ for all $z \in \mathbf{C}$.

Consider the function $g : \mathbf{C} \to \mathbf{C}$ defined by $g(z) = 1/(f(z) - w)$. Clearly, $g(z)$ is entire since $f(z)$ is; moreover, $|g(z)| \leq 1/r$, so $g(z)$ is bounded. By Theorem 18.6, $g(z)$ must be a constant, and hence $f(z)$ must be a constant.

18.18. Consider $\frac{F(t)-F(t_0)}{t-t_0} = \frac{1}{t-t_0} \int_\gamma [f(z,t) - f(z,t_0)]dz$. If f is real-valued, then by the Mean Value Theorem we have $\frac{F(t)-F(t_0)}{t-t_0} = \int_\gamma \frac{\partial f(z,t^*)}{\partial t} dz$, where t^* lies between t_0 and t but depends on z. Thus,

$$\left| \frac{F(t)-F(t_0)}{t-t_0} - \int_\gamma \frac{\partial f(z,t)}{\partial t} dz \right| \leq \int_\gamma \left| \frac{\partial f(z,t^*)}{\partial t} - \frac{\partial f(z,t)}{\partial t} \right| |dz|.$$

Now, using uniform continuity of $\frac{\partial f}{\partial t}$ in $\gamma \times [t_0, t]$ and letting $t \to t_0$, the result follows. If f is complex-valued, we use the same argument for the real and imaginary parts separately. Finally, the continuity of $\frac{\partial f}{\partial t}$ implies that of $\frac{\partial F}{\partial t}$.

18.19. Since $f(z, \lambda)$ is continuous, for every $\epsilon > 0$ there exists a $\delta = \delta(z_0, \lambda)$ such that $|z - z_0| < \delta \Rightarrow |f(z,\lambda) - f(z_0,\lambda)| < \epsilon/3$. Since $f(z,\lambda) \to F(z)$ uniformly, there exists $0 < \rho' \leq \rho$ such that $|\lambda - \lambda_0| < \rho' \Rightarrow |f(z,\lambda) - F(z)| < \epsilon/3$. Thus, $|z - z_0| < \delta$, $|\lambda - \lambda_0| < \rho' \Rightarrow |F(z) - F(z_0)| \leq |F(z) - f(z,\lambda)| + |f(z,\lambda) - f(z_0,\lambda)| + |f(z_0,\lambda) - F(z_0)| < \epsilon$. Again, since $f(z,\lambda) \to F(z)$ uniformly, for every $\epsilon > 0$ there exists $0 < \rho' \leq \rho$ such that $|\lambda - \lambda_0| < \rho' \Rightarrow |f(z,\lambda) - F(z)| < \epsilon/L(\gamma)$, but then $\left| \int_\gamma (f(z,\lambda) - F(z))dz \right| < \epsilon$.

18.20. Similar to that of Problem 18.19.

Lecture 19
The Fundamental Theorem of Algebra

In this lecture, we shall prove the Fundamental Theorem of Algebra, which states that every nonconstant polynomial with complex coefficients has at least one zero. Then, as a consequence of this theorem, we shall establish that every polynomial of degree n has exactly n zeros, counting multiplicities. For a given polynomial, we shall also provide some bounds on its zeros in terms of the coefficients.

A point $z_0 \in \mathbf{C}$ is called a *zero of order* m for the function f if f is analytic at z_0 and $f(z_0) = f'(z_0) = \cdots = f^{(m-1)}(z_0) = 0$, $f^{(m)}(z_0) \neq 0$. A zero of order 1 is called a *simple* zero. The main result of this lecture is the following theorem.

Theorem 19.1 (The Fundamental Theorem of Algebra).
Every nonconstant polynomial

$$P_n(z) = a_n z^n + a_{n-1} z^{n-1} + \cdots + a_1 z + a_0, \quad n \geq 1, \qquad (19.1)$$

where the coefficients a_j, $j = 0, 1, \cdots, n$ are complex and $a_n \neq 0$ has at least one zero in $\mathbf{C}$.

To prove this theorem, we need the following lemma.

Lemma 19.1. Let $A = \max\limits_{0 \leq j \leq n} |a_j|$. Then, for $|z| \geq 2nA/|a_n|$, $|P_n(z)| \geq |a_n| |z|^n / 2$.

Proof. Since

$$P_n(z) = z^n \left(a_n + \frac{a_{n-1}}{z} + \frac{a_{n-2}}{z^2} + \cdots + \frac{a_0}{z^n} \right) \qquad (19.2)$$

by the triangle inequality, we find

$$\left| a_n + \frac{a_{n-1}}{z} + \frac{a_{n-2}}{z^2} + \cdots + \frac{a_0}{z^n} \right| \geq |a_n| - \left| \frac{a_{n-1}}{z} + \cdots + \frac{a_1}{z^{n-1}} + \frac{a_0}{z^n} \right|.$$
$$(19.3)$$

Now, for $|z| \geq 2nA/|a_n|$ (≥ 1, indeed $A \geq |a_n|$, and hence $A/|a_n| \geq 1$ and thus obviously $2nA/|a_n| \geq 1$), we have

$$\left| \frac{a_{n-k}}{z^k} \right| \leq \frac{A}{|z|^k} \leq \frac{A}{|z|} \leq \frac{|a_n|}{2n}.$$

Thus, it follows that

$$
\left| \frac{a_{n-1}}{z} + \cdots + \frac{a_1}{z^{n-1}} + \frac{a_0}{z^n} \right| \leq \left| \frac{a_{n-1}}{z} \right| + \cdots + \left| \frac{a_1}{z^{n-1}} \right| + \left| \frac{a_0}{z^n} \right|
$$

$$
\leq \frac{|a_n|}{2n} + \cdots + \frac{|a_n|}{2n} + \frac{|a_n|}{2n} = \frac{|a_n|}{2}.
$$

(19.4)

Hence, from (19.2)-(19.4), we get

$$
|P_n(z)| = |z|^n \left| a_n + \frac{a_{n-1}}{z} + \frac{a_{n-2}}{z^2} + \cdots + \frac{a_0}{z^n} \right|
$$

$$
\geq |z|^n \left(|a_n| - \frac{|a_n|}{2} \right) = \frac{|a_n||z|^n}{2}. \qquad \blacksquare
$$

From Lemma 19.1, it is clear that $|P_n(z)| \to \infty$ as $|z| \to \infty$.

Proof of Theorem 19.1. Suppose to the contrary that $P_n(z)$ has no zeros. Then, $1/P_n(z)$ is an entire function. Now, for $|z| \geq 2nA/|a_n|$, in view of Lemma 19.1, we have

$$
\left| \frac{1}{P_n(z)} \right| \leq \frac{1}{|a_n||z|^n/2} \leq \frac{2|a_n|^{n-1}}{(2nA)^n}.
$$

Also, for $|z| \leq 2nA/|a_n|$, $1/P_n(z)$ is a continuous function on a closed disk. Hence, it is bounded there. Therefore, $1/P_n(z)$ is bounded on the whole complex plane. But then, by Theorem 18.6 it must be a constant. Thus, $P_n(z)$ itself is a constant, which contradicts our assumption. $\blacksquare$

There are numerous other proofs of the Fundamental Theorem of Algebra. In fact, we shall give another proof in Lecture 37.

Now let $z_1 \in \mathbf{C}$ be a zero of $P_n(z)$; i.e., $P_n(z_1) = 0$. Then, we have

$$
P_n(z) = P_n(z) - P_n(z_1) = a_n(z^n - z_1^n) + a_{n-1}(z^{n-1} - z_1^{n-1}) + \cdots + a_1(z - z_1),
$$

and hence it follows that

$$
P_n(z) = (z - z_1)P_{n-1}(z), \qquad (19.5)
$$

where $P_{n-1}(z)$ is a polynomial of degree $(n-1)$. Repetition of this process finally leads to a complete factorization

$$
P_n(z) = a_n(z - z_1)(z - z_2) \cdots (z - z_n), \qquad (19.6)
$$

where $z_1, z_2, \cdots, z_n$ are not necessarily distinct. From (19.6), we can conclude that $P_n(z)$ does not vanish for any $z \neq z_j$, $i \leq j \leq n$. Furthermore, the factorization (19.5) is unique except for the order of factors. We summarize these considerations in the following corollary.

Corollary 19.1. The polynomial $P_n(z)$ has exactly n zeros, counting multiplicities.

Multiplying the factors in (19.6) and equating the coefficients of identical powers of z on the left- and right-hand sides, we get the following relations between the roots and the coefficients of $P_n(z)$:

$$z_1 + z_2 + \cdots + z_n = -\frac{a_{n-1}}{a_n},$$

$$z_1 z_2 + z_1 z_3 + \cdots + z_{n-1} z_n = \frac{a_{n-2}}{a_n},$$

$$\cdots \quad \cdots \quad \cdots$$

$$z_1 z_2 \cdots z_n = (-1)^n \frac{a_0}{a_n}.$$

$$(19.7)$$

For our next result, we shall need the following lemma.

Lemma 19.2. For any complex number $c \neq 1$, the following identity holds:

$$1 + c + c^2 + \cdots + c^n = \frac{1 - c^{n+1}}{1 - c}.$$
$$(19.8)$$

Proof. It suffices to observe that

$$(1 - c)(1 + c + c^2 + \cdots + c^n)$$
$$= (1 + c + c^2 + \cdots + c^n) - (c + c^2 + \cdots + c^{n+1}) = 1 - c^{n+1}. \quad \blacksquare$$

Theorem 19.2. Let $B = \max_{0 \leq j \leq n-1} |a_j|$. Then, all zeros z_k, $k = 1, 2, \cdots, n$ of $P_n(z)$ lie inside the circle

$$|z| = 1 + \frac{B}{|a_n|} = R.$$
$$(19.9)$$

Proof. If $B = 0$, then the result is obviously true. So, we consider the case when $B > 0$. For $|z| > 1$, from (19.1), the triangle inequality, and Lemma 19.2, we have

$$
\begin{aligned}
|P_n(z)| &\geq |a_n z^n| - \left(|a_{n-1} z^{n-1}| + \cdots + |a_1 z| + |a_0|\right) \\
&\geq |a_n||z|^n - B\left(|z|^{n-1} + \cdots + |z| + 1\right) \\
&= |a_n||z|^n - B\frac{|z|^n - 1}{|z| - 1} \\
&> \left(|a_n| - \frac{B}{|z| - 1}\right)|z|^n.
\end{aligned}
$$

Thus, if

$$|a_n| - \frac{B}{|z| - 1} \geq 0,$$

which is the same as

$$|z| \geq 1 + \frac{B}{|a_n|} = R,$$

then $|P_n(z)| > 0$; i.e., there is no zero on or outside the circle $|z| = R$.   ∎

Corollary 19.2.   Let $a_0 \neq 0$ and $C = \max_{1 \leq j \leq n} |a_j|$. Then, all zeros z_k, $k = 1, 2, \cdots, n$ of $P_n(z)$ lie outside the circle

$$|z| = \frac{1}{1 + \dfrac{C}{|a_0|}} = r. \tag{19.10}$$

Proof.   Set $z = 1/w$, so that $P_n(z) = Q_n(w)/w^n$, where

$$Q_n(w) = a_0 w^n + a_1 w^{n-1} + \cdots + a_{n-1} w + a_n.$$

The zeros of $Q_n(w)$ in view of Theorem 19.2 lie inside the circle

$$|w| = 1 + \frac{C}{|a_0|} = \frac{1}{r}.$$

Thus, the zeros of $P_n(z)$ lie outside the circle $|z| = r$.   ∎

Remark 19.1.   From Theorem 19.2 and Corollary 19.2, it follows that all zeros z_k, $k = 1, 2, \cdots, n$ of $P_n(z)$ lie inside the annulus $r \leq |z| \leq R$; i.e., satisfy the inequality

$$\frac{1}{1 + \dfrac{C}{|a_0|}} < |z_k| < 1 + \frac{B}{|a_n|}. \tag{19.11}$$

Example 19.1.   For the polynomial

$$z^3 - z^2 + z - 1 = (z^2 + 1)(z - 1),$$

we have $B = C = 1$, and hence, from Remark 19.1, all its zeros $i, -i, 1$ must lie inside the annulus $1/2 \leq |z| \leq 2$.

Example 19.2.   For the polynomial

$$z^5 - 4z^4 + z^3 + 4z^2 + 2z - 4,$$

$z_1 = 1$ is an exact zero, and the other approximate zeros are $z_2 = -0.778384 -0.603392i$, $z_3 = -0.778384 + 0.603392i$, $z_4 = 1.245346$, $z_5 = 3.311422$. Since for this polynomial $B = C = 4$, from Remark 19.1 all these zeros must lie inside the annulus $1/2 \leq |z| \leq 5$.

Theorem 19.3. The smallest convex polygon that contains the zeros of a polynomial $P_n(z)$ also contains the zeros of $P_n'(z)$.

Proof. We shall prove that "If all zeros of a polynomial $P_n(z)$ lie in a half-plane, then all zeros of the derivative $P_n'(z)$ lie in the same half-plane." The general result will follow by considering the half-planes below (above) every pair of adjacent vertices of the smallest convex polygon that contains the zeros of $P_n(z)$. From (19.6), it follows that

$$\frac{P_n'(z)}{P_n(z)} = \frac{1}{z - z_1} + \frac{1}{z - z_2} + \cdots + \frac{1}{z - z_n}. \tag{19.12}$$

Suppose that the half-plane S is defined as the part of the plane where $\text{Im}\,(z - a)/b < 0$ (see dilation and translation in Lecture 10). If z_j is in S and z is not, then we have

$$\text{Im}\left(\frac{z - z_j}{b}\right) = \text{Im}\left(\frac{z - a}{b}\right) - \text{Im}\left(\frac{z_j - a}{b}\right) > 0.$$

But the imaginary parts of reciprocal numbers have opposite signs. Therefore, under the same assumption, $\text{Im}\, b(z - z_j)^{-1} < 0$. If this is true for all j, we conclude from (19.12) that

$$\text{Im}\frac{bP_n'(z)}{P_n(z)} = \sum_{j=1}^{n} \text{Im}\frac{b}{z - z_j} < 0,$$

and hence $P_n'(z) \neq 0$. ∎

Problems

19.1. Show that if the coefficients of the polynomial equation $P_n(z) = 0$ are real and if z_0 is a root, then $\bar{z}_0$ is also a root.

19.2. Let $P(z)$ and $Q(z)$ be two polynomials of degree at most n that agree on $n + 1$ distinct points. Show that $P(z) = Q(z)$, $z \in \mathbf{C}$.

19.3. Show that if the coefficients of the polynomial equation

$$P_n(z) = a_n z^n + a_{n-1} z^{n-1} + \cdots + a_1 z + a_0 = 0 \tag{19.13}$$

are positive and nondecreasing; i.e., $0 < a_n \leq a_{n-1} \leq \cdots \leq a_0$, then (19.13) has no root in the circle $|z| \leq 1$, except perhaps at $z = -1$.

19.4. If $P_n(z)$ has distinct roots $z_1, \cdots, z_n$ and if $Q(z)$ is a polynomial of degree $< n$, show that

$$\frac{Q(z)}{P_n(z)} = \sum_{j=1}^{n} \frac{Q(z_j)}{P_n'(z_j)(z - z_j)}.$$

19.5. Show that

$$\frac{1}{2} + \cos\theta + \cos 2\theta + \cdots + \cos n\theta = \frac{\sin(n+1/2)\theta}{2\sin\theta/2},$$

$$\sin\theta + \sin 2\theta + \cdots + \sin n\theta = \frac{\cos\theta/2 - \cos(n+1/2)\theta}{2\sin\theta/2}.$$

The first equality is due to Lagrange.

19.6. Consider the polynomial equation

$$z^4 - iz^3 + 7z^2 - iz + 6 = 0$$

for which the roots are $i, -i, -2i, 3i$.

(a). Verify the relations (19.7).

(b). Use (19.11) to find the annulus region in which the roots lie.

19.7. Show that there are at most n points in the extended w-plane with fewer than n distinct inverse images under the mapping $w = P_n(z)$, $n > 1$.

19.8. Show that there are at most $n + m$ points in the extended w-plane with fewer than $N = \max\{n, m\} > 1$ distinct inverse images under the rational mapping $w = P(z)/Q(z)$, where $P(z)$ and $Q(z)$ are polynomials of degree $n > 1$ and $m > 1$, respectively.

Answers or Hints

19.1. $\overline{P_n(z_0)} = a_n\overline{z}_0^n + a_{n-1}\overline{z}_0^{n-1} + \cdots + a_1\overline{z}_0 + a_0 = 0$.

19.2. The function $\phi(z) = P(z) - Q(z)$ is a polynomial of degree at most n that has $n + 1$ distinct zeros. Now apply Corollary 19.1.

19.3. Obviously, $z = 1$ is not a solution. Consider $|z| \leq 1$ except at $z = \pm 1$. It suffices to show that

$$|(1 - z)(a_n z^n + a_{n-1}z^{n-1} + \cdots + a_1 z + a_0)| > 0. \qquad (19.14)$$

Since $(1 - z)(a_n z^n + a_{n-1}z^{n-1} + \cdots + a_1 z + a_0) = a_0 - [a_n z^{n+1} + (a_{n-1} - a_n)z^n + \cdots + (a_0 - a_1)z]$, it follows that

$$|(1 - z)(a_n z^n + a_{n-1}z^{n-1} + \cdots + a_1 z + a_0)|$$
$$\geq |a_0| - |a_n z^{n+1} + (a_{n-1} - a_n)z^n + \cdots + (a_0 - a_1)z|. \qquad (19.15)$$

Now,

$$|a_n z^{n+1} + (a_{n-1} - a_n)z^n + \cdots + (a_0 - a_1)z|$$
$$\leq a_n|z^{n+1}| + (a_{n-1} - a_n)|z^n| + \cdots + (a_0 - a_1)|z| \qquad (19.16)$$

with equality if and only if $z \in \mathbb{R}$ and $z \geq 0$. However, for such z, $P(z) > 0$ ($a_0 > 0$). Thus, in (19.16) we need to consider only strict inequality. Then, it follows that

$$|a_n z^{n+1} + (a_{n-1} - a_n)z^n + \cdots + (a_0 - a_1)z|$$
$$< a_n + (a_{n-1} - a_n) + \cdots + (a_0 - a_1) = a_0. \tag{19.17}$$

Using (19.17) in (19.15), we get the required inequality (19.14).

19.4. Since $P_n(z) = a_n(z - z_1)(z - z_2) \cdots (z - z_n)$, it suffices to show that

$$\phi(z) = Q(z) - \sum_{j=1}^{n} \frac{(z - z_1) \cdots (z - z_{j-1})(z - z_{j+1}) \cdots (z - z_n)}{(z_j - z_1) \cdots (z_j - z_{j-1})(z_j - z_{j+1}) \cdots (z_j - z_n)} Q(z_j) \equiv 0.$$

Clearly, $\phi(z)$ is a polynomial of degree at most $(n-1)$. Furthermore, $\phi(z_k) = Q(z_k) - Q(z_k) = 0$, $1 \le k \le n$; i.e., $\phi(z)$ has at least n zeros. Now use Problem 19.2.

19.5. In (19.8), let $c = e^{i\theta}$ and separate real and imaginary parts.

19.6. (a). Verify directly, (b). $6/13 \le |z| \le 8$.

19.7. Since $P_n(z) = \infty$ if and only if $z = \infty$, the point ∞ has just one inverse image. If $A \ne \infty$ has fewer than n distinct inverse images, the equation $P_n(z) = A$ must have a multiple root; i.e., must satisfy the equation $P'_n(z) = 0$. But this equation is of degree $n - 1$, and hence can have at most $n - 1$ distinct roots, say $\alpha_1, \cdots, \alpha_r$, $1 \le r \le n - 1$. Thus, $P_n(\alpha_1), \cdots, P_n(\alpha_r), \infty$ are the only values of A for which the equation $P_n(z) = A$ can have a multiple root. Clearly, at most n of these numbers are distinct.

19.8. Arguments are similar to that of Problem 19.7.

Lecture 20
Maximum Modulus Principle

In this lecture, we shall prove that a function analytic in a bounded domain and continuous up to and including its boundary attains its maximum modulus on the boundary. This result has direct applications to harmonic functions.

Let f be a function analytic inside and on the positively oriented circle γ_r of radius r around z_0. Parameterizing γ_r by $z = z_0 + re^{it}$, $0 \le t \le 2\pi$, we can write (18.6) as

$$f^{(n)}(z_0) = \frac{n!}{2\pi i} \int_0^{2\pi} \frac{f(z_0 + re^{it})}{r^{n+1} e^{i(n+1)t}} i r e^{it} dt$$

$$= \frac{n!}{2\pi r^n} \int_0^{2\pi} f(z_0 + re^{it}) e^{-int} dt,$$

which for $n = 0$ is the same as

$$f(z_0) = \frac{1}{2\pi} \int_0^{2\pi} f(z_0 + re^{it}) \, dt, \tag{20.1}$$

i.e., $f(z_0)$ is the average of its value around the circle γ_r. This result is known as *Gauss's mean-value property*. We shall use this result to prove the following lemma.

Lemma 20.1. Suppose that $f(z)$ is analytic in an open disk centered at z_0 and that the maximum value of $|f(z)|$ over this disk is $|f(z_0)|$. Then, $|f(z)|$ is a constant in the disk.

Proof. Let ρ be the radius of the disk. Then, for every r such that $0 < r < \rho$, we have the mean-value property (20.1). Hence, we find

$$|f(z_0)| = \left| \frac{1}{2\pi} \int_0^{2\pi} f(z_0 + re^{it}) \, dt \right| \le \frac{1}{2\pi} \int_0^{2\pi} |f(z_0 + re^{it})| \, dt. \tag{20.2}$$

However, since $|f(z_0)|$ is the maximum value of $|f(z)|$, we have $|f(z)| \le |f(z_0)|$ whenever $|z - z_0| < \rho$. Thus, from (20.2), we get

$$|f(z_0)| \le \frac{1}{2\pi} \int_0^{2\pi} |f(z_0 + re^{it})| \, dt \le \frac{1}{2\pi} \int_0^{2\pi} |f(z_0)| dt = |f(z_0)|. \tag{20.3}$$

From (20.3), it follows that

$$|f(z_0)| = \frac{1}{2\pi} \int_0^{2\pi} |f(z_0 + re^{it})| \, dt$$

or

$$\int_0^{2\pi} [|f(z_0)| - |f(z_0 + re^{it})|] \, dt = 0.$$

The integrand $|f(z_0)| - |f(z_0 + re^{it})|$ is continuous in the variable t, and by our assumption on $|f(z_0)|$ it is greater than or equal to zero on the entire interval $0 \le t \le 2\pi$. Because the value of the integral is zero, it follows that the integrand is identically zero. i.e.,

$$|f(z_0 + re^{it})| = |f(z_0)|, \quad 0 < r < \rho, \quad 0 \le t \le 2\pi$$

or

$$|f(z)| = |f(z_0)| \quad \text{whenever} \quad |z - z_0| < \rho.$$

Hence, $|f(z)|$ must be constant. ∎

Now we shall use Lemma 20.1 to prove the following important theorem.

Theorem 20.1 (Maximum Modulus Principle). If f is analytic in a domain S and $|f(z)|$ achieves its maximum value at a point z_0 in S, then f is constant in S.

Proof. We shall prove that $|f|$ is constant in S. Then, by Theorem 7.3, we can conclude that f is constant.

Let w be a point in S. Since S is connected, there is a polygonal path joining z_0 to w. Using topological considerations, we can always cover the path by a sequence of disks $\{B_0, B_1, \cdots, B_n\}$ centered at $z_0, z_1, \cdots, z_n = w$, respectively, that satisfy the following properties:

(i). B_i is contained in the domain S for every i.

(ii). z_i lies on the path for every i.

(iii). B_i contains the point z_{i+1} for $i = 0, 1, \cdots, n - 1$.

Since $|f(z_0)|$ is the maximum value of $|f(z)|$ over S, it is the maximum value of $|f(z)|$ over B_0. By Lemma 20.1, $|f(z)|$ is constant in the disk B_0. In particular, $|f(z_0)| = |f(z_1)|$ since $z_1 \in B_0$. Therefore, $|f(z_1)|$ is the maximum value of $|f(z)|$ over the disk B_1. By Lemma 20.1 again, $|f(z)|$ is constant in B_1. By using the same argument repeatedly, we deduce that $|f(z)|$ is constant over the disks $B_0, B_1, \cdots, B_n$. In particular, $|f(z_0)| = |f(w)|$. Since w is arbitrary, we conclude that $|f|$ is constant in S. Hence, f is constant in S. ∎

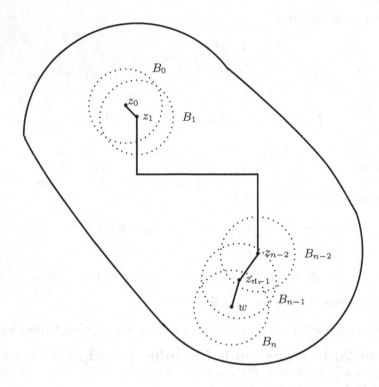

Figure 20.1

The following version of Theorem 20.1 is directly applicable in applications.

Theorem 20.2. A function analytic in a bounded domain and continuous up to and including its boundary attains its maximum modulus on the boundary.

Example 20.1. Find the maximum value of $|z^2 + 2z - 3|$ in the disk $\overline{B}(0,1)$. Clearly, $z^2 + 2z - 3$ is analytic in the disk $B(0,1)$ and continuous on the closed disk $\overline{B}(0,1)$. By Theorem 20.2, the maximum of $|z^2 + 2z - 3|$ must occur on the boundary of the disk; i.e., on the circle $|z| = 1$. Parameterizing the circle by $z(t) = e^{it}$, $0 \le t \le 2\pi$, we get

$$
\begin{aligned}
|z^2 + 2z - 3|^2 &= (z^2 + 2z - 3)\overline{(z^2 + 2z - 3)} \\
&= \left(e^{2it} + 2e^{it} - 3\right)\left(e^{-2it} + 2e^{-it} - 3\right) \\
&= (14 - 6\cos 2t - 8\cos t).
\end{aligned}
$$

The function $14 - 6\cos 2t - 8\cos t$ attains its maximum when $t = \cos^{-1}(1/3)$.

Thus, the maximum value of $|z^2 + 2z - 3|$ is $8/\sqrt{3}$.

Example 20.2. Find the maximum of $|\sin z|$ on the square $\{x + iy : 0 \le x \le 2\pi, \ 0 \le y \le 2\pi\}$. Since $\sin z$ is entire, we can apply Theorem 20.2. Now, since $|\sin(x + iy)|^2 = \sin^2 x + \sinh^2 y$ (see Example 8.2) on the boundary $y = 0$, $|\sin^2 z|$ has maximum 1; for $x = 0$, the maximum is $\sinh^2 2\pi$ since $\sinh y$ increases with y; for $x = 2\pi$, the maximum is again $\sinh^2 2\pi$; and for $y = 2\pi$, the maximum is $\sinh^2 2\pi + 1$. Thus, the maximum of $|\sin z|^2$ occurs at $x = \pi/2$, $y = 2\pi$ and is $\sinh^2 2\pi + 1 = \cosh^2 2\pi$. Therefore, the maximum of $|\sin z|$ on the square is $\cosh 2\pi$.

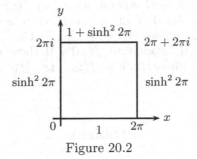

Figure 20.2

Example 20.3. Find the maximum of $|e^{z^3}|$ on $\overline{B}(0,1)$. Since e^{z^3} is an entire function, it is analytic in the disk $B(0,1)$ and continuous on the closed disk $\overline{B}(0,1)$. By Theorem 20.2, the maximum of $|e^{z^3}|$ occurs on the boundary of the disk; i.e., the circle $|z| = 1$. Let $z = x + iy$. Then

$$|e^{z^3}| = |e^{x^3 + 3ix^2 y - 3xy^2 - iy^3}| = e^{x^3 - 3xy^2}.$$

Since $|z| = 1$, we have $x^2 + y^2 = 1$. Thus,

$$|e^{z^3}| = e^{4x^3 - 3x}, \quad -1 \le x \le 1.$$

The maximum value of $e^{4x^3 - 3x}$, $-1 \le x \le 1$, is equal to e and is attained at $x = -1/2$ and $x = 1$. Hence, the maximum of $|e^{z^3}|$ is e, which occurs at $x = -1/2$, $y = \pm\sqrt{3}/2$, and $x = 1$, $y = 0$.

Example 20.4. Consider the entire function $e^{-iz^2} = e^{-i(x^2 - y^2)} e^{2xy}$ in the positive quadrant $S = \{z = x + iy : x \ge 0, \ y \ge 0\}$. Since $|f(z)| = e^{2xy}$, on the boundary of S, $|f(z)| = 1$; however, $|f(z)| \to \infty$ as $y = x \to \infty$. Thus, the maximum modulus principle fails on unbounded domains.

An immediate application of Theorem 20.2 applied to the function $g = 1/f$ gives the following result.

Theorem 20.3 (Minimum Modulus Principle). A function f analytic in a bounded domain S, continuous up to and including its boundary, and $f(z) \neq 0$ for all $z \in S$, attains its minimum modulus on the boundary.

A combination of Theorems 20.1-20.3 leads to the following result.

Theorem 20.4. Suppose that f is analytic in a bounded domain S, continuous up to and including its boundary, and $f(z) \neq 0$ for all $z \in S$. Then, $|f|$ attains a maximum M and a minimum m on the boundary, and either f is a constant or $m < |f(z)| < M$ for all $z \in S$.

Finally, we note that the Maximum Modulus Principle and Minimum Modulus Principle have local versions also: Let $f(z)$ be analytic and not constant in a neighborhood N of z_0. Then, there are points in N lying arbitrarily close to z_0, where $|f(z)| > |f(z_0)|$; i.e., $|f(z)|$ cannot have a local maximum at z_0. If, in addition, $f(z_0) \neq 0$, there are points in N lying arbitrarily close to z_0, where $|f(z)| < |f(z_0)|$; i.e., $|f(z)|$ cannot have a local minimum at z_0.

Problems

20.1. Find the maximum of $|ze^z + z^2|$ and $|iz^2 - 2z|$ on the set $\{z : |z| \leq 1 \text{ and } \operatorname{Im} z \geq 0\}$.

20.2. Find the maximum of $|\cos z|$ on the square $\{x + iy : 0 \leq x \leq 2\pi, \ 0 \leq y \leq 2\pi\}$. (Observe that $|\cos(0 + 2\pi i)| > 1$).

20.3. Consider the function $g(z)$ defined in (18.7) on the rectangle with vertices at 0, π, i, $\pi + i$. Find the maximum and minimum values of $|g(z)|$ and determine where these values occur.

20.4. Consider the function $f(z) = z + 1$. Find the maximum and minimum of $|f|$ in the closed triangle with vertices at $z = 0$, $z = 2$, and $z = i$.

20.5. Suppose $f(z)$ and $g(z)$ are continuous in $\overline{B}(0, r)$ and analytic in $B(0, r)$ and that $f(z) \neq 0$ and $g(z) \neq 0$ for all $z \in \overline{B}(0, r)$. If $|f(z)| = |g(z)|$ for $|z| = r$, show that there exists a constant c such that $|c| = 1$ and $f(z) = cg(z)$ for all z in $\overline{B}(0, r)$.

Answers or Hints

20.1. $|ze^z + z^2| = |z||e^z + z| \leq |e^z| + |z| \leq e + 1$ if $|z| \leq 1$ (note that when $z = 1$, $|1 \cdot e^1 + 1| = e + 1$), $|iz^2 - 2z| = |z||iz - 2| \leq |iz| + 2 \leq 3$ if $|z| \leq 1$ (note that when $z = i$, $|i \cdot i^2 - 2i| = |-3i| = 3$).

20.2. We need only to consider either $x = 0, 2\pi$ or $y = 0, 2\pi$.

$$|\cos(x+iy)| = |\cos x \cos iy - \sin x \sin iy|$$

$$= \begin{cases} |\cos iy| = \left|\frac{e^{-y}+e^{y}}{2}\right| & \text{if } x = 0, 2\pi \\[2mm] |\cos x| = \left|\frac{e^{ix}+e^{-ix}}{2}\right| & \text{if } y = 0 \\[2mm] \left|\frac{e^{ix-2\pi}+e^{-ix+2\pi}}{2}\right| & \text{if } y = 2\pi. \end{cases}$$

Note that $\left|\frac{e^{ix-2\pi}+e^{-ix+2\pi}}{2}\right| \leq \frac{e^{-2\pi}+e^{2\pi}}{2} = \cos(0 + 2\pi i).$

20.3. $\sinh 1$, $z = i$; $0, z = \pi$.

20.4. The maximum is attained at $z = 2$ and equals 3, whereas the minimum is attained at $z = 0$ and equals 1.

20.5. The function $h(z) = f(z)/g(z)$ satisfies the conditions of Theorem 20.4, and hence there exist $u, v \in \partial B(0, r)$ such that $|h(u)| \leq |h(z)| \leq |h(v)|$ for all $z \in \overline{B}(0, r)$. But, $|f(z)| = |g(z)|$ for $z \in \delta B(0, r)$, and hence $|h(z)| = 1$ for all $z \in \overline{B}(0, r)$. The result now follows from Theorem 7.3.

Lecture 21
Sequences and Series
of Numbers

In this lecture, we shall collect several results for complex sequences and series of numbers. Their proofs require essentially the same arguments as in calculus.

A sequence of complex numbers is a function whose domain is the set of nonnegative integers and whose range is a subset of complex numbers. We denote a complex sequence as $\{z_n\}_{n=0}^{\infty}$. This sequence is said to have the *limit z*, and we write $\lim_{n \to \infty} z_n = z$, or equivalently $z_n \to z$ as $n \to \infty$, if for any $\epsilon > 0$ there exists an integer $N = N(\epsilon)$ such that $|z_n - z| < \epsilon$ for all $n > N$. Intuitively, this means that as n increases the distance between z_n and z gets smaller and smaller. If $\{z_n\}$ has no limit, we say it diverges.

Example 21.1. (i) $\lim_{n \to \infty} i/n = 0$, (ii) $\{i^n\}$ diverges, (iii) if $z \in \mathbf{C}$, then

$$\lim_{n \to \infty} z^n = \begin{cases} 0 & \text{if } |z| < 1 \\ 1 & \text{if } z = 1 \\ \text{divergent} & \text{otherwise.} \end{cases} \tag{21.1}$$

Example 21.2. Consider $z_n = 1 + (-1)^n (i/n)$, $n \geq 1$. Clearly, $|z_n| \to 1$; however, if the principal values of the arguments are chosen, then

$$\arg z_{2k} = \tan^{-1} \frac{1}{2k} \to 0 \quad \text{and} \quad \arg z_{2k+1} = 2\pi - \tan^{-1} \frac{1}{2k+1} \to 2\pi.$$

Thus, $\lim_{n \to \infty} \arg z_n$ does not exist.

We list the principal properties of complex sequences as follows:

P1. A convergent sequence has a *unique limit*.

P2. A convergent sequence is bounded.

P3. If $\lim_{n \to \infty} z_n = z$ and $\lim_{n \to \infty} w_n = w$, and α and β are complex numbers, then $\lim_{n \to \infty} (\alpha z_n + \beta w_n) = \alpha z + \beta w$.

P4. If $\lim_{n \to \infty} z_n = z$ and $\lim_{n \to \infty} w_n = w$, then $\lim_{n \to \infty} (z_n w_n) = \lim_{n \to \infty} z_n \times \lim_{n \to \infty} w_n = zw$.

P5. If $\lim_{n \to \infty} z_n = z$ and $\lim_{n \to \infty} w_n = w$, then $\lim_{n \to \infty} (z_n/w_n) = \lim_{n \to \infty} z_n / \lim_{n \to \infty} w_n = z/w$, provided $w \neq 0$.

P6. If $\lim_{n\to\infty} z_n = z$, then $\lim_{n\to\infty} \bar{z}_n = \bar{z}$ and $\lim_{n\to\infty} |z_n| = |\lim_{n\to\infty} z_n| = |z|$.

P7. If $z_n = x_n + iy_n$, $n \geq 0$ and $z = x + iy$, then $\lim_{n\to\infty} z_n = z$ if and only if $x_n \to x$ and $y_n \to y$.

P8. If $\lim_{n\to\infty} z_n = 0$ and $|w_n| \leq |z_n|$, $n \geq 0$, then $\lim_{n\to\infty} w_n = 0$.

P9. If $\lim_{n\to\infty} z_n = 0$ and $\{w_n\}$ is a bounded sequence, then $\lim_{n\to\infty} w_n z_n = 0$.

P10. From any bounded sequence, it is possible to extract a convergent subsequence.

P11. A sequence $\{z_n\}$ is called a *Cauchy sequence* if for any $\epsilon > 0$ there exists an integer $N > 0$ such that $n, m \geq N \Rightarrow |z_n - z_m| < \epsilon$. The sequence $\{z_n\}$ is convergent if and only if it is a Cauchy sequence.

P12. (*Heine's Criterion*). Let f be a function defined in a neighborhood of z. The function f is continuous at z if and only if for any sequence $\{z_n\}$ converging to z the condition $\lim_{n\to\infty} f(z_n) = f(z)$ holds.

A point z is called a *limit point* (or an *accumulation point*) of the sequence $\{z_n\}$ if for every neighborhood $N(z)$ there exists a subsequence $\{z_{n_k}\}$ for which all terms belong to $N(z)$. Alternatively, z is said to be a limit point of $\{z_n\}$ if every neighborhood $N(z)$ contains infinitely many terms of $\{z_n\}$. From P10, in particular, it follows that every bounded real sequence $\{a_n\}$ has at least one limit point. For example, for the sequence $\{(-1)^n\}$, -1 and 1 are the limit points. An unbounded sequence may not have a limit point; e.g. $\{\sqrt{n}\}$; however, an unbounded real sequence $\{a_n\}$ has at least one limit point if we allow the values $+\infty$ and $-\infty$. This means that the real sequence $\{a_n\}$ has $+\infty$ as a limit point if it contains arbitrarily large positive terms, and $-\infty$ as a limit point if it contains negative terms of arbitrarily large absolute value. Clearly, for the sequence $\{1, 1, 2, 1, 2, 3, 1, 2, 3, 4, \cdots\}$, each natural number is a limit point. The largest (smallest) limit point of the (bounded or unbounded) real sequence $\{a_n\}$ is called the *limit superior* or *upper limit* (*limit inferior* or *lower limit*) and is denoted as $\limsup_{n\to\infty} a_n$ ($\liminf_{n\to\infty} a_n$). For example, the sequence $\{a_n\}$ defined by $u_n = (-1)^n(3 + 1/n)$ is bounded with lower bound -4 and upper bound $7/2$. For this sequence, the limit points are -3 and 3, and hence $\limsup_{n\to\infty} a_n = 3$ and $\liminf_{n\to\infty} a_n = -3$. For the sequence $\{1, 1, 2, 1, 2, 3, 1, 2, 3, 4, \cdots\}$, $\limsup_{n\to\infty} a_n = \infty$ and $\liminf_{n\to\infty} a_n = 1$. For a real bounded sequence $\{a_n\}$, it follows that $\lim_{n\to\infty} a_n = a$ if and only if $\liminf_{n\to\infty} a_n = a = \limsup_{n\to\infty} a_n$. For an unbounded real sequence $\{a_n\}$, we have $\lim_{n\to\infty} a_n = \infty$ if and only if $\liminf_{n\to\infty} a_n = \infty = \limsup_{n\to\infty} a_n$, and $\lim_{n\to\infty} a_n = -\infty$ if and only if $\liminf_{n\to\infty} a_n = -\infty = \limsup_{n\to\infty} a_n$. Now let $\{a_n\}$ and $\{b_n\}$ be bounded sequences of real numbers. For these sequences, it can easily be shown that:

(1). If $a_n \leq b_n$, $n \in \mathcal{N}$, then $\limsup_{n\to\infty} a_n \leq \limsup_{n\to\infty} b_n$ and $\liminf_{n\to\infty} a_n \leq \liminf_{n\to\infty} b_n$.

(2). $\limsup_{n\to\infty}(a_n + b_n) \leq \limsup_{n\to\infty} a_n + \limsup_{n\to\infty} b_n$.

(3). $\liminf_{n\to\infty}(a_n + b_n) \geq \liminf_{n\to\infty} a_n + \liminf_{n\to\infty} b_n$.

In (2) and (3), the inequality sign cannot be replaced by equality. For this it suffices to consider $a_n = (-1)^n$ and $b_n = (-1)^{n+1}$.

A *series* is a formal expression of the form $z_0 + z_1 + \cdots$ or, equivalently, $\sum_{j=0}^{\infty} z_j$, where the terms z_j are complex numbers. The nth *partial sum* of the series, denoted by s_n, is the sum of the first $n + 1$ terms; i.e., $s_n = \sum_{j=0}^{n} z_j$. If the sequence of partial sums $\{s_n\}_{n=0}^{\infty}$ has a limit s, the series is said to *converge*, or *sum*, to s, and we write $s = \sum_{j=0}^{\infty} z_j$. A series that does not converge is said to *diverge*. The series $\sum_{j=0}^{\infty} z_j$ is said to be *absolutely convergent* provided that the series of magnitudes $\sum_{j=0}^{\infty} |z_j|$ converges. For example, the series $\sum_{j=0}^{\infty}(-i)^j/(j+1)^2$ converges absolutely. A complex series that is convergent but not absolutely convergent is called *conditionally convergent*. For example, the series $\sum_{j=0}^{\infty}(-1)^j/(j+1)$ converges conditionally.

Example 21.3. Consider the geometric series $\sum_{j=0}^{\infty} ac^j$, where $|c| < 1$. From Lemma 19.2, we have

$$\frac{a}{1-c} - (a + ac + ac^2 + \cdots + ac^n) = \frac{ac^{n+1}}{1-c}.$$

Since $|c| < 1$, the term $|a||c|^{n+1}/|1-c|$ converges to 0 as $n \to \infty$. Thus, the geometric series converges to $a/(1-c)$.

The following properties of complex series are similar to those of real series:

Q1. If $\sum_{j=0}^{\infty} z_j$ converges, then $\lim_{n\to\infty} z_n = 0$.

Q2. If $\sum_{j=0}^{\infty} z_j$ converges, then $\lim_{m\to\infty} \sum_{n=m+1}^{\infty} z_n = 0$.

Q3. If $\sum_{j=0}^{\infty} z_j$ converges, then there exists a real constant M such that $|z_j| \leq M$ for all j.

Q4. If $\sum_{j=0}^{\infty} z_j$ and $\sum_{j=0}^{\infty} w_j$ converge, and α and β are complex numbers, then $\sum_{j=0}^{\infty}(\alpha z_j + \beta w_j) = \alpha \sum_{j=0}^{\infty} z_j + \beta \sum_{j=0}^{\infty} w_j$.

Q5. If $z_j = x_j + iy_j$ and $s = u + iv$, then $\sum_{j=0}^{\infty} z_j$ converges to s if and only if $\sum_{j=0}^{\infty} x_j = u$ and $\sum_{j=0}^{\infty} y_j = v$.

Q6. If $\sum_{j=0}^{\infty} z_j$ converges absolutely, then it converges; however, the converse is not true.

Q7. If $\sum_{j=0}^{\infty} z_j$ converges absolutely, then its every rearrangement is absolutely convergent and converges to the same limit.

Q8. The series $\sum_{j=0}^{\infty} z_j$ converges if and only if for any given $\epsilon > 0$ there exists an $N > 0$ such that $m > n \geq N$ implies $|s_m - s_n| = |z_{n+1} +$

$z_{n+2} + \cdots + z_m| < \epsilon$. This is known as *Cauchy's criterion*.

Q9. (*Comparison Test*). If $|z_j| \leq a_j$ for all j and $\sum_{j=0}^{\infty} a_j$ converges, then $\sum_{j=0}^{\infty} z_j$ converges.

Q10. (*Cauchy's Root Test*). If $\rho = \lim_{j \to \infty} |z_j|^{1/j}$ either exists or is infinite, then the series $\sum_{j=0}^{\infty} z_j$ converges absolutely if $\rho < 1$ and diverges if $\rho > 1$. If $\rho = 1$ the test is inconclusive.

Q11. (*d'Alembert's Ratio Test*). If $\lim_{j \to \infty} |c_{j+1}/c_j| = \rho$, then the series $\sum_{j=0}^{\infty} z_j$ converges absolutely if $\rho < 1$ and diverges if $\rho > 1$. If $\rho = 1$ the test is inconclusive.

Now let $\sum_{j=0}^{\infty} z_j$ and $\sum_{j=0}^{\infty} w_j$ be given series. The *Cauchy product* of these series is defined to be the new series $\sum_{j=0}^{\infty} t_j$, where

$$t_j = z_0 w_j + z_1 w_{j-1} + \cdots + z_{j-1} w_1 + z_j w_0 = \sum_{k=0}^{j} z_k w_{j-k} = \sum_{k=0}^{j} z_{j-k} w_k.$$
(21.2)

The following example shows that the Cauchy product of two convergent series need not be convergent.

Example 21.4. The series $\sum_{j=0}^{\infty} z_j$ and $\sum_{j=0}^{\infty} w_j$, where $z_j = w_j = (-1)^j (j+1)^{-1/2}$, are convergent. The Cauchy product of these series is $\sum_{j=0}^{\infty} t_j$, where

$$t_j = (-1)^j \sum_{k=0}^{j} (k+1)^{-1/2} (j+1-k)^{-1/2}.$$

Now since from the arithmetic-geometric mean inequality

$$\sqrt{(k+1)(j+1-k)} \leq \frac{k+1+j+1-k}{2} = \frac{j+2}{2},$$

we find

$$(k+1)^{-1/2}(j+1-k)^{-1/2} \geq \frac{2}{j+2}.$$

Thus, it follows that

$$(-1)^j t_j \geq (j+1)\frac{2}{j+2} > 1 \quad \text{for all} \quad j > 0.$$

Hence, t_j does not tend to zero as $j \to \infty$. This shows that the Cauchy product of these series diverges.

Q12. If $\sum_{j=0}^{\infty} z_j$ and $\sum_{j=0}^{\infty} w_j$ converge to z and w, respectively, and their Cauchy product converges, then

$$\sum_{j=0}^{\infty} t_j = \left(\sum_{j=0}^{\infty} z_j\right)\left(\sum_{j=0}^{\infty} w_j\right) = zw. \qquad (21.3)$$

However, if $\sum_{j=0}^{\infty} z_j$ and $\sum_{j=0}^{\infty} w_j$ converge absolutely to z and w, respectively, then their Cauchy product also converges absolutely, and (21.3) holds.

Problems

21.1. Discuss the convergence of the following sequences:

(a). $\left\{\dfrac{3n + 2(i)^n}{n}\right\}$, (b). $\left\{\left(\dfrac{1+i}{\sqrt{2}}\right)^n\right\}$, (c). $\left\{\dfrac{3n + 7ni}{2n + 5i}\right\}$.

21.2. Discuss the convergence of the following series:

(a). $\displaystyle\sum_{j=0}^{\infty} \dfrac{i^j}{(j+1)^2}$,   (b). $\displaystyle\sum_{j=0}^{\infty} \dfrac{(1+3i)^j}{5^j}$,

(c). $\displaystyle\sum_{j=0}^{\infty} \left(\cos\dfrac{j\pi}{5} + i\sin\dfrac{j\pi}{5}\right)$,   (d). $\displaystyle\sum_{j=0}^{\infty} \dfrac{2i^j}{5 + ij^2}$.

21.3. Examine the series $\sum_{j=0}^{\infty} z_j$ for its convergence or divergence, where z_j are recursively defined as

$$z_0 = 1 + i, \quad z_{j+1} = \dfrac{(2+3i)j}{7+5ij^2} z_j.$$

21.4. If $0 \le r < 1$, show that

(a). $\displaystyle\sum_{j=0}^{\infty} r^j \cos j\theta = \dfrac{1 - r\cos\theta}{1 + r^2 - 2r\cos\theta}$, (b). $\displaystyle\sum_{j=0}^{\infty} r^j \sin j\theta = \dfrac{r\sin\theta}{1 + r^2 - 2r\cos\theta}$.

21.5. If $\lim_{n\to\infty} z_n = z$, show that $\lim_{n\to\infty}(z_0 + z_1 + \cdots + z_n)/(n+1) = z$.

21.6. Prove that if $\sum_{j=0}^{\infty} z_j$ converges and $|\arg z_j| \le \theta < \pi/2$, then it converges absolutely.

21.7. Suppose that $\sum_{j=0}^{\infty} z_j$ and $\sum_{j=0}^{\infty} z_j^2$ are convergent and $\operatorname{Re} z_j \ge 0$. Show that $\sum_{j=0}^{\infty} |z_j|^2$ converges.

21.8. A metric space (S, d) is said to be *complete* if every Cauchy sequence in S converges. Show that $(\mathbf{C}, d)$ with any metric d is complete.

21.9. The function $f : S \to \mathbf{C}$ is said to be *uniformly continuous* on S if for every given $\epsilon > 0$ there exists a $\delta = \delta(\epsilon) > 0$ such that $|z - w| < \delta$ implies $|f(z) - f(w)| < \epsilon$ for all $z, w \in S$. Show that:

(a). The function $f(z) = 1/z$ is continuous on $S = \{z : 0 < |z| < 1\}$, but not uniformly continuous.

(b). If S is compact and $f(z)$ is continuous on S, then it is uniformly continuous.

Answers or Hints

21.1. (a). Converges to 3, (b). in polar form $\lim_{n \to \infty} e^{i\, n\pi/4}$ oscillates, and hence diverges, (c). converges to $(3 + 7i)/2$.

21.2. (a). Absolutely convergent, (b). geometric series $5/(4 - 3i)$, (c). divergent, (d). absolutely convergent.

21.3. Absolutely convergent.

21.4. In Example 21.3, let $a = 1$, $c = re^{i\theta}$.

21.5. Writing $w_n = z_n - z$, this is equivalent to showing that if $\lim w_n = 0$, then $\lim(w_0 + w_1 + \cdots + w_n)/(n + 1) = 0$. Given $\varepsilon > 0$, there are $N_1 \geq 1$ and N_2 such that $n \geq N_1$ implies $|w_n| < \varepsilon/2$ and

$$n \geq N_2 \Rightarrow \frac{|w_0 + w_1 + \cdots + w_{N_1 - 1}|}{n + 1} < \frac{\varepsilon}{2}.$$

Then, by the triangle inequality,

$$n \geq \max\{N_1, N_2\} \quad \Rightarrow \quad \frac{|w_0 + w_1 + \cdots + w_n|}{n + 1} \leq \frac{|w_0 + w_1 + \cdots + w_{N_1 - 1}|}{n + 1}$$
$$+ \frac{|w_{N_1}| + \cdots + |w_n|}{n + 1} < \frac{\varepsilon}{2} + \frac{n - N_1 + 1}{n + 1} \frac{\varepsilon}{2} < \varepsilon.$$

21.6. Writing $z_j = x_j + iy_j = r_j e^{i\theta_j}$, since $x_j > 0$, $|y_j| = x_j |\tan\theta_j| \leq x_j \tan\theta$, and $\sum x_j$ converges, $\sum |y_j|$ also converges. Since $|z_j| \leq |x_j| + |y_j|$ by the triangle inequality, then $\sum |z_j|$ converges.

21.7. Writing $z_j = x_j + iy_j$, since $\sum x_j$ converges and $x_j \geq 0$, $\{x_j\}$ is bounded, say, $x_j \leq M$. Then $0 \leq x_j^2 \leq Mx_j$, so $\sum x_j^2$ also converges. Since $z_j^2 = x_j^2 - y_j^2 + 2ixy$, $\sum(x_j^2 - y_j^2)$ converges. Then so does $\sum y_j^2$ since $y_j^2 = x_j^2 - (x_j^2 - y_j^2)$. Since $|z_j|^2 = x_j^2 + y_j^2$, then $\sum |z_j|^2$ also converges.

21.8. We will show that $\mathbf{C}$ with the Euclidean metric is complete. Let $\{z_n\}$ be a Cauchy sequence in $\mathbf{C}$, so that for $\epsilon > 0$ there is an N such that $|z_n - z_m| < \epsilon$ for $n, m \geq N$. Let $z_n = x_n + iy_n$. Since $|x_n - x_m| < |z_n - z_m|$ and $|y_n - y_m| < |z_n - z_m|$ for all n, m, we find that $\{x_n\}$ and $\{y_n\}$ are real Cauchy sequences. They both converge since $\mathbb{R}$ is complete, say to p and q, respectively. Then, it follows that $\{z_n\}$ converges to $p + iq \in \mathbf{C}$.

21.9. (a). Let $z = \delta$ ($0 < \delta < 1$) and $\epsilon = \delta/(1 + \epsilon)$. Then, $|z - w| < \delta$, but $|f(z) - f(w)| = \epsilon/\delta > \epsilon$. (b). Suppose not. Then there exists a z_0, and for each n there are points z_n and w_n in S such that $|z_n - w_n| < 1/n$ and $|f(z_n) - f(w_n)| \geq \epsilon_0$. Now, since S is compact, there exists a subsequence $z_{n_1}, z_{n_2}, \cdots$ and a point $z_* \in S$ such that $z_* = \lim_{k \to \infty} z_{n_k}$. Since $|z_* - w_{n_k}| \leq |z_* - z_{n_k}| + 1/n_k$, it is clear that $z_* = \lim_{k \to \infty} w_{n_k}$. Thus, from the continuity of $f(z)$, we have $f(z_*) = \lim_{k \to \infty} f(z_{n_k}) = \lim_{k \to \infty} f(w_{n_k})$. Now notice that in $\epsilon_0 \leq |f(z_{n_k}) - f(w_{n_k})| \leq |f(z_{n_k}) - f(z_*)| + |f(z_*) - f(w_{n_k})|$ the right-hand side tends to zero.

Lecture 22
Sequences and Series
of Functions

In this lecture, we shall prove some results for complex sequences and series of functions. These results will be needed repeatedly in later lectures.

Let $\{f_n(z)\}_{n=0}^{\infty}$ be a sequence of complex functions defined on a domain $S \subseteq \mathbf{C}$. Suppose that for any $z \in S$ the complex sequence $\{f_n(z)\}$ converges. Define $f(z) = \lim_{n \to \infty} f_n(z)$ for any $z \in S$. We say $\{f_n(z)\}$ converges to $f(z)$ *pointwise* on S. The sequence $\{f_n(z)\}$ is said to converge to $f(z)$ *uniformly* on S if for every $\epsilon > 0$ there exists an integer $N > 0$ such that $n \geq N$ implies $|f_n(z) - f(z)| < \epsilon$ for all $z \in S$; i.e., the same N works for all points $z \in S$. The series $\sum_{j=0}^{\infty} f_j(z)$ converges uniformly to $s(z)$ on S if the sequence of its partial sums; i.e., $s_n(z) = \sum_{j=0}^{n} f_j(z)$, $n \geq 0$ converges uniformly to $s(z)$ there. For the sequences and series of functions we shall prove the following results.

Theorem 22.1 (Cauchy's Criterion). The sequence $\{f_n(z)\}$ converges uniformly on a domain S to $f(z)$ if and only if for any $\epsilon > 0$ there exists an $N(\epsilon)$ such that, for all $z \in S$, $n \geq N$, and any natural number m, the inequality

$$|f_{n+m}(z) - f_n(z)| < \epsilon \tag{22.1}$$

holds.

Proof. If $\{f_n(z)\}$ converges uniformly on S, then for any $\epsilon > 0$ there exists an $N = N(\epsilon)$ such that, for all $z \in S$, $n \geq N$, and any natural number m,

$$|f_n(z) - f(z)| < \frac{\epsilon}{2} \quad \text{and} \quad |f_{n+m}(z) - f(z)| < \frac{\epsilon}{2}.$$

Thus, from the triangle inequality, we find

$$|f_{n+m}(z) - f_n(z)| \leq |f_{n+m}(z) - f(z)| + |f(z) - f_n(z)| < \frac{\epsilon}{2} + \frac{\epsilon}{2} = \epsilon.$$

Conversely, if (22.1) holds, then in view of P11 (Lecture 21), for any fixed $z \in S$, the sequence $\{f_n(z)\}$ converges. Define $f(z) = \lim_{n \to \infty} f_n(z)$. Then, we have

$$\lim_{m \to \infty} |f_{n+m}(z) - f_n(z)| = |f(z) - f_n(z)| \leq \epsilon$$

145

for $n \geq N$ at all points on S; i.e., the convergence is uniform. ∎

Example 22.1. Let $f_n(z) = z^n$, $f(z) \equiv 0$. Then,

(i). $f_n \to f$ pointwise on the open disk $B(0,1)$,

(ii). $f_n \to f$ uniformly on the closed disk $\overline{B}(0,r)$, $0 < r < 1$, and

(iii). $f_n \not\to f$ uniformly on $\overline{B}(0,1)$.

Example 22.2. Let $f_n(z) = 1/(1 + nz)$, $f(z) \equiv 0$. Then, for $|z| \geq 2$, we find

$$\left| \frac{1}{1 + nz} \right| \leq \frac{1}{n|z| - 1} \leq \frac{1}{2n - 1} \leq \frac{1}{n},$$

i.e., the convergence is uniform. We also note that this sequence for $|z| \leq 1$ and $\operatorname{Re} z \geq 0$ converges pointwise but not uniformly. For this, clearly $f(0) = 1$ and $f(z) = 0$, $z \neq 0$ is the pointwise limit of this sequence.

Corollary 22.1. The series $\sum_{j=0}^{\infty} f_j(z)$ converges uniformly on a domain S to $s(z)$; i.e., $\sum_{j=0}^{\infty} f_j(z) = s(z)$ if and only if for any $\epsilon > 0$ there exists an $N(\epsilon)$ such that, for all $z \in S$, $n \geq N$, and any natural number m, the inequality $|s_{n+m}(z) - s_n(z)| < \epsilon$ holds.

Example 22.3. The series $\sum_{j=0}^{\infty} z^j$ does not converge uniformly on $\overline{B}(0,1)$. In fact, since

$$
\begin{aligned}
|s_{n+m}(z) - s_n(z)| &= |z^{n+1}(1 + z + \cdots + z^{m-1})| \\
&= \frac{|z|^{n+1}|1 - z^m|}{|1 - z|} \geq \frac{|z|^{n+1}(1 - |z|^m)}{|1 - z|},
\end{aligned}
$$

letting $m = n$ and $z_n = (n - 1)/n$, and recalling that $\lim_{n \to \infty}(1 - 1/n)^n = 1/e$, we find

$$|s_{2n}(z_n) - s_n(z_n)| \geq n \left(1 - \frac{1}{n} \right)^{n+1} \left[1 - \left(1 - \frac{1}{n} \right)^n \right] \to \infty \quad \text{as} \quad n \to \infty.$$

Theorem 22.2. If the sequence $\{f_n(z)\}$ converges uniformly on a domain S to $f(z)$ and each $f_n(z)$ is continuous on S, then $f(z)$ is continuous on S.

Proof. Let z_0 be any point in S. Then, for $z \neq z_0$, we have

$$|f(z) - f(z_0)| \leq |f_n(z) - f(z)| + |f_n(z) - f_n(z_0)| + |f_n(z_0) - f(z_0)|. \quad (22.2)$$

Since $\{f_n(z)\}$ is uniformly convergent on S, given any $\epsilon > 0$, there is an $N = N(\epsilon)$ such that $|f_n(z) - f(z)| < \epsilon/3$ for all $n \geq N$ and $z \in S$. Thus, in particular, $|f_n(z_0) - f(z_0)| < \epsilon/3$ for all $n \geq N$. Furthermore, since $f_n(z)$ is continuous at $z_0 \in S$, we can choose $\delta > 0$ such that $|f_n(z) - f_n(z_0)| < \epsilon/3$

if $|z - z_0| < \delta$. Using these bounds in (22.2), we find that $|f(z) - f(z_0)| < \epsilon$ if $|z - z_0| < \delta$; i.e., $f(z)$ is continuous at z_0. ∎

Corollary 22.2. If the series $\sum_{j=0}^{\infty} f_j(z)$ converges uniformly on a domain S to $s(z)$ and each $f_n(z)$ is continuous on S, then $s(z)$ is continuous on S.

Theorem 22.3. Let $\{f_n(z)\}$ be a sequence of functions continuous on a domain S containing the contour γ, and suppose that $\{f_n(z)\}$ converges uniformly to $f(z)$ on S. Then, the following holds:

$$\lim_{n \to \infty} \int_{\gamma} f_n(z)dz = \int_{\gamma} \lim_{n \to \infty} f_n(z)dz = \int_{\gamma} f(z)dz.$$

Proof. Let L be the length of γ. Choose N large enough so that $|f(z) - f_n(z)| < \epsilon/L$ for any $n \geq N$ and for all z on γ. Then, from Theorem 13.1, we have

$$\left| \int_{\gamma} f(z)dz - \int_{\gamma} f_n(z)dz \right| = \left| \int_{\gamma} [f(z) - f_n(z)]dz \right| < \frac{\epsilon}{L}L = \epsilon. \quad ∎$$

Theorem 22.4 (Weierstrass's M-Test). Let $\sum_{j=0}^{\infty} M_j$ be a convergent series of positive numbers. Suppose that $|f_j(z)| \leq M_j$ for all z on a domain S and $j \geq 0$. Then, $\sum_{j=0}^{\infty} f_j(z)$ converges uniformly and absolutely on S.

Proof. Since $\sum_{j=0}^{\infty} M_j$ converges, from Q8 (Lecture 21) it follows that for any given $\epsilon > 0$ there exists an $N = N(\epsilon)$ such that, for all $n \geq N$ and positive integer m, $|M_{n+1} + M_{n+2} + \cdots + M_{n+m}| < \epsilon$. But then $|s_{n+m}(z) - s_n(z)| \leq |f_{n+1}(z)| + |f_{n+2}(z)| + \cdots + |f_{n+m}(z)| \leq M_{n+1} + M_{n+2} + \cdots + M_{n+m} < \epsilon$ for all $n \geq N$, $m > 0$, and $z \in S$. The result now follows from Corollary 22.1. ∎

Theorem 22.5. If $\sum_{j=0}^{\infty} f_j(z)$ converges uniformly to $s(z)$ on a domain S, then for any contour γ in S the following holds

$$\int_{\gamma} \sum_{j=0}^{\infty} f_j(z)dz = \sum_{j=0}^{\infty} \int_{\gamma} f_j(z)dz. \tag{22.3}$$

Proof. Since the partial sums $s_n(z) = \sum_{j=0}^{n} f_j(z)$, $n \geq 0$ converge to $s(z)$ uniformly on S, from Theorem 22.3 it follows that

$$\lim_{n \to \infty} \int_{\gamma} s_n(z)dz = \int_{\gamma} s(z)dz.$$

Now it suffices to note that

$$\lim_{n\to\infty}\int_\gamma s_n(z)dz = \lim_{n\to\infty}\int_\gamma \sum_{j=0}^{n} f_j(z)dz = \lim_{n\to\infty}\sum_{j=0}^{n}\int_\gamma f_j(z)dz = \sum_{j=0}^{\infty}\int_\gamma f_j(z)dz$$

and

$$\int_\gamma s(z)dz = \int_\gamma \sum_{j=0}^{\infty} f_j(z)dz. \quad\blacksquare$$

Theorem 22.6. Let $\{f_n(z)\}$ be a sequence of analytic functions on a domain S. If $f_n(z) \to f(z)$ uniformly on S, then $f(z)$ is analytic on S.

Proof. Let $z_0 \in S$. Since S is open, there exists an $r > 0$ such that the open disk $B(z_0, r) \subseteq S$. Let γ be a closed contour in $B(z_0, r)$. Then, by Theorem 22.3, we have

$$\lim_{n\to\infty}\int_\gamma f_n(z)dz = \int_\gamma f(z)dz.$$

Since, for each n, $f_n(z)$ is analytic in S, $\int_\gamma f_n(z)dz = 0$. Hence, $\int_\gamma f(z)dz = 0$. By Theorem 22.2, $f(z)$ is continuous on S. Therefore, by Theorem 18.4, $f(z)$ is analytic in $B(z_0, r)$. Since $z_0 \in S$ is arbitrary, $f(z)$ is analytic in S. $\blacksquare$

Corollary 22.3. Let $\{f_n(z)\}$ be a sequence of analytic functions on a domain S. If $\sum_{j=0}^{\infty} f_j(z)$ converges to $s(z)$ uniformly on S, then $s(z)$ is analytic on S.

To prove our next result, we need the following elementary lemma.

Lemma 22.1. Suppose that the sequence $\{f_n(z)\}$ (series $\sum_{j=0}^{\infty} f_j(z)$) converges uniformly to $f(z)$ ($s(z)$) on a compact set S, and $g(z)$ is a continuous function on S. Then, $\{g(z)f_n(z)\}$ ($\sum_{j=0}^{\infty} g(z)f_j(z)$) converges to $g(z)f(z)$ ($g(z)s(z)$) uniformly on S.

Theorem 22.7. Let $\{f_n(z)\}$ be a sequence of analytic functions on a domain S. If $f_n(z) \to f(z)$ uniformly on every compact subset of S, then, for any $k \geq 1$, $f_n^{(k)}(z) \to f^{(k)}(z)$ for all $z \in S$; i.e., the limit of the kth derivative is the kth derivative of the limit. Moreover, for each $k \geq 1$, the differentiated sequence $\{f_n^{(k)}(z)\}$ converges to $f^{(k)}(z)$ uniformly on every compact subset of S.

Proof. Let $z_0 \in S$ and γ be a positively oriented closed contour in S. From (18.6), we have

$$f^{(k)}(z_0) = \frac{k!}{2\pi i}\int_\gamma \frac{f(z)}{(z-z_0)^{k+1}}dz \quad \text{and} \quad f_n^{(k)}(z_0) = \frac{k!}{2\pi i}\int_\gamma \frac{f_n(z)}{(z-z_0)^{k+1}}dz.$$

Now, since $f_n(z) \to f(z)$ uniformly on γ and $1/(z - z_0)^{k+1}$ is continuous on γ, from Lemma 22.1 it follows that

$$\frac{f_n(z)}{(z - z_0)^{k+1}} \longrightarrow \frac{f(z)}{(z - z_0)^{k+1}}$$

uniformly on γ. Thus, we can apply Theorem 22.3 to obtain

$$f_n^{(k)}(z_0) = \frac{k!}{2\pi i} \int_\gamma \frac{f_n(z)}{(z - z_0)^{k+1}} dz \longrightarrow f^{(k)}(z_0) = \frac{k!}{2\pi i} \int_\gamma \frac{f(z)}{(z - z_0)^{k+1}} dz.$$

Now let S_1 be a compact subset of S, and in S let γ be a positively oriented closed contour containing S_1 so that the shortest distance d between S_1 and γ is positive; i.e., for all $\xi \in \gamma$ and $z \in S_1$, $|\xi - z| \geq d > 0$. Then, from (18.4), for all $z \in S_1$ it follows that

$$
\begin{aligned}
|f_n^{(k)}(z) - f^{(k)}(z)| &= \left| \frac{k!}{2\pi i} \int_\gamma \frac{f_n(\xi) - f(\xi)}{(\xi - z)^{k+1}} d\xi \right| \\
&\leq \frac{k!}{2\pi d^{k+1}} L(\gamma) \max_{\xi \in \{\gamma\}} |f_n(\xi) - f(\xi)|.
\end{aligned}
$$

Finally, since $f_n(\xi) \to f(\xi)$ uniformly on every compact subset of S, from the inequality above it follows that $f_n^{(k)}(z) \to f^{(k)}(z)$ uniformly on S_1. ∎

Corollary 22.4. Let $\{f_n(z)\}$ be a sequence of analytic functions on a domain S. If $\sum_{j=0}^\infty f_j(z)$ converges to $s(z)$ uniformly on every compact subset of S, then for any $k \geq 1$ and all $z \in S$,

$$\sum_{j=0}^\infty f_j^{(k)}(z) = s^{(k)}(z), \tag{22.4}$$

i.e., the series may be differentiated term-by-term. Moreover, for each k, the convergence in (22.4) is uniform on every compact subset of S.

Problems

22.1. Check for the pointwise and uniform convergence of the series $\sum_{j=0}^\infty f_j(z)$ on the given domain, where $f_j(z)$ is

(a). $\dfrac{z^j}{(j+1)(j+2)}$, $|z| \leq 1$,   (b). $\dfrac{1}{(z+j+1)^2}$, $\operatorname{Re} z > 0$,

(c). $\dfrac{1}{(1+j^2+z^2)}$, $1 < |z| < 2$,   (d). $\dfrac{1}{(1+j^2 z^2)}$, $|z| \leq 1$.

22.2. Does the sequence of functions

$$f_n(z) = \begin{cases} n|z| & \text{if } |z| \leq 1/n \\ 1 & \text{if } 1/n \leq |z| \leq 1 \end{cases}$$

converge uniformly on the unit disk $\overline{B}(0,1)$?

22.3. Prove that Theorem 22.7 does not hold if S is assumed to be an arbitrary set instead of a domain.

22.4. Suppose that $\{f_n(z)\}$ converges to $f(z)$ pointwise on a domain S. For each $n \geq 0$, let

$$\sup\{|f_n(z) - f(z)| : z \in S\} = M_n.$$

Show that $\{f_n(z)\}$ converges to $f(z)$ uniformly on S if and only if $\lim_{n \to \infty} M_n = 0$.

22.5. Let the functions $f_n(z)$ be analytic on a domain S and continuous on $\overline{S}$, and let the series $\sum_{j=0}^{\infty} f_j(z)$ converge uniformly on the boundary γ of S. Show that the series $\sum_{j=0}^{\infty} f_j(z)$ converges uniformly on $\overline{S}$.

Answers or Hints

22.1. (a). For all $|z| \leq 1$, $|f_j(z)| \leq 1/(j+1)^2$, and hence the convergence is uniform. (b). For all $\mathrm{Re}\, z > 0$, $|x+iy+j+1|^2 = (x+j+1)^2 + y^2 > (j+1)^2$. (c). For $j \geq 4$, $|1 + j^2 + z^2| > 1 + j^2 - 4 > (j+1)^2/2$, and hence the convergence is uniform. (d). $f_j(z) \to 0$ if $z \neq 0$ and 1 if $z = 0$, and hence from Theorem 22.2 the convergence is only pointwise.

22.2. Each $f_n(z)$ is continuous on $|z| \leq 1$; however, the limiting function $\lim_{n \to \infty} f_n(z) = \begin{cases} 1 & \text{if } 0 < |z| \leq 1 \\ 0 & \text{if } z = 0 \end{cases}$ is not continuous. Hence, in view of Theorem 22.2, the convergence is not uniform.

22.3. The sequence $\{(\sin nx)/n\}$ converges uniformly to zero on the real axis; however, the sequence of its derivative $\{\cos nx\}$ converges only at $x = 0$. Thus, the sequence $\{(\sin nz)/n\}$ cannot converge uniformly on any domain containing points of the real axis.

22.4. See the definition of uniform convergence.

22.5. Clearly, the function $s_{n+m}(z) - s_n(z)$, being a finite sum of analytic functions, is analytic on S and continuous on $\overline{S}$. From the uniform convergence on γ, Corollary 22.1 gives $|s_{n+m}(z) - s_n(z)| = |f_{n+m}(z) + \cdots + f_{n+1}(z)| < \epsilon$ for $n \geq N$ for any natural number m and all $z \in \{\gamma\}$. Now, by Theorem 20.2, it follows that $|s_{n+m}(z) - s_n(z)| < \epsilon$ for $n \geq N$ for any natural number m and all $z \in \overline{S}$.

Lecture 23
Power Series

Power series are a special type of series of functions that are of fundamental importance. For a given power series we shall introduce and show how to compute its radius of convergence. We shall also show that within its radius of convergence a power series can be integrated and differentiated term-by-term.

An infinite series of the form

$$\sum_{j=0}^{\infty} a_j(z - z_0)^j = a_0 + a_1(z - z_0) + \cdots + a_j(z - z_0)^j + \cdots \qquad (23.1)$$

is called a *power series*, with z_0 as the point of expansion. The constants a_j are called the *coefficients* of the power series. It is clear that this series converges at $z = z_0$; it may converge for all z, or it may converge for some values of z and not for others. If it converges absolutely for $|z - z_0| < R$ and diverges for $|z - z_0| > R$, then R is called the *radius of convergence*. It is clear that this R is unique. If (23.1) converges nowhere except at z_0, we define R to be zero; if (23.1), converges for all z, we say R is infinite. On the circle of convergence; i.e., $|z - z_0| = R$, the series (23.1) may converge at some, all, or none of the points. The following result determines the domain of convergence of (23.1).

Theorem 23.1. If (23.1) converges at $z = z_1$ ($\neq z_0$), then it converges absolutely and uniformly in the closed disk $\overline{B}(z_0, r)$, where $r < |z_1 - z_0|$.

Proof. Since the series (23.1) converges at $z = z_1$, $a_j(z_1 - z_0)^j \to 0$ as $j \to \infty$. Consequently, there exists a constant M such that $|a_j||z_1 - z_0|^j \leq M$, $j \geq 0$. Let z be an arbitrary point such that $|z - z_0| \leq r < |z_1 - z_0|$; i.e.,

$$\frac{|z - z_0|}{|z_1 - z_0|} = \rho < 1.$$

But then

$$|a_j||z - z_0|^j \leq M\frac{|z - z_0|^j}{|z_1 - z_0|^j} = M\rho^j, \quad j \geq 0.$$

Now, since in view of Example 21.3 the geometric series $\sum_{j=0}^{\infty} M\rho^j$ converges, from Weierstrass's M-test it follows that (23.1) converges absolutely and uniformly in the closed disk $\overline{B}(z_0, r)$. ∎

Corollary 23.1. If (23.1) diverges at some point $z = z_1$, then it diverges at all points z that satisfy the inequality $|z - z_0| > |z_1 - z_0|$.

Corollary 23.2. Let R be the radius of convergence of (23.1). For each $0 < R_1 < R$, the series converges uniformly on the closed disk $\bar{B}(z_0, R_1)$.

Example 23.1. From Example 21.3 and Theorem 23.1, it follows that the geometric series

$$\sum_{j=0}^{\infty} z^j = 1 + z + z^2 + \cdots + z^j + \cdots, \qquad (23.2)$$

which is in fact a power series with $z_0 = 0$, $a_j = 1$, converges absolutely and uniformly on $|z| \le r < 1$ to the analytic function $1/(1-z)$. Since at $z = 1$ the series (23.2) diverges, from Corollary 23.1 we find that it diverges on $|z| > 1$. Thus, the radius of convergence of (23.2) is $R = 1$. If $|z| = 1$, then the terms of (23.2) do not tend to zero, and hence it also diverges on $|z| = 1$.

From Theorem 23.1 and Corollary 23.1, it is clear that the radius of convergence R of (23.1) is the least upper bound of the distances $|z - z_0|$ from the point z_0 to the point z at which the series (23.1) converges. However, for the computation of R, one of the following methods is usually employed:

d'Alembert's Ratio Test. $R = \left(\lim_{j \to \infty} |a_{j+1}/a_j|\right)^{-1}$, provided the limit exists.

Cauchy's Root Test. $R = \left(\lim_{j \to \infty} |a_j|^{1/j}\right)^{-1}$, provided the limit exists.

Cauchy-Hadamard Formula. $R = \left(\limsup_{j \to \infty} |a_j|^{1/j}\right)^{-1}$; this limit always exists.

Example 23.2. Since in (23.2), $a_j = 1$, the ratio test confirms its radius of convergence $R = 1$. Similarly, for the series $\sum_{j=0}^{\infty} z^{j+2}/(j+1)$, we have $|a_{j+1}/a_j| = |(j-1)/j| \to 1$ as $j \to \infty$. Thus, for this series also, the radius of convergence $R = 1$. This series diverges at $z = 1$ and converges at $z = -1$. For the series $\sum_{j=0}^{\infty} (-1)^j (z - 2 - 3i)^{j+1}/(j+1)!$, we have $|a_{j+1}/a_j| = 1/(j+2) \to 0$ as $j \to \infty$. Thus, the radius of convergence $R = \infty$.

Example 23.3. For the series $\sum_{j=0}^{\infty} (j+1)^j (z+i)^j$, we have $|a_j|^{1/j} = (j+1) \to \infty$ as $j \to \infty$, and hence by the root test its radius of convergence $R = 0$; i.e., it converges only at $z = -i$. For the series $\sum_{j=0}^{\infty} (z - 3 - 2i)^j/(j+1)^j$, the root test gives $R = \infty$, whereas for the series $\sum_{j=0}^{\infty} [(j+1)/(2j+3)]^j (z - 2 - i)^j$, $R = 2$.

Example 23.4. In the series $\sum_{j=0}^{\infty} z^{2j}/2^j$, $a_{2m} = 1/2^m$ and $a_{2m+1} = 0$. Thus, $|a_j|^{1/j} = 0$ if j is odd and $|a_j|^{1/j} = (1/2^m)^{1/2m} = 1/\sqrt{2}$ if $j = 2m$.

Hence, $\limsup_{j\to\infty} |a_j|^{1/j} = 1/\sqrt{2}$. Therefore, by the Cauchy-Hadamard formula, $R = \sqrt{2}$.

We shall now prove the following theorem.

Theorem 23.2. Let R be the radius of convergence of (23.1).

(I). $s(z) = \sum_{j=0}^{\infty} a_j(z - z_0)^j$ is an analytic function on $B(z_0, R)$.

(II). (*Term-by-Term Integration*). If γ is a contour in $B(z_0, R)$ and $g(z)$ is a continuous function on γ, then

$$\int_{\gamma} g(z)s(z)dz \;=\; \int_{\gamma} g(z) \sum_{j=0}^{\infty} a_j(z - z_0)^j dz \;=\; \sum_{j=0}^{\infty} a_j \int_{\gamma} g(z)(z - z_0)^j dz.$$

(23.3)

In particular, if $g(z) \equiv 1$, then

$$\int_{\gamma} \sum_{j=0}^{\infty} a_j(z - z_0)^j dz \;=\; \sum_{j=0}^{\infty} a_j \int_{\gamma} (z - z_0)^j dz. \tag{23.4}$$

(III). (*Term-by-Term Differentiation*).

$$s'(z) \;=\; \frac{d}{dz} \sum_{j=0}^{\infty} a_j(z - z_0)^j \;=\; \sum_{j=1}^{\infty} ja_j(z - z_0)^{j-1}. \tag{23.5}$$

The radius of convergence of the series $\sum_{j=1}^{\infty} ja_j(z - z_0)^{j-1}$ is also R.

Proof. (I). Let $z_1 \in B(z_0, R)$. Choose any r such that $|z_1 - z_0| < r < R$. Then, $s_n(z) = \sum_{j=0}^{n} a_j(z - z_0)^j \to s(z)$ uniformly on $\overline{B}(z_0, r)$ by Theorem 23.1. Now, since each term of the series $a_j(z - z_0)^j$ is entire, the function $s(z)$ is analytic on $\overline{B}(z_0, r)$ by Corollary 22.3. Hence, in particular, $s(z)$ is analytic at z_1.

(II). Let $0 < r < R$ be such that $\{\gamma\} \subset \overline{B}(z_0, r)$. Then, by Theorem 23.1, $s_n(z) \to s(z)$ uniformly on $\overline{B}(z_0, r)$. Since $\{\gamma\}$ is a closed subset of $\overline{B}(z_0, r)$, from Lemma 22.1 it follows that $g(z)s_n(z) \to g(z)s(z)$ uniformly on $\{\gamma\}$. Now, from Theorem 22.3, we have

$$\lim_{n\to\infty} \int_{\gamma} g(z)s_n(z)dz \;=\; \int_{\gamma} \lim_{n\to\infty} g(z)s_n(z)dz,$$

and hence

$$\sum_{j=0}^{\infty} a_j \int_{\gamma} g(z)(z - z_0)^j dz \;=\; \int_{\gamma} g(z)s(z)dz.$$

(III). Let $z \in B(z_0, R)$, and let $\gamma_r = \{\xi : |\xi - z| = r\} \subset B(z_0, R)$. Then, from part (II) with $g(\xi) = 1/[2\pi i(\xi - z)^2]$, it follows that

$$\frac{1}{2\pi i} \int_{\gamma_r} \frac{s(\xi)}{(\xi - z)^2} d\xi = \sum_{j=0}^{\infty} a_j \frac{1}{2\pi i} \int_{\gamma_r} \frac{(\xi - z_0)^j}{(\xi - z)^2} d\xi.$$

Now, from (18.3), we have

$$s'(z) = \sum_{j=0}^{\infty} a_j \frac{d}{dz}(z - z_0)^j = \sum_{j=1}^{\infty} ja_j(z - z_0)^{j-1}.$$

Finally, since

$$\lim_{j \to \infty} \left| \frac{(j+1)a_{j+1}}{ja_j} \right| = \lim_{j \to \infty} \left| \frac{j+1}{j} \right| \lim_{j \to \infty} \left| \frac{a_{j+1}}{a_j} \right| = \lim_{j \to \infty} \left| \frac{a_{j+1}}{a_j} \right|,$$

the radius of convergence of the series $\sum_{j=1}^{\infty} ja_j(z - z_0)^{j-1}$ is also R. ∎

Corollary 23.3. Let R be the radius of convergence of (23.1). Then, $s(z) = \sum_{j=0}^{\infty} a_j(z - z_0)^j$ is infinitely differentiable for all $z \in B(z_0, R)$. In fact, for any k,

$$s^{(k)}(z) = \sum_{j=k}^{\infty} j(j-1) \cdots (j-k+1)a_j(z - z_0)^{j-k}. \tag{23.6}$$

Remark 23.1. From (23.6), it immediately follows that $a_k = s^{(k)}(z_0)/k!$, $k = 0, 1, \cdots$.

Example 23.5. From Example 23.1, we have $\sum_{j=0}^{\infty} z^j = 1/(1 - z)$, $z \in B(0, 1)$. Applying Theorem 23.2 (III) to differentiate the series term-by-term, it follows that

$$1 + 2z + 3z^2 + \cdots = \sum_{j=0}^{\infty} (j+1)z^j = \frac{1}{(1-z)^2}.$$

Example 23.6. We shall show that on $B(0, 1)$,

$$\text{Log}(1 + z) = \sum_{j=0}^{\infty} (-1)^j \frac{z^{j+1}}{j+1} = z - \frac{z^2}{2} + \frac{z^3}{3} + \cdots. \tag{23.7}$$

For this, let $s(z)$ be the right-hand side of (23.7). Since $\lim_{j \to \infty} |a_{j+1}/a_j| = \lim_{j \to \infty} |j/(j+1)| = 1$, the radius of convergence of $s(z)$ is 1. Using Theorem 23.2 (III), we find

$$s'(z) = 1 - z + z^2 - z^3 - \cdots = (1 + z)^{-1} = \frac{d}{dz} \text{Log}(1 + z) \tag{23.8}$$

on $B(0,1)$; i.e., $[s(z) - \text{Log}\,(1+z)]' = 0$. The relation (23.7) now follows from Theorem 7.2 and the fact that $s(0) - \text{Log}\,1 = 0$. Clearly, (23.7) also follows if we begin with (23.8) and use Theorem 23.2 (II) to integrate the series term-by-term along any path connecting 0 and $z \in B(0,1)$.

We conclude this lecture by stating the following theorem.

Theorem 23.3. Assume that the power series $f(z) = \sum\limits_{j=0}^{\infty} a_j(z - z_0)^j$

and $g(z) = \sum\limits_{j=0}^{\infty} b_j(z - z_0)^j$ have the radius of convergence R_1 and R_2, respectively. Then,

(i). $f(z) \pm g(z) = \sum\limits_{j=0}^{\infty}(a_j \pm b_j)(z - z_0)^j$ and $f(z)g(z) = \sum\limits_{j=0}^{\infty} c_j(z - z_0)^j$,

where $c_j = \sum\limits_{k=0}^{j} a_k b_{j-k} = \sum\limits_{k=0}^{j} a_{j-k} b_k$, have the radius of convergence $R = \min\{R_1, R_2\}$, and

(ii). if $g(z) \neq 0$ in the disk $B(z_0, r)$ (necessarily $b_0 \neq 0$), then the quotient $f(z)/g(z) = \sum\limits_{j=0}^{\infty} c_j(z - z_0)^j$, where the coefficients satisfy the equation $a_j = \sum\limits_{k=0}^{j} b_{j-k} c_k$, has the radius of convergence $R = \min\{r, R_1, R_2\}$.

Problems

23.1. Use the ratio test to compute the radius of convergence of the following series:

(a). $\sum\limits_{j=0}^{\infty} \dfrac{1}{(1 + 3i)^{j+1}}(z - 7i)^j$, (b). $\sum\limits_{j=0}^{\infty} \dfrac{(j!)^2}{(2j)!}(z - 3 - 2i)^j$.

23.2. Use the root test to compute the radius of convergence of the following series:

(a). $\sum\limits_{j=0}^{\infty} \left(\dfrac{11j + 9}{2j + 5}\right)^j (z - 2 - 5i)^j$, (b). $\sum\limits_{j=0}^{\infty} \left(\dfrac{4j^2}{2j + 1} - \dfrac{6j^2}{3j + 4}\right)(z - 3i)^j$.

23.3. Use the Cauchy-Hadamard formula to compute the radius of convergence of the series $\sum_{j=0}^{\infty} a_j z^j$, where (a). $a_j = \left(\sum\limits_{k=0}^{j} \dfrac{1}{k!} \right)^j$,   (b). $a_j = (8 - (-3)^j)^j$.

23.4. Suppose that (23.1) has radius of convergence R. Show that $\sum_{j=0}^{\infty} a_j^2 (z - z_0)^j$ has radius of convergence R^2.

23.5. Show that

$$\lim_{n \to \infty} \left(\frac{n!}{n^n} \right)^{1/n} = \frac{1}{e}$$

and use it to find the radius of convergence of the series

$$1 + z + \frac{2^2}{2!} z^2 + \cdots + \frac{j^j}{j!} z^j + \cdots .$$

23.6. The *Bessel function of order n* is defined by

$$J_n(z) = \sum_{j=0}^{\infty} \frac{(-1)^j}{j!(j+n)!} \left(\frac{z}{2} \right)^{2j+n} .$$

(a). Show that $J_n(z)$ is entire.

(b). Verify the identity $[z^n J_n(z)]' = z^n J_{n-1}(z)$.

23.7. Consider the function

$$f(z) = 1 + \sum_{j=1}^{\infty} \frac{a(a-1)(a-2) \cdots (a-j+1)}{j!} z^j .$$

(a). Show that its radius of convergence is 1.

(b). For all $|z| < 1$, $f'(z) = af(z)/(1 + z)$. Hence, deduce the *binomial expansion* $(1 + z)^a = f(z)$.

23.8. Use power series $f(z) = \sum_{j=0}^{\infty} a_j z^j$ to solve the *functional equation* $f(z) = z + f(z^2)$.

23.9. If the sum of two power series in a neighborhood of the point of expansion z_0 is the same, then show that the identical powers of $(z - z_0)$ have identical coefficients; i.e., there is a unique power series that has a given sum in a neighborhood of z_0.

23.10. For the power series $s(z) = \sum_{j=0}^{\infty} a_j z^j$, let the radius of convergence be R. Show that

(a). the coefficients of the odd powers of z vanish if $s(z)$ is even; i.e., if $s(-z) = s(z)$, and (b). the coefficients of the even powers of z vanish if $s(z)$ is odd; i.e., if $s(-z) = -s(z)$.

23.11. For the power series $s(z) = \sum_{j=0}^{\infty} a_j z^j$, let the radius of convergence be R.

(a). Show that $\overline{s(\overline{z})}$ is a power series in z with the same radius of convergence R.

(b). Suppose $R > 1$ and $|s(e^{i\theta})| \le 4$ for $0 \le \theta \le \pi$ and $|s(e^{i\theta})| \le 9$ for $\pi < \theta \le 2\pi$. Show that $|s(0)| \le 6$.

23.12. Let $f(z) = \sum_{j=0}^{\infty} a_j z^j = 1 + z + 2z^2 + 3z^3 + 5z^4 + 8z^5 + 13z^6 + 21z^7 + \cdots$, where the coefficients a_j are the *Fibonacci numbers* defined by $a_0 = 1$, $a_1 = 1$, $a_j = a_{j-1} + a_{j-2}$, $j \ge 2$. Show that $f(z) = 1/(1 - z - z^2)$, $z \in B(0, R)$, where $R = (\sqrt{5} - 1)/2$.

23.13. Let $f(z) = \sum_{j=0}^{\infty} a_j z^j$, where the coefficients a_j are the *Lucas numbers* defined by $a_0 = 1$, $a_1 = 3$, $a_j = a_{j-1} + a_{j-2}$, $j \ge 2$. Show that $f(z) = (1 + 2z)/(1 - z - z^2)$, $z \in B(0, R)$, where $R = (\sqrt{5} - 1)/2$.

23.14 (Weierstrass Double Series Theorem). Suppose that for each $k = 0, 1, 2, \cdots$ the power series $f_k(z) = \sum_{j=0}^{\infty} a_j^{(k)} (z - z_0)^j$ converges in the disk $B(z_0, R)$, $R \le \infty$; i.e., each power series defines the function $f_k(z)$ in the disk $B(z_0, R)$. Suppose that $F(z) = \sum_{k=0}^{\infty} f_k(z)$ converges uniformly for $|z - z_0| \le \rho$ for every $\rho < R$. Show that $F(z)$ is analytic on $|z - z_0| \le \rho$ and has a power series expansion, $F(z) = \sum_{j=0}^{\infty} A_j (z - z_0)^j$, that converges for all $|z - z_0| < R$; here, $A_j = \sum_{k=0}^{\infty} a_j^{(k)}$. Furthermore, show that

(a). $\displaystyle\sum_{j=1}^{\infty} \frac{z^j}{1 + z^{2j}}$ converges uniformly for $|z| \le r < 1$, and

(b). $\displaystyle\sum_{j=1}^{\infty} \frac{z^j}{1 + z^{2j}} = \sum_{k=0}^{\infty} (-1)^k \frac{z^{2k+1}}{1 - z^{2k+1}}$ for $|z| < 1$.

Answers or Hints

23.1. (a). $\sqrt{10}$, (b). 4.

23.2. (a). 2/11, (b). 3/5.

23.3. (a). $1/e$, (b). 1/11.

23.4. $\limsup_{j \to \infty} |a_j^2|^{1/j} = \left(\limsup_{j \to \infty} |a_j|^{1/j}\right)^2$.

23.5. $e^n = 1 + \frac{n}{1!} + \cdots + \frac{n^{n-1}}{(n-1)!} + \frac{n^n}{n!} \left[1 + \frac{n}{n+1} + \frac{n^2}{(n+1)(n+2)} + \cdots\right]$ implies that $\frac{n^n}{n!} < e^n < n\frac{n^n}{n!} + \frac{n^n}{n!}\left[1 + \frac{n}{n+1} + \left(\frac{n}{n+1}\right)^2 + \cdots\right] = (2n+1)\frac{n^n}{n!}$. Hence,

$\frac{1}{e^n} < \frac{n!}{n^n} < \frac{2n+1}{e^n}$. $R = 1/e$.

23.6. (a). $a_j = \begin{cases} 0 & \text{if } j \text{ is odd} \\ (-1)^{j/2}/[2^{j+n}\,(j/2)!\,(j/2+n)!] & \text{if } j \text{ is even.} \end{cases}$ We will

show that $\limsup_{j\to\infty} |a_j|^{1/j} = 0$. Clearly, $|a_j|^{1/j} = 0$ if j is odd, and for j
even we have $|a_j|^{1/j} \le 1/[(j/2)!]^{2/j}$. Now use $[(k)!]^{1/k} \to \infty$ as $k \to \infty$.

(b). $[z^n J_n(z)]' = \sum_{j=0}^{\infty} \frac{(-1)^j 2^n (j+n)}{j!(j+n)!}\left(\frac{z}{2}\right)^{2j+2n-1} = z^n \sum_{j=0}^{\infty} \frac{(-1)^j}{j!(j+n-1)!}$
$\times \left(\frac{z}{2}\right)^{2j+n-1}$.

23.7. (a). $\lim_{j\to\infty} |a_{j+1}/a_j| = \lim_{j\to\infty} |(a-j)/(j+1)| = 1$. (b). Compute
$(1+z)f'(z)$ directly and show that it is the same as $af(z)$.

23.8. $f(z) = a_0 + \sum_{j=0}^{\infty} z^{2^j}$, $|z| < 1$.

23.9. If $\sum_{j=0}^{\infty} a_j(z-z_0)^j$ and $\sum_{j=0}^{\infty} b_j(z-z_0)^j$ have the same sum $s(z)$ in
a neighborhood of z_0, then from Corollary 23.3 and Remark 23.1 it follows
that $a_j = b_j = s^{(j)}(z_0)/j!$, $j = 0, 1, \cdots$.

23.10. (a). $s^{(2k+1)}(0) = 0$, $k = 0, 1, \cdots$, (b). $s^{(2k)}(0) = 0$, $k = 0, 1, \cdots$.

23.11. (a). $s(z) = \sum_{j=0}^{\infty} a_j z^j$, so $\overline{s(\overline{z})} = \sum_{j=0}^{\infty} \overline{a}_j z^j$, and since $\overline{\lim}|\overline{a}_j|^{1/j} = \overline{\lim}|a_j|^{1/j}$, they have the same radius of convergence. (b). Both $s(z)$ and
$\overline{s(\overline{z})}$ converge absolutely and uniformly on $B(0,R)$, $R > 1$, so they are
analytic on $\overline{B(0,1)}$. Hence, $g(z) = s(z)\overline{s(\overline{z})}$ is also analytic on $\overline{B(0,1)}$.
Since $|s(e^{i\theta})| \le 4$ for $0 \le \theta \le \pi$ and $|s(e^{i\theta})| \le 9$ for $\pi < \theta \le 2\pi$, $|g(e^{i\theta})| = |s(e^{i\theta})\overline{s(e^{-i\theta})}| = |s(e^{i\theta})||s(e^{i(2\pi-\theta)})| \le 4 \times 9$ or 9×4 according to whether
$0 \le \theta \le \pi$ or $\pi < \theta \le 2\pi$. Thus, $|g(e^{i\theta})| \le 36$ for all $0 \le \theta \le 2\pi$. Now, by the
Maximum Modulus Principle, $|g(0)| \le |g(e^{i\theta})|$ for some θ, so $|g(0)| \le 36$;
i.e., $|s(0)|^2 \le 36$, and hence $|s(0)| \le 6$.

23.12. $1 + zf(z) + z^2 f(z) = 1 + \sum_{j=0}^{\infty} a_j z^{j+1} + \sum_{j=0}^{\infty} a_j z^{j+2} = 1 + z + \sum_{j=2}^{\infty}(a_{j-1}+a_{j-2})z^j = 1+z+\sum_{j=2}^{\infty} a_j z^j$. Hence, $1+zf(z)+z^2 f(z) = f(z)$.

23.13. Show that $f(z) = 1 + 2z + zf(z) + z^2 f(z)$.

23.14. Use results of Lecture 22. For (a) and (b), expand each function in
the power series.

Lecture 24
Taylor's Series

In this lecture, we shall prove Taylor's Theorem, which expands a given analytic function in an infinite power series at each of its points of analyticity. The novelty of the proof comes from the fact that it requires only Cauchy's integral formula for derivatives.

Theorem 24.1 (Taylor's Theorem). Let f be analytic in a domain S and let at $z_0 \in S$, $B(z_0, R)$ be the largest open disk in S. Then, f has the series representation

$$f(z) = f(z_0) + \frac{f'(z_0)}{1!}(z-z_0) + \frac{f''(z_0)}{2!}(z-z_0)^2 + \cdots = \sum_{j=0}^{\infty} \frac{f^{(j)}(z_0)}{j!}(z-z_0)^j,$$

$$(24.1)$$

which converges for all $z \in B(z_0, R)$. Furthermore, for any $0 \le r < R$, the convergence is uniform on the closed disk $\overline{B}(z_0, r)$. The series (24.1) is called the *Taylor series of f at z_0*.

Proof. Let $z \in B(z_0, R)$, and let μ denote the distance between z and z_0; i.e., $|z - z_0| = \mu$. Clearly, $0 \le \mu < R$. Let ν be such that $0 \le \mu < \nu < R$, and let γ be the positively oriented circle $|\xi - z_0| = \nu$. For a point ζ on γ, we define $w = (z - z_0)/(\zeta - z_0)$ so that $|w| = |z - z_0|/|\xi - z_0| = \mu/\nu < 1$. Thus, in view of Example 23.1, we have

$$\frac{1}{\zeta - z} = \frac{1}{(\zeta - z_0) - (z - z_0)} = \frac{1}{\zeta - z_0} \frac{1}{1 - w} = \frac{1}{\zeta - z_0} \sum_{j=0}^{\infty} w^j,$$

which is the same as

$$\frac{f(\zeta)}{\zeta - z} = \sum_{j=0}^{\infty} f(\zeta) \frac{(z - z_0)^j}{(\zeta - z_0)^{j+1}}$$

uniformly in ζ on γ.

Now, from Theorem 22.5 and Cauchy's integral formula for derivatives

(18.4), we find

$$f(z) = \frac{1}{2\pi i} \int_\gamma \frac{f(\zeta)}{\zeta - z} d\zeta = \frac{1}{2\pi i} \int_\gamma \sum_{j=0}^\infty \frac{f(\zeta)}{(\zeta - z_0)^{j+1}} (z - z_0)^j d\zeta$$

$$= \sum_{j=0}^\infty \left[\frac{1}{2\pi i} \int_\gamma \frac{f(\zeta)}{(\zeta - z_0)^{j+1}} d\zeta \right] (z - z_0)^j$$

$$= \sum_{j=0}^\infty \frac{f^{(j)}(z_0)}{j!} (z - z_0)^j.$$

It is clear that the radius of convergence of (24.1) is at least R. This in turn implies that the power series converges uniformly on every closed disk $\overline{B}(z_0, r)$, where $0 \le r < R$. ∎

Remark 24.1. Taylor's series (24.1) with $z_0 = 0$ reduces to

$$f(z) = f(0) + \frac{f'(0)}{1!} z + \frac{f''(0)}{2!} z^2 + \cdots = \sum_{j=0}^\infty \frac{f^{(j)}(0)}{j!} z^j. \qquad (24.2)$$

This series is called the *Maclaurin series of f*.

Remark 24.2. In view of Theorem 23.2, Taylor's series (24.1) in $B(z_0, R)$ can be integrated as well as differentiated any number of times term-by-term. The radius of convergence of each differentiated series is also R.

Theorem 24.2 (Uniqueness of Taylor Series). If $f(z) = \sum_{j=0}^\infty a_j (z - z_0)^j$ for all $z \in B(z_0, R)$, then the series is the Taylor series of f at z_0.

Proof. From Corollary 23.3 and Remark 23.1, it follows that $a_j = f^{(j)}(z_0)/j!$, $j = 0, 1, \cdots$. ∎

Remark 24.3. The function $f(z)$ is analytic at z_0 if and only if it can be expanded in Taylor's series at z_0.

Theorem 24.3. Let f and g have Taylor's series expansions $f(z) = \sum_{j=0}^\infty a_j (z - z_0)^j$, $|z - z_0| < R_1$ and $g(z) = \sum_{j=0}^\infty b_j (z - z_0)^j$, $|z - z_0| < R_2$; here $a_j = f^{(j)}(z_0)/j!$ and $b_j = g^{(j)}(z_0)/j!$. Then, the following hold:

(i). $\alpha f(z) = \sum_{j=0}^\infty \alpha a_j (z - z_0)^j$, $|z - z_0| < R_1$, where $\alpha \in \mathbf{C}$ is a constant.

(ii). $f(z) \pm g(z) = \sum_{j=0}^\infty (a_j \pm b_j)(z - z_0)^j$, $|z - z_0| < R = \min\{R_1, R_2\}$.

(iii). $f(z)g(z) = \sum_{j=0}^\infty c_j (z - z_0)^j$, $|z - z_0| < R = \min\{R_1, R_2\}$, where $c_j = \sum_{k=0}^j a_{j-k} b_k = \sum_{k=0}^j a_k b_{j-k}$.

Example 24.1. Calculating the derivatives of all orders at $z_0 = 0$ of the entire functions e^z, $\cos z$, $\sin z$, $\cosh z$, $\sinh z$, we obtain the following Maclaurin series expansions, which are valid on $|z| < \infty$:

$$e^z = 1 + z + \frac{z^2}{2!} + \frac{z^3}{3!} + \cdots = \sum_{j=0}^{\infty} \frac{z^j}{j!}, \tag{24.3}$$

$$\cos z = 1 - \frac{z^2}{2!} + \frac{z^4}{4!} - \frac{z^6}{6!} + \cdots = \sum_{j=0}^{\infty} (-1)^j \frac{z^{2j}}{(2j)!}, \tag{24.4}$$

$$\sin z = z - \frac{z^3}{3!} + \frac{z^5}{5!} - \frac{z^7}{7!} + \cdots = \sum_{j=0}^{\infty} (-1)^j \frac{z^{2j+1}}{(2j+1)!}, \tag{24.5}$$

$$\cosh z = 1 + \frac{z^2}{2!} + \frac{z^4}{4!} + \frac{z^6}{6!} + \cdots = \sum_{j=0}^{\infty} \frac{z^{2j}}{(2j)!}, \tag{24.6}$$

$$\sinh z = z + \frac{z^3}{3!} + \frac{z^5}{5!} + \frac{z^7}{7!} + \cdots = \sum_{j=0}^{\infty} \frac{z^{2j+1}}{(2j+1)!}. \tag{24.7}$$

Example 24.2. Clearly, $\dfrac{d^j}{dz^j}(1-z)^{-1} = j!(1-z)^{-j-1}$. Evaluating these at $z = 0$ gives the Maclaurin series expansion

$$\frac{1}{1-z} = 1 + z + \frac{2!z^2}{2!} + \frac{3!z^3}{3!} + \cdots = \sum_{j=0}^{\infty} z^j. \tag{24.8}$$

This is the geometric series, which is valid for $|z| < 1$.

Example 24.3. The consecutive derivatives of $\text{Log}\, z$ are $1/z, -1/z^2$, $2/z^3, \cdots$; in general,

$$\frac{d^j \text{Log}\, z}{dz^j} = (-1)^{j+1}(j-1)! z^{-j}, \quad j = 1, 2, \cdots.$$

Evaluating these at $z = 1$, we find the Taylor series expansion

$$\begin{aligned}
\text{Log}\, z &= 0 + (z-1) - \frac{(z-1)^2}{2!} + 2!\frac{(z-1)^3}{3!} - 3!\frac{(z-1)^4}{4!} + \cdots \\
&= \sum_{j=1}^{\infty} \frac{(-1)^{j+1}(z-1)^j}{j}.
\end{aligned}$$

$$\tag{24.9}$$

This is valid for $|z - 1| < 1$, the largest open disk centered at 1 over which $\text{Log}\, z$ is analytic.

Example 24.4. From Example 24.1, the following expansions follow on $|z| < \infty$:

$$\cos z + i \sin z = \left(1 - \frac{z^2}{2!} + \frac{z^4}{4!} - \cdots\right) + i\left(z - \frac{z^3}{3!} + \frac{z^5}{5!} - \cdots\right)$$

$$= 1 + iz - \frac{z^2}{2!} - i\frac{z^3}{3!} + \frac{z^4}{4!} + i\frac{z^5}{5!} - \cdots$$

$$= 1 + iz + \frac{(iz)^2}{2!} + \frac{(iz)^3}{3!} + \frac{(iz)^4}{4!} + \cdots = e^{iz},$$

$$\sin z \cdot \cos z = \left(z - \frac{z^3}{3!} + \frac{z^5}{5!} - \frac{z^7}{7!} + \cdots\right)\left(1 - \frac{z^2}{2!} + \frac{z^4}{4!} - \frac{z^6}{6!} + \cdots\right)$$

$$= z - \left(\frac{1}{3!} + \frac{1}{2!}\right)z^3 + \left(\frac{1}{5!} + \frac{1}{3!\,2!} + \frac{1}{4!}\right)z^5$$

$$- \left(\frac{1}{7!} + \frac{1}{5!\,2!} + \frac{1}{3!\,4!} + \frac{1}{6!}\right)z^7 + \cdots$$

$$= z - \frac{4}{3!}z^3 + \frac{16}{5!}z^5 - \frac{64}{7!}z^7 + \cdots = \frac{1}{2}\sin 2z.$$

Example 24.5. (a). Differentiate (24.8) term-by-term to get

$$\frac{1}{(1-z)^2} = 1 + 2z + 3z^2 + \cdots = \sum_{j=0}^{\infty}(j+1)z^j, \quad |z| < 1.$$

(b). Write $1/(9 + z^2) = 1/[9(1 - w)]$, where $w = -z^2/9$. Thus, as long as $|w| < 1$; i.e., $|-z^2/9| < 1$ or $|z| < 3$, we have

$$\frac{1}{9 + z^2} = \frac{1}{9}\sum_{j=0}^{\infty} w^j = \frac{1}{9}\sum_{j=0}^{\infty}\left(-\frac{z^2}{9}\right)^n.$$

(c). Use partial fractions to obtain

$$\frac{1}{z^2 - 5z + 6} = \frac{1}{(z-3)(z-2)} = \frac{1}{z-3} - \frac{1}{z-2}$$

$$= \frac{1}{2}\left(\frac{1}{1 - z/2}\right) - \frac{1}{3}\left(\frac{1}{1 - z/3}\right)$$

$$= \frac{1}{2}\sum_{j=0}^{\infty}\left(\frac{z}{2}\right)^j - \frac{1}{3}\sum_{j=0}^{\infty}\left(\frac{z}{3}\right)^j$$

$$= \sum_{j=0}^{\infty}\left[\left(\frac{1}{2}\right)^{j+1} - \left(\frac{1}{3}\right)^{j+1}\right]z^j, \quad |z| < 2.$$

Example 24.6. Find the first few terms of the Maclaurin series for $\tan z$. Let $\tan z = \dfrac{\sin z}{\cos z} = \sum\limits_{j=0}^{\infty} c_j z^j$. In view of Theorem 23.3 (ii), for $|z| < \pi/2$, we have

$$z - \frac{z^3}{3!} + \frac{z^5}{5!} - \cdots$$
$$= \left(1 - \frac{z^2}{2!} + \frac{z^4}{4!} - \cdots\right)\left(c_0 + c_1 z + c_2 z^2 + c_3 z^3 + \cdots\right)$$
$$= c_0 + c_1 z + \left(c_2 - \frac{c_0}{2!}\right) z^2 + \left(c_3 - \frac{c_1}{2!}\right) z^3 + \left(c_4 - \frac{c_2}{2!} + \frac{c_0}{4!}\right) z^4 + \cdots.$$

Thus, $c_0 = 0$, $c_1 = 1$, $c_2 - c_0/2 = 0$, and hence $c_2 = 0$, $c_3 - c_1/2 = -1/6$, and hence $c_3 = 1/3$, $c_4 - (c_2/2) + (c_0/24) = 0$, and hence $c_4 = 0, \cdots$. Therefore, it follows that

$$\tan z = z + \frac{1}{3} z^3 + \frac{2}{15} z^5 + \cdots.$$

Example 24.7. Consider the function $f(z) = (1 + 2z)/(z^3 + z^4)$. We cannot find a Maclaurin series for $f(z)$ since it is not analytic at $z = 0$. However, we can write $f(z)$ as $f(z) = \dfrac{1}{z^3}\left(2 - \dfrac{1}{1+z}\right)$. Now $1/(1+z)$ has a Taylor series expansion around the point $z = 0$. Thus, when $0 < |z| < 1$, it follows that

$$f(z) = \frac{1}{z^3}(2 - 1 + z - z^2 + z^3 - z^4 + z^5 - z^6 + \cdots)$$
$$= \frac{1}{z^3} + \frac{1}{z^2} - \frac{1}{z} + 1 - z + z^2 - z^3 + \cdots.$$

Problems

24.1. Find the first four terms of the power series expansions about $z = 0$ of the following functions:

(a). $\dfrac{z}{z^2 + 3}$, (b). $\dfrac{1}{z^2 - 2}$, (c). $\dfrac{z}{2 - z^2}$, (d). $\dfrac{1}{z^2 - 2z + 2}$, (e). $\sec z$.

24.2. Show that, for any constant $z_0 \in \mathbf{C}$,

(a). $e^z = e^{z_0} \sum\limits_{j=0}^{\infty} \dfrac{(z - z_0)^j}{j!}$, $|z| < \infty$. Hence, deduce that $e^{z_1 + z_2} = e^{z_1} e^{z_2}$.

(b). $\sin z = \sin z_0 \cos(z - z_0) + \cos z_0 \sin(z - z_0)$.

(c). $\cos z = \cos z_0 \cos(z - z_0) - \sin z_0 \sin(z - z_0)$.

24.3. Show that

(a). $\dfrac{1}{z^2} = \sum\limits_{j=0}^{\infty} (j+1)(z+1)^j, \quad |z+1| < 1,$

(b). $\sin^3 z = \sum\limits_{j=0}^{\infty} (-1)^j \dfrac{3(1-9^j)}{4\,(2j+1)!} z^{2j+1}, \quad |z| < \infty,$

(c). $\dfrac{e^z}{1-z} = 1 + 2z + \dfrac{5}{2}z^2 + \dfrac{8}{3}z^3 + \cdots, \quad |z| < 1.$

24.4. For the function $g(z)$ defined in (18.7), show that

$$g(z) = \sum\limits_{j=0}^{\infty} (-1)^j \dfrac{z^{2j}}{(2j+1)!}, \quad z \in \mathbf{C}.$$

Hence, deduce that the function $g(z)$ is entire.

24.5. Expand $\sinh z$ in Taylor's series at $z_0 = \pi i$, and show that

$$\lim_{z \to \pi i} \dfrac{\sinh z}{z - \pi i} = -1.$$

23.6. The *error function* $\mathrm{erf}(z)$ is defined by the integral $\mathrm{erf}(z) = \dfrac{2}{\sqrt{\pi}} \displaystyle\int_0^z e^{-t^2}\, dt$. Find the Maclaurin series for $\mathrm{erf}(z)$.

23.7. The *Fresnel integrals* $C(z)$ and $S(z)$ are defined by

$$C(z) = \int_0^z \cos(\xi^2)\,d\xi \quad \text{and} \quad S(z) = \int_0^z \sin(\xi^2)\,d\xi.$$

Use $f(z) = C(z) + iS(z) = \displaystyle\int_0^z e^{i\xi^2}\, d\xi$ to compute the Maclaurin series for $f(z)$.

24.8. Suppose that an entire function $f(z)$ satisfies $|f(z)| \le k|z|^j$ for sufficiently large $|z|$, where j is a positive integer and k is a positive constant. Show that f is a polynomial of degree at most j.

24.9. Let $f(z)$ be an entire function such that $\lim_{z \to \infty} f(z)/z = 0$. Show that $f(z)$ is a constant.

24.10. Bernoulli's numbers B_k are defined as follows:

$$B_0 = 1, \quad B_1 = -\dfrac{1}{2}, \quad B_{2j+1} = 0, \quad j = 1, 2, \cdots,$$

$$\frac{2^{2j}}{1}\binom{2j}{0}B_{2j} + \frac{2^{2j-2}}{3}\binom{2j}{2}B_{2j-2} + \cdots + \frac{1}{2j+1}\binom{2j}{2j}B_0 = 1,$$
$$j = 1, 2, \cdots.$$

Extend the method of Example 24.6 to show that

(a). $\dfrac{z}{e^z - 1} = B_0 + B_1 z + \displaystyle\sum_{j=1}^{\infty} \frac{B_{2j}}{(2j)!} z^{2j}, \quad |z| < 2\pi,$

(b). $z \cot z = B_0 + \displaystyle\sum_{j=1}^{\infty} (-1)^j \frac{2^{2j} B_{2j}}{(2j)!} z^{2j}, \quad |z| < \pi,$

(c). $\tan z = \displaystyle\sum_{j=1}^{\infty} (-1)^{j-1} \frac{2^{2j}(2^{2j} - 1) B_{2j}}{(2j)!} z^{2j-1}, \quad |z| < \pi/2,$

(d). $z \sec z = B_0 + \displaystyle\sum_{j=1}^{\infty} (-1)^{j-1} \frac{(2^{2j} - 2) B_{2j}}{(2j)!} z^{2j}, \quad |z| < \pi.$

24.11. Euler's numbers E_k are defined as

$$E_0 = 1, \quad E_{2j+1} = 0, \quad j = 0, 1, 2, \cdots,$$

$$\binom{2j}{0} E_{2j} + \binom{2j}{2} E_{2j-2} + \cdots + \binom{2j}{2j} E_0 = 0, \quad j = 1, 2, \cdots.$$

Extend the method of Example 24.6 to show that

$$\operatorname{sech} z = E_0 + \sum_{j=1}^{\infty} \frac{E_{2j}}{(2j)!} z^{2j}, \quad |z| < \pi/2.$$

24.12. The *Legendre polynomials* $P_j(\xi)$, $j = 0, 1, 2, \cdots$ are defined as the coefficients of z^j in the Maclaurin expansion of

$$(1 - 2\xi z + z^2)^{-1/2} = \sum_{j=0}^{\infty} P_j(\xi) z^j.$$

Show that $P_j(\xi)$ is a polynomial of degree j, and compute $P_j(\xi)$, $j = 0, 1, 2, 3, 4, 5$.

24.13 (Parseval's Formula). Let $f(z) = \sum_{j=0}^{\infty} a_j (z - z_0)^j$ be Taylor's expansion of the analytic function $f(z)$ on $B(z_0, R)$. Show that, for any $0 < r < R$,

$$\sum_{j=0}^{\infty} |a_j|^2 r^{2j} = \frac{1}{2\pi} \int_0^{2\pi} |f(z_0 + re^{it})|^2 dt.$$

24.14. Suppose that $f(z)$ is analytic in an open set containing the closed unit disk $|z| \leq 1$, and suppose that $|f(z)| \leq 1$ for all $|z| = 1$. If $f(1/2) = 0$, show that $|f(3/4)| \leq 2/5$.

24.15. Let $a_i(z)$, $i = 1, 2, \cdots, n$ be analytic functions in a disk $B(z_0, R)$. It is known that the initial value problem

$$\frac{d^n f}{dz^n} + a_1(z)\frac{d^{n-1}f}{dz^{n-1}} + \cdots + a_{n-1}(z)\frac{df}{dz} + a_n(z)f(z) = 0,$$

$$f(z_0) = \alpha_0, \quad f'(z_0) = \alpha_1, \cdots, f^{(n-1)}(z_0) = \alpha_{n-1},$$

has a unique solution, which is analytic in $B(z_0, R)$. Use Taylor's series to compute the solution of the following initial value problems:

(a). $f' - if = 0$, $f(0) = 1$,

(b). $f'' + f = 0$, $f(0) = 0$, $f'(0) = 1$,

(c). $f'' + f = 0$, $f(0) = 1$, $f'(0) = 0$.

24.16. Let $g(t)$ be a continuous complex-valued function of a real variable on $[0, \pi]$. Show that the function $f(z) = \int_0^\pi g(t) \sin(tz) dt$ is entire, and find its power series around the origin.

Answers or Hints

24.1. (a). $\frac{z}{z^2+3} = \frac{z}{3}\left(1 + \frac{z^2}{3}\right)^{-1} = \frac{z}{3} - \left(\frac{z^2}{3}\right) + \frac{z}{3}\left(\frac{z^2}{3}\right)^2 - \frac{z}{3}\left(\frac{z^2}{3}\right)^3 + \cdots$,

(b). $\frac{1}{z^2-2} = -\frac{1}{2}\left(1 - \frac{z^2}{2}\right)^{-1} = -\frac{1}{2}\left(1 + \frac{z^2}{2} + \left(\frac{z^2}{2}\right)^2 + \left(\frac{z^2}{2}\right)^3 + \cdots\right)$,

(c). $\frac{z}{2-z^2} = \frac{1}{2}z\left(1 - \frac{z^2}{2}\right)^{-1} = \frac{1}{2}\left(z + \frac{z^3}{2} + \frac{z^5}{4} + \frac{z^7}{8} + \cdots\right)$,

(d). $\frac{1}{z^2-2z+2} = \frac{1/2i}{z-(1+i)} - \frac{1/2i}{z-(1-i)} = -\frac{i}{2(1+i)}\frac{1}{z/(1+i)-1} + \frac{i}{2(1-i)}\frac{1}{z/(1-i)-1}$

$= \frac{1+i}{4}\left(1 + \frac{z}{1+i} + \left(\frac{z}{1+i}\right)^2 + \cdots\right) + \frac{1-i}{4}\left(1 + \frac{z}{1-i} + \left(\frac{z}{1-i}\right)^2 + \cdots\right)$

$= \frac{1}{2} + \frac{z}{2} + \frac{z^2}{4} + 0 \, z^3 + \cdots$, (e). $1 + \frac{1}{2}z^2 + \frac{5}{24}z^4 + \cdots$.

24.2. (a). Let $f(z) = e^z$. Then $f^{(j)}(z) = e^z$, and hence $f(z) = f(z_0) + \frac{f'(z_0)}{1!}(z - z_0) + \frac{f''(z_0)}{2!}(z - z_0)^2 + \cdots = e^{z_0} + e^{z_0}(z - z_0) + e^{z_0}\frac{(z-z_0)^2}{2!} + \cdots$.

Now, since $e^{z_1+z_2} = 1 + (z_1 + z_2) + \frac{(z_1+z_2)^2}{2!} + \cdots = \sum_{j=0}^\infty \frac{(z_1+z_2)^j}{j!}$ and

$e^{z_1} = \sum_{j=0}^\infty \frac{z_1^j}{j!}$, $e^{z_2} = \sum_{j=0}^\infty \frac{z_2^j}{j!}$, it follows that $e^{z_1}e^{z_2} =$

$\sum_{j=0}^\infty \left(\sum_{k=0}^j \frac{1}{(j-k)!}\frac{1}{k!}z_1^{j-k}z_2^k\right) = \sum_{j=0}^\infty \frac{1}{j!}\left(\sum_{k=0}^j \frac{j!}{(j-k)!k!}z_1^{j-k}z_2^k\right)$

$= \sum_{j=0}^\infty \frac{(z_1+z_2)^j}{j!}$.

(b). Let $f(z) = \sin z$. Then $f'(z) = \cos z$, $f''(z) = -\sin z, \cdots$, and hence
$\sin z = \sin z_0 + \frac{\cos z_0}{1!}(z - z_0) - \frac{\sin z_0}{2!}(z - z_0)^2 + \frac{\cos z_0}{3!}(z - z_0)^3 - \cdots$
$= \sin z_0 \left[1 - \frac{(z-z_0)^2}{2!} + \frac{(z-z_0)^4}{4!} - \cdots \right] + \cos z_0 \left[(z - z_0) - \frac{(z-z_0)^3}{3!} + \cdots \right]$
$= \sin z_0 \cos(z - z_0) + \cos z_0 \sin(z - z_0)$.

(c). $\cos z = \cos z_0 - \frac{(z-z_0)}{1!} \sin z_0 - \frac{(z-z_0)^2}{2!} \cos z_0 + \frac{(z-z_0)^3}{3!} \sin z_0 - \cdots$
$= \cos z_0 \cos(z - z_0) - \sin z_0 \sin(z - z_0)$.

24.3. (a). $\frac{1}{z} = -\frac{1}{1-(z+1)} = -1 - (z+1) - (z+1)^2 - \cdots$, (b). $\sin^3 z = \frac{3}{4} \sin z - \frac{1}{4} \sin 3z$, (c). $\frac{e^z}{1-z} = (1 + z + z^2 + z^3 + \cdots)(1 + z + \frac{z^2}{2!} + \frac{z^3}{3!} + \cdots) = 1 + (1+1)z + (1 + \frac{1}{2} + 1)z^2 + (1 + \frac{1}{2} + \frac{1}{6} + 1)z^3 + \cdots = 1 + 2z + \frac{5}{2}z^2 + \frac{8}{3}z^3 + \cdots$.

24.4. Use (24.5). Since $g(z)$ is represented by a convergent power series for $z \in \mathbf{C}$ it is entire.

24.5. For even j, $f^{(j)}(\pi i) = \sinh \pi i = i \sin \pi = 0$, and for odd j, $f^{(j)}(\pi i) = \cosh \pi i = \cos \pi = -1$, so

$$\sinh z = \sum_{j=0}^{\infty} \left[-\frac{(z-\pi i)^{2j+1}}{(2j+1)!} \right], \quad |z| < \infty,$$

in particular,

$$\lim_{z \to \pi i} \frac{\sinh z}{z - \pi i} = \sum_{j=0}^{\infty} \left[\lim_{z \to \pi i} -\frac{(z-\pi i)^{2j}}{(2j+1)!} \right] = -1$$

since the convergence is uniform on $\overline{B}(\pi i, r)$ for any $r \in (0, \infty)$.

24.6. Since $e^{-z^2} = \sum_{j=0}^{\infty} \frac{(-z^2)^j}{j!}$, $|z| < \infty$, and the convergence is uniform on $\overline{B}(\pi i, r)$ for any $r \in (0, \infty)$, term-by-term integration gives

$$\operatorname{erf}(z) = \frac{2}{\sqrt{\pi}} \sum_{j=0}^{\infty} \frac{(-1)^j z^{2j+1}}{j! (2j+1)}, \quad |z| < \infty.$$

24.7. Since $e^{iz^2} = \sum_{j=0}^{\infty} \frac{(iz^2)^j}{j!}$, $|z| < \infty$, and the convergence is uniform on $\overline{B}(\pi i, r)$ for any $r \in (0, \infty)$, term-by-term integration gives

$$f(z) = \sum_{j=0}^{\infty} \frac{i^j z^{2j+1}}{j! (2j+1)}, \quad |z| < \infty.$$

24.8. Taylor's expansion of f at 0; i.e., $f(z) = \sum_{j=0}^{\infty} a_j z^j$ has infinite radius of convergence; here $a_j = \frac{1}{j!} f^{(j)}(0) = \frac{1}{j!} \frac{j!}{2\pi i} \int_{\gamma_R} \frac{f(z)}{z^{j+1}} dz$ and γ_R is a positively oriented circle of radius R around 0. Clearly, for large R, $|a_m| \leq \frac{1}{2\pi} \int_{C_R} \frac{|f(z)|}{|z|^{m+1}} dz \leq \frac{1}{2\pi} k \frac{R^j}{R^{m+1}} 2\pi R$. Thus, $|a_m| \leq \frac{k}{R^{m-j}} \to 0$ as $R \to \infty$ for $m > j$, and hence $a_m = 0$ for $m > j$. So, f is a polynomial of degree $\leq j$.

24.9. Observe that $|f(z)| \leq k|z|$ for large $|z|$. Now use Problem 24.8.

24.10. Similar to 24.11 below. In part (a), first note that $z/(e^z - 1) - B_1 z = z(e^z + 1)/(e^z - 1)$ is an even function.

24.11. Since $\cosh z$ is an even entire function whose zeros are $(2k+1)\pi i/2$, $k \in \mathbf{Z}$, the function $f(z) = \operatorname{sech} z = 1/\cosh z$ is analytic in $B(0, \pi/2)$ and even, so it has the Taylor series expansion $f(z) = \sum_{j=0}^{\infty} \frac{b_j}{(2j)!} z^{2j}$, $|z| < \pi/2$, where $b_j = f^{(2j)}(0)$. Then

$$1 = f(z)\cosh z = \sum_{j=0}^{\infty} \frac{b_j}{(2j)!} z^{2j} \sum_{j=0}^{\infty} \frac{z^{2j}}{(2j)!}$$

$$= \sum_{j=0}^{\infty} \left[\sum_{k=0}^{j} \frac{b_{j-k}}{[2(j-k)]!\,(2k)!} \right] z^{2j}$$

$$= \sum_{j=0}^{\infty} \left[\sum_{k=0}^{j} \binom{2j}{2k} b_{j-k} \right] \frac{z^{2j}}{(2j)!},$$

and hence $b_0 = 1$, $\sum_{k=0}^{j} \binom{2j}{2k} b_{j-k} = 0$, $j \geq 1$, so $b_j = E_{2j}$.

24.12. If $|\xi| \leq r$, where r is arbitrary, and $|z| < (1 + r^2)^{1/2} - r$, then it follows that $|2\xi z - z^2| \leq 2|\xi||z| + |z^2| < 2r(1 + r^2)^{1/2} - 2r^2 + 1 + r^2 + r^2 - 2r(1 + r^2)^{1/2} = 1$, and hence we can expand $(1 - 2\xi z + z^2)^{-1/2}$ binomially (see Problem 23.7) to obtain $[1 - z(2\xi - z)]^{-1/2} = 1 + \frac{1}{2}z(2\xi - z) + \frac{1}{2}\frac{3}{4}z^2(2\xi - z)^2 + \cdots + \frac{1.3\cdots(2j-1)}{2.4\cdots(2j)} z^j(2\xi - z)^j + \cdots$. The coefficient of z^j in this expansion is

$$\frac{1.3\cdots(2j-1)}{2.4\cdots(2j)}(2\xi)^j - \frac{1.3\cdots(2j-3)}{2.4\cdots(2j-2)}\frac{(j-1)}{1!}(2\xi)^{j-2} + \frac{1.3\cdots(2j-5)}{2.4\cdots(2j-4)}\frac{(j-2)(j-3)}{2!}(2\xi)^{j-4} - \cdots$$

$$= \frac{1.3\cdots(2j-1)}{j!}\left[\xi^j - \frac{j(j-1)}{(2j-1)1.2}\xi^{j-2} + \frac{j(j-1)(j-2)(j-3)}{(2j-1)(2j-3)\,2.4}\xi^{j-4} - \cdots \right] = P_j(\xi).$$

$P_0(\xi) = 1$, $P_1(\xi) = \xi$, $P_2(\xi) = \frac{1}{2}(3\xi^2 - 1)$, $P_3(\xi) = \frac{1}{2}(5\xi^3 - 3\xi)$, $P_4(\xi) = \frac{1}{8}(35\xi^4 - 30\xi^2 + 3)$, $P_5(\xi) = \frac{1}{8}(63\xi^5 - 70\xi^3 + 15\xi)$.

24.13. Let z be on the circle of radius r with center at z_0. Since $z = z_0 + re^{it}$, where $0 \leq t \leq 2\pi$, $f(z) = \sum_{j=0}^{\infty} a_j r^j e^{ijt}$. Thus, $|f(z)|^2 = f(z)\overline{f(z)} = \sum_{j=1}^{\infty}\sum_{k=1}^{\infty} a_j \overline{a_k} r^{j+k} e^{i(j-k)t}$. Now integrate this term-by-term.

24.14. Clearly, $\lim_{z\to 1/2} \frac{(2-z)f(z)}{2z-1} = \lim_{z\to 1/2} \frac{(2-z)(f(z)-f(1/2))}{2(z-1/2)} = (3/4)\times f'(1/2)$. Consider the function $g(z) = \frac{(2-z)f(z)}{2z-1}$, $z \neq 1/2$ and $g(z) = (3/4)f'(1/2)$, $z = 1/2$. Since $f(z)$ is analytic, in a small neighborhood around $z = 1/2$ we have $f(z) = f'(1/2)(z - 1/2) + \sum_{n\geq 2} a_n(z - 1/2)^n$. Hence, $\lim_{z\to 1/2} \frac{g(z)-g(1/2)}{z-1/2} = \lim_{z\to 1/2} \frac{[(1-2z)/4]f'(1/2) + \sum_{n\geq 2} a_n(z-1/2)^{n-1}}{z-1/2}$ $= -(1/2)f'(1/2) + a_2$. So, $g'(1/2)$ exists and clearly $g'(z)$ exists at $z \neq 1/2$. Thus, $g(z)$ is analytic on $|z| \leq 1$. Applying Theorem 20.2 to $g(z)$ on $|z| \leq 1$, we have $|g(z)| \leq \max_{|\alpha|=1} |g(\alpha)| = \frac{|2-\alpha||f(\alpha)|}{|2\alpha-1|} \leq \frac{|\alpha-2|}{|2\alpha-1|} = 1$ (see Problem 3.5 with $a = 1$, $b = -2$). Therefore, $|g(z)| \leq 1$ for $|z| \leq 1$. In particular, $|g(3/4)| \leq 1$, which implies $|f(3/4)| \leq 2/5$.

24.15. (a). e^{iz}, (b). $\sin z$, (c). $\cos z$.

24.16. For fixed $z \in \mathbf{C}$, Lemma 22.1 ensures that the series $g(t)\sin(tz) = \sum_{j=0}^{\infty}(-1)^j \frac{(tz)^{2j+1}}{(2j+1)!} g(t)$ converges uniformly on $[0, \pi]$. Now, by Theorem 22.5, we can use term-by-term integration to obtain

$$f(z) = \sum_{j=0}^{\infty} \left(\int_0^{\pi} (-1)^j \frac{(t)^{2j+1}}{(2j+1)!} g(t)dt \right) z^{2j+1},$$

which is a power series in z. Clearly, this series converges for all z, and hence in view of Theorem 23.1, $f(z)$ is entire.

Lecture 25
Laurent's Series

In this lecture, we shall expand a function that is analytic in an annulus domain. The resulting expansion, although it resembles a power series, involves positive as well as negative integral powers of $(z - z_0)$. From an applications point of view, such an expansion is very useful.

Theorem 25.1 (Laurent's Theorem). Let $f(z)$ be analytic in an annulus domain $A = \{z : R_1 < |z - z_0| < R_2\}$. Then, $f(z)$ can be represented by the *Laurent series*

$$f(z) = \sum_{j=0}^{\infty} a_j(z - z_0)^j + \sum_{j=1}^{\infty} \frac{b_j}{(z - z_0)^j}, \quad z \in A, \qquad (25.1)$$

where

$$a_j = \frac{1}{2\pi i} \int_\gamma \frac{f(\xi)}{(\xi - z_0)^{j+1}} d\xi, \quad j = 0, 1, 2, \cdots \qquad (25.2)$$

and

$$b_j = \frac{1}{2\pi i} \int_\gamma f(\xi)(\xi - z_0)^{j-1} d\xi, \quad j = 1, 2, \cdots, \qquad (25.3)$$

and γ is any positively oriented, simple, closed contour around z_0 lying in A (see Figure 25.1). The second sum in (25.1) is called the *principal part* of the Laurent series.

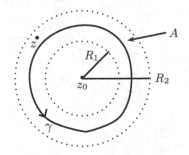

Figure 25.1

Proof. From Cauchy's integral formula (17.1), we have

$$f(z) = \frac{1}{2\pi i} \left[\int_{\gamma_2} - \int_{\gamma_1} \right] \frac{f(\xi)}{\xi - z} d\xi, \qquad (25.4)$$

where γ_1 and γ_2 are circles centered at z_0 contained in A, as indicated in Figure 25.2.

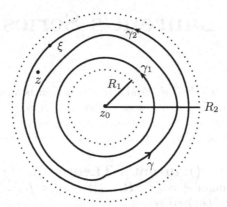

Figure 25.2

For $\xi \in \gamma_2$, $\left| \dfrac{z - z_0}{\xi - z_0} \right| < 1$ and

$$\frac{1}{\xi - z} = \frac{1}{(\xi - z_0) - (z - z_0)} = \frac{1}{\xi - z_0} \frac{1}{1 - [(z - z_0)/(\xi - z_0)]}$$

$$= \frac{1}{\xi - z_0} \sum_{j=0}^{\infty} \left(\frac{z - z_0}{\xi - z_0} \right)^j$$

uniformly on γ_2. Hence, we get

$$\frac{1}{2\pi i} \int_{\gamma_2} \frac{f(\xi)}{\xi - z} d\xi = \frac{1}{2\pi i} \int_{\gamma_2} f(\xi) \sum_{j=0}^{\infty} \frac{(z - z_0)^j}{(\xi - z_0)^{j+1}} d\xi$$

$$= \sum_{j=0}^{\infty} \left[\frac{1}{2\pi i} \int_{\gamma} \frac{f(\xi)}{(\xi - z_0)^{j+1}} d\xi \right] (z - z_0)^j \qquad (25.5)$$

$$= \sum_{j=0}^{\infty} a_j (z - z_0)^j.$$

Now, for $\xi \in \gamma_1$, $\left| \dfrac{\xi - z_0}{z - z_0} \right| < 1$ and

$$\frac{1}{z - \xi} = \frac{1}{(z - z_0) - (\xi - z_0)} = \frac{1}{z - z_0} \frac{1}{1 - [(\xi - z_0)/(z - z_0)]}$$

$$= \frac{1}{z - z_0} \sum_{j=0}^{\infty} \left(\frac{\xi - z_0}{z - z_0} \right)^j$$

uniformly on γ_1. Thus, we have

$$-\frac{1}{2\pi i}\int_{\gamma_1}\frac{f(\xi)}{\xi - z}d\xi = \frac{1}{2\pi i}\int_{\gamma_1}f(\xi)\sum_{j=0}^{\infty}\frac{(\xi - z_0)^j}{(z - z_0)^{j+1}}d\xi$$

$$= \sum_{j=0}^{\infty}\left[\frac{1}{2\pi i}\int_{\gamma_1}f(\xi)(\xi - z_0)^j d\xi\right]\frac{1}{(z - z_0)^{j+1}}$$

$$= \sum_{j=1}^{\infty}\left[\frac{1}{2\pi i}\int_{\gamma}f(\xi)(\xi - z_0)^{j-1}d\xi\right]\frac{1}{(z - z_0)^j}$$

$$= \sum_{j=1}^{\infty}\frac{b_j}{(z - z_0)^j}. \tag{25.6}$$

Combining (25.4)-(25.6) we immediately get (25.1). ∎

Remark 25.1. 1. If $f(z)$ is analytic everywhere in $B(z_0, R_2)$ except at z_0, then the Laurent series is valid in $0 < |z - z_0| < R_2$; i.e., we can take $R_1 = 0$.

2. If $f(z)$ is analytic in $B(z_0, R_2)$, then $f(\xi)(\xi - z_0)^{j-1}$ is analytic in $B(z_0, R_2)$, so that by the Cauchy-Goursat Theorem (Theorem 15.2),

$$b_j = \frac{1}{2\pi i}\int_{\gamma}f(\xi)(\xi - z_0)^{j-1}d\xi = 0.$$

Hence, the Laurent series for $f(z)$ reduces to the Taylor series for $f(z)$.

3. We can write

$$f(z) = \sum_{j=-\infty}^{\infty}a_j(z - z_0)^j, \quad z \in A,$$

where

$$a_j = \frac{1}{2\pi i}\int_{\gamma}\frac{f(\xi)}{(\xi - z_0)^{j+1}}d\xi \quad \text{for all} \quad j \in \mathbf{Z}.$$

4. If a series $\sum_{j=-\infty}^{\infty}a_j(z - z_0)^j$ converges to $f(z)$ in some annular domain, then it must be the Laurent series for $f(z)$ at z_0; i.e., Laurent's series for $f(z)$ in a given annulus is unique.

5. Laurent's series of an analytic function in an annular region can be differentiated term-by-term. As a consequence, since $\text{Log } z$ is not analytic in any annulus around 0, it cannot be represented by a Laurent series around 0.

Example 25.1. Expand $e^{1/z}$ in the Laurent series around $z = 0$. Since

$$e^{\xi} = 1 + \xi + \frac{\xi^2}{2!} + \frac{\xi^3}{3!} + \cdots$$

for all (finite) ξ, if $z \neq 0$, we let $\xi = 1/z$ and find

$$e^{1/z} = 1 + \frac{1}{z} + \frac{1}{2!z^2} + \frac{1}{3!z^3} + \cdots, \quad 0 < |z| < \infty.$$

Similarly, we have

$$\frac{1}{1-z} = -\frac{1}{z}\left(1 - \frac{1}{z}\right)^{-1} = -\frac{1}{z} - \frac{1}{z^2} - \frac{1}{z^3} - \cdots, \quad 1 < |z| < \infty,$$

and by term-by-term differentiation,

$$\frac{1}{(1-z)^2} = \frac{1}{z^2} + \frac{2}{z^3} + \frac{3}{z^4} + \cdots, \quad 1 < |z| < \infty.$$

Example 25.2. Find the Laurent series for the function $(z^2 - 2z + 7)/(z - 2)$ in the domain $|z - 1| > 1$. We have

$$\begin{aligned}
\frac{z^2 - 2z + 7}{z - 2} &= [(z-1)^2 + 6]\frac{1}{(z-1) - 1} = \frac{[(z-1)^2 + 6]}{(z-1)}\frac{1}{1 - \frac{1}{(z-1)}} \\
&= \frac{[(z-1)^2 + 6]}{(z-1)}\left[1 + \frac{1}{(z-1)} + \frac{1}{(z-1)^2} + \frac{1}{(z-1)^3} + \cdots\right] \\
&= (z-1) + 1 + \frac{7}{(z-1)} + \frac{7}{(z-1)^2} + \cdots.
\end{aligned}$$

Example 25.3. Find the Laurent series of $e^{2z}/(z-1)^3$ around $z = 1$. We have

$$\begin{aligned}
\frac{e^{2z}}{(z-1)^3} &= \frac{e^{2+2(z-1)}}{(z-1)^3} = \frac{e^2}{(z-1)^3}e^{2(z-1)} \\
&= \frac{e^2}{(z-1)^3}\left[1 + 2(z-1) + \frac{[2(z-1)]^2}{2!} + \frac{[2(z-1)]^3}{3!} + \cdots\right] \\
&= \frac{e^2}{(z-1)^3} + \frac{2e^2}{(z-1)^2} + \frac{2e^2}{(z-1)} + \frac{4e^2}{3} + \frac{2e^2}{3}(z-1) + \cdots.
\end{aligned}$$

Example 25.4. Find the Laurent series of $(z-3)\sin 1/(z+2)$ around $z = -2$. We have

$$\begin{aligned}
(z-3)\sin\frac{1}{z+2} &= [(z+2) - 5]\left(\frac{1}{(z+2)} - \frac{1}{3!(z+2)^3} + \frac{1}{5!(z+2)^5} - \cdots\right) \\
&= 1 - \frac{5}{(z+2)} - \frac{1}{3!(z+2)^2} + \frac{5}{3!(z+2)^3} + \frac{1}{5!(z+2)^4} - \cdots.
\end{aligned}$$

Example 25.5. Expand $f(z) = 1/(z^2 + 1)$ in a Laurent series in a punctured ball centered at $z = i$ (in powers of $z - i$). We have that

$$f(z) = \frac{1}{(z+i)(z-i)} = \frac{1}{z-i}\frac{1}{(z-i)+2i} = \frac{1}{z-i}\frac{1}{2i}\frac{1}{1+(z-i)/2i}$$

$$= -\frac{i}{2}\frac{1}{z-i}\frac{1}{1-(i/2)(z-i)} = -\frac{i}{2}\frac{1}{z-i}\sum_{j=0}^{\infty}\left[\frac{i}{2}(z-i)\right]^j$$

(converges if $|i(z-i)/2| < 1$)

$$= -\sum_{j=0}^{\infty}\left(\frac{i}{2}\right)^{j+1}(z-i)^{j-1} = -\frac{i/2}{z-i} - \sum_{j=0}^{\infty}\left(\frac{i}{2}\right)^{j+2}(z-i)^j$$

converges if $|i(z-i)/2| < 1$ and $z \neq i$; i.e., $0 < |z-i| < 2$.

Example 25.6. For the function $(b-a)/(z-a)(z-b)$, $0 < |a| < |b|$, find the Laurent series expansion in (i) the domain $|z| < |a|$, (ii) the domain $|a| < |z| < |b|$, and (iii) the domain $|z| > |b|$. Note that

$$\frac{b-a}{(z-a)(z-b)} = \frac{1}{z-b} - \frac{1}{z-a}.$$

(i). For $|z| < |a|$, we have $|z/a| < 1$ and $|z/b| < 1$. Thus,

$$\frac{1}{z-b} - \frac{1}{z-a} = -\frac{1}{b}\frac{1}{1-z/b} + \frac{1}{a}\frac{1}{1-z/a}$$

$$= -\frac{1}{b}\sum_{j=0}^{\infty}\left(\frac{z}{b}\right)^j + \frac{1}{a}\sum_{j=0}^{\infty}\left(\frac{z}{a}\right)^j$$

$$= \sum_{j=0}^{\infty}\left(\frac{1}{a^{j+1}} - \frac{1}{b^{j+1}}\right)z^j.$$

(ii). For $|a| < |z| < |b|$, we have $|a/z| < 1$ and $|z/b| < 1$. Thus,

$$\frac{1}{(z-b)} - \frac{1}{(z-a)} = -\frac{1}{b}\frac{1}{1-z/b} - \frac{1}{z}\frac{1}{1-a/z}$$

$$= -\sum_{j=0}^{\infty}\left(\frac{z^j}{b^{j+1}} + \frac{a^j}{z^{j+1}}\right).$$

(iii). For $|z| > |b|$, we have $|a/z| < 1$ and $|b/z| < 1$. Thus,

$$\frac{1}{(z-b)} - \frac{1}{(z-a)} = \frac{1}{z}\frac{1}{1-b/z} - \frac{1}{z}\frac{1}{1-a/z}$$

$$= \sum_{j=0}^{\infty}\left(\frac{b^j}{z^{j+1}} - \frac{a^j}{z^{j+1}}\right).$$

Problems

25.1. Expand each of the following functions $f(z)$ in a Laurent series on the indicated domain:

(a). $\dfrac{z^2 - 2z + 5}{(z-2)(z^2+1)}$, $\quad 1 < |z| < 2$, $\quad$ (b). $\dfrac{(z-1)}{z^2}$, $\quad |z-1| > 1$,

(c). $\text{Log}\left(\dfrac{z-a}{z-b}\right)$, where $b > a > 1$ are real, $|z| > b$.

25.2. Find the Laurent series for the function $1/[z(z-1)]$ in the following domains: (a). $0 < |z| < 1$, (b). $1 < |z|$, (c). $0 < |z-1| < 1$, (d). $1 < |z-1|$, (e). $1 < |z-2| < 2$.

25.3. Find the Laurent series for the function $z/[(z^2+1)(z^2+4)]$ in the following domains (a). $0 < |z| < 1$, (b). $1 < |z| < 2$, (c). $|z| > 2$.

25.4. (a). Show that when $0 < |z| < 4$,

$$\frac{1}{4z - z^2} = \frac{1}{4z} + \sum_{n=0}^{\infty} \frac{z^n}{4^{n+2}}.$$

(b). Show that, when $0 < |z-1| < 2$,

$$\frac{z}{(z-1)(z-3)} = -\frac{1}{2(z-1)} - 3\sum_{n=0}^{\infty} \frac{(z-1)^n}{2^{n+2}}.$$

(c). Show that, when $2 < |z| < \infty$,

$$\frac{1}{z^4 + 4z^2} = \frac{1}{z^4} \sum_{n=0}^{\infty} (-1)^n \left(\frac{4}{z^2}\right)^n.$$

25.5. Find the Laurent series for the function $1/[(z-1)(z-2)(z-3)]$ in the following domains: (a). $0 < |z| < 1$, (b). $1 < |z| < 2$, (c). $2 < |z| < 3$, (d). $|z| > 3$.

25.6. For the series $\sum_{j=0}^{\infty} b_j (z-z_0)^{-j}$, let $r = \limsup_{j \to \infty} |b_j|^{1/j}$. Show that

(a). if $r = 0$, the series is absolutely convergent for all z in the extended plane except at $z = z_0$,

(b). if $0 < r < \infty$, the series is absolutely convergent for all $|z - z_0| > r$ and divergent for all $|z - z_0| < r$, and

(c). if $r = \infty$, the series is divergent for all finite z.

25.7. Show that if the Laurent series $\sum_{j=-\infty}^{\infty} a_j(z-z_0)^j$ represents an even function, then $a_{2j+1} = 0$, $j = 0, \pm 1, \pm 2, \cdots$ and if it represents an odd function, then $a_{2j} = 0$, $j = 0, \pm 1, \pm 2, \cdots$.

25.8. Let $\sum_{j=0}^{\infty} a_j(z-z_0)^j$ converge for $|z-z_0| < R_2$ and $\sum_{j=1}^{\infty} a_{-j}(z-z_0)^{-j}$ converge for $|z-z_0| > R_1$, and $R_1 < R_2$. Show that there exists an analytic function $f(z)$, $R_1 < |z-z_0| < R_2$ whose Laurent series is $\sum_{j=-\infty}^{\infty} a_j(z-z_0)^j$.

25.9. Determine conditions on a_j, $j = 0, \pm 1, \pm 2, \cdots$ such that the Laurent series $\sum_{j=-\infty}^{\infty} a_j z^j$ is analytic on the punctured disk $0 < |z| < 1$.

25.10. Let $f(z)$ be analytic in an annulus domain $\overline{A} = \{z : R_1 \leq |z-z_0| \leq R_2\}$ and $|f(z)| \leq M$, $z \in \overline{A}$. Show that the coefficients a_j and b_j in the Laurent expansion of $f(z)$ in the annulus $\overline{A}$ satisfy

$$|a_j| \leq \frac{M}{R_2^j}, \quad j = 0, 1, 2, \cdots \quad \text{and} \quad |b_j| \leq MR_1^j, \quad j = 1, 2, \cdots.$$

25.11. Show that

$$\exp\left[\frac{\xi}{2}\left(z - \frac{1}{z}\right)\right] = \sum_{n=-\infty}^{\infty} J_n(\xi)z^n, \quad |z| > 0 \qquad (25.7)$$

where

$$J_n(\xi) = (-1)^n J_{-n}(\xi) = \frac{1}{2\pi} \int_0^{2\pi} \cos(\xi \sin\theta - n\theta)d\theta. \qquad (25.8)$$

The $J_n(\xi)$ are the *Bessel functions* of the first kind (see Problem 23.6). From (25.8), deduce that for $z = x$ a real number $|J_n(x)| \leq 1$.

25.12. Show that

$$\cosh\left(z + \frac{1}{z}\right) = a_0 + \sum_{j=1}^{\infty} a_j\left(z^j + \frac{1}{z^j}\right),$$

where

$$a_j = \frac{1}{2\pi} \int_0^{2\pi} \cos j\theta \cosh(2\cos\theta)d\theta.$$

Answers or Hints

25.1. (a). $-\sum_{j=0}^{\infty} z^j/2^{j+1} + 2\sum_{j=1}^{\infty}(-1)^j/2^{2j}$, (b). $\sum_{j=1}^{\infty} j(-1)^{j-1}(z-1)^{-j}$, (c). $\text{Log}\left(\frac{z-a}{z-b}\right) = \text{Log}\left(1 - \frac{a}{z}\right) - \text{Log}\left(1 - \frac{b}{z}\right) = \sum_{j=1}^{\infty}(b^j - a^j)/(jz^j)$.

25.2. (a). $-\frac{1}{z} - \sum_{j=0}^{\infty} z^j$, (b). $\sum_{j=2}^{\infty} 1/z^j$, (c). $\frac{1}{z-1} - \sum_{j=0}^{\infty} (-1)^j (z-1)^j$, (d). $\sum_{j=2}^{\infty} (-1)^j/(z-1)^j$, (e). $\sum_{j=1}^{\infty} (-1)^{j+1}(z-2)^{-j} + \sum_{j=0}^{\infty} (-1)^{j+1}(z-2)^j/2^{j+1}$.

25.3. $\frac{z}{(z^2+1)(z^2+4)} = \frac{z}{3}\left(\frac{1}{z^2+1} - \frac{1}{z^2+4}\right)$.

25.4. (a). $\frac{1}{4z-z^2} = \frac{1}{4z}\left(1 - \frac{z}{4}\right) = \frac{1}{4z}\sum_{n=0}^{\infty}\left(\frac{z}{4}\right)^n = \frac{1}{4z} + \sum_{n=0}^{\infty}\frac{z^n}{4^{n+2}}$.

(b). $\frac{z}{(z-1)(z-3)} = -\frac{1}{2(z-1)} + \frac{3}{2}\frac{1}{(z-1)-2} = -\frac{1}{2(z-1)} + \frac{3}{2}\left(-\frac{1}{2}\right)\left(1 - \frac{z-1}{2}\right)^{-1}$

$= -\frac{1}{2(z-1)} - 3\sum_{n=0}^{\infty}\frac{(z-1)^n}{2^{n+2}}$.

(c). $\frac{1}{z^4+4z^2} = \frac{1}{z^4}\frac{1}{1+4/z^2} = \frac{1}{z^4}\sum_{n=0}^{\infty}(-1)^n\left(\frac{4}{z^2}\right)^n$.

25.5. $\frac{1}{(z-1)(z-2)(z-3)} = \frac{1}{2}\frac{1}{z-1} - \frac{1}{z-2} + \frac{1}{2}\frac{1}{z-3}$.

25.6. Use the substitution $\xi = 1/(z - z_0)$ to obtain the series $\sum_{j=0}^{\infty} b_j \xi^j$. For this new series, the radius of convergence is $1/r$.

25.7. See Problem 23.10.

25.8. Define $f(z)$ as the sum of two series. The result now follows from the uniqueness of the Laurent series.

25.9. $\sum_{j=-\infty}^{-1} a_j z^j$ must converge for all $z \neq 0$, and $\sum_{j=0}^{\infty} a_j z^j$ must converge on the disk $|z| < 1$, $\limsup_{j\to\infty} |a_{-j}|^{1/j} = 0$, and $\limsup_{j\to\infty} |a_j|^{1/j} \leq 1$.

25.10. Let $\gamma = R_2$ and R_1 in (25.2) and (25.3), respectively, and follow as in Theorem 18.5.

25.11. To show $J_n(\xi) = (-1)^n J_{-n}(\xi)$, replace $z = -1/s$ in (25.7). Let $\gamma : |z| = 1$. Then, from Examples 14.6 and 16.1, it follows that $J_n(\xi) = \frac{1}{2\pi i}\int_\gamma \exp\left[\frac{\xi}{2}\left(z - \frac{1}{z}\right)\right]\frac{dz}{z^{n+1}}$. Now let $z = e^{i\theta}$.

25.12. Use Laurent expansion to obtain $\cosh(z + z^{-1}) = \sum_{j=-\infty}^{\infty} a_j z^j$, and note that $a_j = a_{-j}$.

Lecture 26
Zeros of Analytic Functions

In this lecture, we shall use Taylor's series to study zeros of analytic functions. We shall show that unless a function is identically zero, about each point where the function is analytic there is a neighborhood throughout which the function has no zero except possibly at the point itself; i.e., the zeros of an analytic function are isolated. We begin by proving the following theorem.

Theorem 26.1. Let $f(z)$ be analytic at z_0. Then, $f(z)$ has a zero of order m at z_0 if and only if $f(z)$ can be written as $f(z) = (z - z_0)^m g(z)$, where $g(z)$ is analytic at z_0 and $g(z_0) \neq 0$.

Proof. Since $f(z)$ is analytic at z_0, it can be expanded in a Taylor series; i.e., $f(z) = \sum_{j=0}^{\infty} a_j(z - z_0)^j$ on some disk $B(z_0, R)$; here $a_j = f^{(j)}(z_0)/j!$, $j = 0, 1, 2, \cdots$. If $f(z)$ has a zero of order m at z_0, then $f^{(k)}(z_0) = 0$, $k = 0, 1, \cdots, m - 1$ and $f^{(m)}(z_0) \neq 0$ (recall the definition in Lecture 19). Thus, it follows that $a_0 = a_1 = \cdots = a_{m-1} = 0$, $a_m \neq 0$, and hence the Taylor series reduces to

$$
\begin{aligned}
f(z) &= a_m(z - z_0)^m + a_{m+1}(z - z_0)^{m+1} + \cdots \\
&= (z - z_0)^m \left[a_m + a_{m+1}(z - z_0) + a_{m+2}(z - z_0)^2 + \cdots \right],
\end{aligned}
$$

which is the same as $f(z) = (z - z_0)^m g(z)$, where $g(z) = \sum_{j=m}^{\infty} a_j(z - z_0)^{j-m}$ is analytic on $|z - z_0| < R$ and $g(z_0) = a_m \neq 0$. Conversely, let $f(z) = (z - z_0)^m g(z)$. Since $g(z)$ is analytic at z_0, it has a Taylor series expansion $g(z) = \sum_{j=0}^{\infty} b_j(z - z_0)^j$, and since $g(z_0) \neq 0$ it follows that $b_0 \neq 0$. Thus, we have

$$
f(z) = b_0(z - z_0)^m + b_1(z - z_0)^{m+1} + \cdots = \sum_{j=0}^{\infty} b_j(z - z_0)^{m+j},
$$

which implies that $f(z)$ has a zero of order m at z_0. ∎

Corollary 26.1. Let $f(z)$ and $g(z)$ be analytic at z_0. If $f(z)$ has a zero of order m at z_0 and $g(z)$ has a zero of order n at z_0, then $f(z)g(z)$ has a zero of order $m + n$ at z_0.

Example 26.1. (i). The zeros of the function $\sin z$, which occur at integer multiples of π, are all simple (at such points, the first derivative, $\cos z$, is nonzero).

(ii).   In view of Problem 24.4, the function $g(z)$ defined in (18.7) has no zero at 0.

(iii).   Since

$$
\begin{aligned}
f_1(z) &= z^2 \sin z &= z^2 \left(z - \frac{z^3}{3!} + \frac{z^5}{5!} - \cdots \right) \\
&&= z^3 \left(1 - \frac{z^2}{3!} + \frac{z^4}{5!} - \cdots \right),
\end{aligned}
$$

$f_1(z)$ has a zero of order 3 at $z = 0$.

(iv).   Since

$$
\begin{aligned}
f_2(z) &= z \sin z^2 &= z \left(z^2 - \frac{z^6}{3!} + \frac{z^{10}}{5!} - \cdots \right) \\
&&= z^3 \left(1 - \frac{z^4}{3!} + \frac{z^8}{5!} - \cdots \right),
\end{aligned}
$$

$f_2(z)$ has a zero of order 3 at $z = 0$.

Corollary 26.2.   If $f(z)$ is analytic at z_0 and $f(z_0) = 0$, then either $f(z)$ is identically zero in a neighborhood of z_0 or there is a punctured disk about z_0 in which $f(z)$ has no zeros.

Proof.   Let $\sum_{j=0}^{\infty} a_j (z - z_0)^j$ be the Taylor series for $f(z)$ about z_0. This series converges to $f(z)$ in some neighborhood of z_0. So, if all the Taylor coefficients a_j are zero, then $f(z)$ must be identically zero in this neighborhood. Otherwise, let $m \geq 1$ be the smallest subscript such that $a_m \neq 0$. Then, $f(z)$ has a zero of order m at z_0, and so the representation $f(z) = (z - z_0)^m g(z)$ by Theorem 26.1 is valid. Since $g(z_0) \neq 0$ and $g(z)$ is continuous at z_0, there exists a disk $B(z_0, \delta)$ throughout which $g(z)$ is nonzero. Consequently, $f(z) \neq 0$ for $0 < |z - z_0| < \delta$.   ∎

Thus, for a function $f(z)$ that is analytic and has a zero at a point z_0 but is not identically equal to zero in any neighborhood of z_0, the point z_0 must be a zero of some finite order; and conversely, if z_0 is a zero of finite order of $f(z)$, then $f(z) \neq 0$ throughout some punctured neighborhood $N_\epsilon(z_0) = \{z : 0 < |z - z_0| < \epsilon\}$ of z_0. Thus, a finite-order zero of $f(z)$ is *isolated* from other zeros of $f(z)$.

Corollary 26.3.   Suppose $f(z)$ is analytic on a region S, and $\{z_n\}$ is an infinite sequence of distinct points in S converging to $\alpha \in S$. If $f(z_n) = 0$ for all n, then $f(z)$ is identically zero on S.

Proof.   Since $f(z)$ is continuous, $0 = \lim_{n \to \infty} f(z_n) = f(\alpha)$. We claim that $f(z)$ is identically zero in some neighborhood of α. In fact, if $f(z) \neq 0$ in some punctured neighborhood $N_\epsilon(\alpha)$, then, from the definition of the limit, for sufficiently large n, there exists $z_n \in N_\epsilon(\alpha)$ such that $f(z_n) = 0$. The rest of the proof is exactly the same as that of Theorem 20.1 (recall Figure 20.1). ∎

Corollary 26.4. If a function $f(z)$ is analytic on a domain S and $f(z) = 0$ at each point z of a subdomain of S or an arc contained in S, then $f(z) \equiv 0$ in S.

Now suppose $f(z)$ and $g(z)$ are analytic on the same domain S and that $f(z) = g(z)$ at each point z of a subdomain of S or an arc contained in S. Then, the function $h(z)$ defined by $h(z) = f(z) - g(z)$ is also analytic on S and $h(z) = 0$ throughout the subdomain or along the arc. But then, in view of Corollary 26.4, $h(z) = 0$ throughout S; i.e., $f(z) = g(z)$, $z \in S$. This proves the following corollary.

Corollary 26.5. A function that is analytic on a domain S is uniquely determined over S by its values over a subdomain of S or along an arc contained in S.

Example 26.2. (i). Since $\sin^2 x + \cos^2 x = 1$, the entire function $f(z) = \sin^2 z + \cos^2 z - 1$ has zero values along the real axis. Thus, Corollary 26.4 implies that $f(z) = 0$ throughout the complex plane; i.e., $\sin^2 z + \cos^2 z = 1$ for all z. Furthermore, Corollary 26.5 tells us that $\sin z$ and $\cos z$ are the only entire functions that can assume the values $\sin x$ and $\cos x$, respectively, along the real axis or any segment of it.

(ii). We can use $\cosh^2 x - \sinh^2 x = 1$ to show that $\cosh^2 z - \sinh^2 z = 1$ for all complex z.

(iii). We can use $\sin(x_1 + x_2) = \sin x_1 \cos x_2 + \cos x_1 \sin x_2$ to show that for all complex z_1, $\sin(z_1 + x_2) = \sin z_1 \cos x_2 + \cos z_1 \sin x_2$, and from this deduce that $\sin(z_1 + z_2) = \sin z_1 \cos z_2 + \cos z_1 \sin z_2$ for all complex numbers z_1 and z_2.

(iv). We can use $e^{x_1 + x_2} = e^{x_1} e^{x_2}$ to show that $e^{z_1 + z_2} = e^{z_1} e^{z_2}$ for all complex numbers z_1 and z_2.

Corollary 26.6. If a function $f(z)$ is analytic on a bounded domain S and continuous and nonvanishing on the boundary of S, then $f(z)$ can have at most finitely many zeros inside S.

Proof. Suppose that $f(z)$ has an infinite number of zeros inside S. Since $\overline{S}$ is closed and bounded, in view of Theorem 4.2, there is an infinite sequence of zeros $\{z_n\}$ that converges to a point $\alpha \in \overline{S}$. Now, since $f(z)$ is continuous on $\overline{S}$, $f(\alpha) = \lim_{n \to \infty} f(z_n) = 0$, and since $f(z)$ is nonvanishing on the boundary, α must be inside S. But then, by Corollary 26.3, $f(z)$ is identically zero on S, and since $f(z)$ is continuous it must be zero on the boundary of S also. This contradiction completes the proof. ∎

From Corollary 26.6, it follows that an analytic function can have an infinite number of zeros only in an open or unbounded domain. From Corollary 26.6, it also follows that an entire function in any bounded part of the complex plane can have only a finite number of zeros. Thus, all the zeros of an entire function can be arranged in some kind of order, for example in order of increasing absolute value. In the extended complex plane, an entire function can have only a countable set of zeros, and the *limit point* of this set is the point at infinity of the complex plane.

Extension of known equalities from real to complex variables as in Example 26.2 can be generalized to a broader class of identities. In the following result we state such a generalization for an important class of identities that involve only polynomials of functions.

Theorem 26.2. Let $P(f_1, \cdots, f_n)$ be a polynomial in the n variables f_j, $j = 1, \cdots, n$, where each f_j is an analytic function of z in a domain S that contains some interval $a < x < b$ of the real axis. If for all $x \in (a, b)$

$$P(f_1(x), \cdots, f_n(x)) = 0,$$

then for all $z \in S$

$$P(f_1(z), \cdots, f_n(z)) = 0.$$

Proof. Clearly, the function $P(f_1(z), \cdots, f_n(z))$ is analytic in S and vanishes over an arc in S. The result now follows from Corollary 26.4. ∎

We conclude this lecture by proving the following result.

Theorem 26.3 (Counting Zeros). Let $f(z)$ be analytic inside and on a positively oriented contour γ. Furthermore, let $f(z) \neq 0$ on γ. Then,

$$\frac{1}{2\pi i} \int_\gamma \frac{f'(z)}{f(z)} dz = Z_f \tag{26.1}$$

holds, where Z_f is the number of zeros (counting with multiplicities) of $f(z)$ that lie inside γ.

Proof. The function $f'(z)/f(z)$ is analytic inside and on γ except at the zeros $a_1, \cdots, a_\ell$ of $f(z)$ lying inside γ. Suppose $m_1, \cdots, m_\ell$ are the respective multiplicities of these zeros. Then $Z_f = m_1 + \cdots + m_\ell$. In view of Theorem 26.1, we can find disjoint open disks $B(a_k, r_k)$, $k = 1, \cdots, \ell$, and analytic nonzero functions $g_k(z)$ in $B(a_k, r_k)$ such that

$$f(z) = (z - a_k)^{m_k} g_k(z), \quad z \in B(a_k, r_k). \tag{26.2}$$

Thus, we have

$$\frac{f'(z)}{f(z)} = \frac{m_k}{z - a_k} + \frac{g_k'(z)}{g_k(z)}, \quad z \in B(a_k, r_k) \backslash \{a_k\}. \tag{26.3}$$

Consider the function

$$
F(z) = \begin{cases} \dfrac{f'(z)}{f(z)} - \displaystyle\sum_{j=1}^{\ell} \dfrac{m_j}{z - a_j}, & z \notin \displaystyle\bigcup_{j=1}^{\ell} B(a_j, r_j), \\[4mm] \dfrac{g_k'(z)}{g_k(z)} - \displaystyle\sum_{j=1, j\neq k}^{\ell} \dfrac{m_j}{z - a_j}, & z \in B(a_k, r_k), \quad k = 1, \cdots, \ell. \end{cases}
$$

Clearly, $F(z)$ is analytic inside and on γ, and hence $\displaystyle\int_{\gamma} F(z)dz = 0$. Therefore, from Theorem 16.2 and Example 16.1, it follows that

$$
\int_{\gamma} \frac{f'(z)}{f(z)} dz \;=\; 2\pi i \sum_{j=1}^{\ell} m_j \;=\; 2\pi i Z_f,
$$

which is the same as (26.1). ∎

Problems

26.1. Locate and determine the order of zeros of the following functions:
(a). $e^{2z} - e^z$,   (b). $z^2 \sinh z$,   (c). $z^4 \cos^2 z$,   (d). $z^3 \cos z^2$.

26.2. Suppose that $f(z)$ is analytic and has a zero of order m at z_0.
Show that

(a). $f'(z)$ has a zero of order $m - 1$ at z_0.

(b). $f'(z)/f(z)$ has a simple pole at z_0.

26.3. Find an entire function $f(z)$ with prescribed distinct zeros $z_1, \cdots, z_k$
with multiplicities $m_1, \cdots, m_k$, respectively. Is $f(z)$ uniquely determined?

26.4. If $f(z)$ is analytic on a domain S and has distinct zeros $z_1, \cdots, z_k$
with multiplicities $m_1, \cdots, m_k$, respectively, show that there exists an analytic function $g(z)$ on S such that $f(z) = (z - z_1)^{m_1} \cdots (z - z_k)^{m_k} g(z)$.

26.5. Suppose that $f(z)$ and $g(z)$ are analytic in a region S and $f(z)g(z)$
is identically zero in S. Show that either $f(z)$ or $g(z)$ is identically zero in
S.

Answers or Hints

26.1. (a). $z = 2k\pi i$, $k = 0, \pm 1, \pm 2, \cdots$ simple zeros, (b). $z = 0$ zero of order 3, $z = ki\pi$, $k = \pm 1, \pm 2, \cdots$ simple zeros, (c). $z = 0$ zero of order 4, $z = (2k + 1)\pi/2$, $k = 0, \pm 1, \pm 2, \cdots$ simple zeros, (d). $z = 0$ zero of order 3, $z = \pm\sqrt{(2k + 1)\pi/2}$, $\pm i\sqrt{(2k + 1)\pi/2}$, $k = 0, 1, 2, \cdots$ simple zeros.

26.2. By Theorem 26.1, $f(z) = (z - z_0)^m g(z)$, where g is analytic at z_0 and $g(z_0) \neq 0$. (a). $f'(z) = (z - z_0)^{m-1} h(z)$, where $h(z) = mg(z) + (z - z_0)g'(z)$ is analytic at z_0 and $h(z_0) = mg(z_0) \neq 0$, so f' has a zero of order $m - 1$ at z_0 by Theorem 26.1 again. (b). $f'(z)/f(z) = p(z)/(z - z_0)$, where $p = h/g$ is analytic at z_0 and $p(z_0) = m \neq 0$, so f'/f has a simple pole at z_0.

26.3. We can take, for example, $f(z) = \prod_{j=1}^{k} (z - z_j)^{m_j}$; it is not uniquely determined since multiplying it by an entire function without zeros gives another such function.

26.4. Taylor's expansion of $f(z)$ at z_1 yields an analytic function $f_1(z)$ such that $f(z) = (z - z_1)^{m_1} f_1(z)$, where z_1 is not a zero of $f_1(z)$. Now expand $f_1(z)$ at z_2.

26.5. If neither f nor g is identically zero in S, then they both have isolated zeros, so the zeros of $h = fg$ are also isolated, and hence h is not identically zero in S.

Lecture 27
Analytic Continuation

The uniqueness result established in Corollary 26.5 will be used here to discuss an important technique in complex function theory known as analytic continuation. The principal task of this technique is to extend the domain of a given analytic function.

Let S_1 and S_2 be two domains with $S_1 \cap S_2 \neq \emptyset$, and let $f_1(z)$ be a function that is analytic in S_1. There may exist a function $f_2(z)$ that is analytic in S_2 such that $f_2(z) = f_1(z)$ for each z in $S_1 \cap S_2$. If so, we call $f_2(z)$ a *direct analytic continuation* of $f_1(z)$ into the second domain S_2. In view of Corollary 26.5, it is clear that if analytic continuation exists, then it is unique; i.e., no more than one function can be analytic in S_2 and also assume the value $f_1(z)$ at each point $z \in S_1 \cap S_2$. However, if there is an analytic continuation $f_3(z)$ of $f_2(z)$ from S_2 into a domain S_3 that intersects S_1, it is not necessary that $f_3(z) = f_1(z)$ for each $z \in S_1 \cap S_3$. We shall illustrate this in Example 27.4.

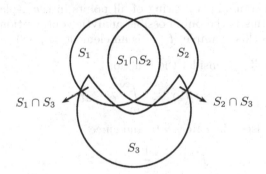

Figure 27.1

If $f_2(z)$ is the analytic continuation of $f_1(z)$ from a domain S_1 into a domain S_2, then the single-valued function $F(z)$ defined by

$$F(z) = \begin{cases} f_1(z), & z \in S_1 \\ f_2(z), & z \in S_2 \end{cases} \tag{27.1}$$

is analytic in the domain $S_1 \cup S_2$. Clearly, the function F is the analytic continuation into $S_1 \cup S_2$ of either $f_1(z)$ or $f_2(z)$. Here $f_1(z)$ and $f_2(z)$ are called elements of F.

Example 27.1. Let $f(x)$, $x \in (a, b)$ be a continuous function. Then in some domain $S \subseteq \mathbf{C}$ such that $(a, b) \subset S$ there may exist a unique analytic function $f(z)$, $z \in S$ that coincides with $f(x)$ on (a, b). This function $f(z)$ is also called an analytic continuation of the function $f(x)$ of the real variable into the complex domain S. Various such analytic continuations were given in Lecture 26.

Example 27.2. Consider the function

$$f_1(z) = \sum_{j=0}^{\infty} z^j.$$

Clearly, this function converges to $1/(1 - z)$ when $|z| < 1$. Hence,

$$f_1(z) = \frac{1}{1 - z} \quad \text{when} \quad |z| < 1,$$

and $f_1(z)$ is not defined when $|z| \geq 1$.

Now the function

$$f_2(z) = \frac{1}{1 - z}, \quad z \neq 1,$$

is defined and analytic everywhere except at the point $z = 1$. Since $f_2(z) = f_1(z)$ inside the circle $|z| = 1$, the function $f_2(z)$ is the analytic continuation of $f_1(z)$ into the domain consisting of all points in the z-plane except for $z = 1$. Clearly, this is the only possible analytic continuation of $f_1(z)$ into that domain. In this example, $f_1(z)$ is an element of $f_2(z)$.

Example 27.3. Consider the function

$$f_1(z) = \int_0^{\infty} e^{-zt} dt.$$

This function exists when $\operatorname{Re} z > 0$, and since

$$f_1(z) = \frac{1}{z}, \quad \operatorname{Re} z > 0,$$

it is analytic in $S_1 : \operatorname{Re} z > 0$.

Let $f_2(z)$ be defined by the geometric series

$$f_2(z) = i \sum_{j=0}^{\infty} \left(\frac{z + i}{i} \right)^j, \quad |z + i| < 1.$$

Within its circle of convergence, which is the unit circle centered at the point $z = -i$, the series is convergent. Clearly,

$$f_2(z) = \frac{i}{1 - (z + i)/i} = \frac{1}{z},$$

where z is in the domain $S_2 : |z + i| < 1$.

Evidently, $f_2(z) = f_1(z)$ for each $z \in S_1 \cap S_2$, and $f_2(z)$ is the analytic continuation of $f_1(z)$ into S_2.

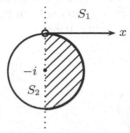

Figure 27.2

The function $F(z) = 1/z$, $S_3 : z \neq 0$ is the analytic continuation of both $f_1(z)$ and $f_2(z)$ into the domain S_3 consisting of all points in the z-plane except the origin. The functions $f_1(z)$ and $f_2(z)$ are elements of $F(z)$.

Example 27.4. Consider the branch of $z^{1/2}$

$$f_1(z) = \sqrt{r}e^{i\theta/2}, \quad r > 0, \quad 0 < \theta < \pi.$$

An analytic continuation of $f_1(z)$ across the negative real axis into the lower half-plane is

$$f_2(z) = \sqrt{r}e^{i\theta/2}, \quad r > 0, \quad \pi/2 < \theta < 2\pi.$$

An analytic continuation of $f_2(z)$ across the positive real axis into the first quadrant is

$$f_3(z) = \sqrt{r}e^{i\theta/2}, \quad \pi < \theta < 5\pi/2.$$

Note that $f_3(z) \neq f_1(z)$ in the first quadrant. In fact, $f_3(z) = -f_1(z)$ there.

In general, it may not be obvious whether a given analytic function can be extended. However, a standard method of analytic continuation is by means of power series, which can be described as follows: Let the function $f_1(z)$ be analytic in the domain S_1. Choose an arbitrary point $z_0 \in S_1$, and consider the Taylor series expansion

$$f_1(z) = \sum_{j=0}^{\infty} \frac{f_1^{(j)}(z_0)}{j!}(z - z_0)^j. \tag{27.2}$$

For (27.2), the radius of convergence R_0 may not exceed the distance from z_0 to the boundary γ_1 of S_1 (see Figure 27.3(a)). In this case, the expansion (27.2) does not provide continuation of $f_1(z)$ beyond the boundary of S_1. Another possibility is that R_0 exceeds the distance from z_0 to the boundary

γ_1 of S_1 (see Figure 27.3(b)). In this case, the domain $S_2 : |z - z_0| < R_0$ is not a subdomain of S_1 but has a common overlapping portion, say S_{12}. In S_2, the series (27.2) defines an analytic function $f_2(z)$, which coincides with $f_1(z)$ in S_{12}. Clearly, $f_2(z)$ is the analytic continuation of $f_1(z)$ into the domain S_2. Thus, the function $F(z)$ defined as in (27.1) is analytic in the domain $S = S_1 \cup S_2$. Now, starting from S_2, the process can be repeated to obtain a chain of overlapping disks.

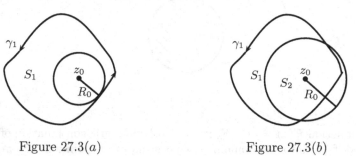

Figure 27.3(a)   Figure 27.3(b)

Example 27.5.   Consider the function $f_1(z) = \text{Log}\, z$. We can expand this function in Taylor's series around the point $z_0 = -1 + i$ to obtain

$$f_2(z) = \text{Log}\,(-1 + i) + \sum_{j=1}^{\infty} \frac{(-1)^{j-1}(z + 1 - i)^j}{j(-1 + i)^j}.$$

The radius of convergence of this series is $\sqrt{2}$, and hence $f_2(z)$ is analytic in the disk $|z + 1 - i| < \sqrt{2}$ (see Figure 27.4). The function $f_2(z)$ crosses the branch cut of $f_1(z)$ but not the branch point $(0,0)$. In the second quadrant $f_2(z)$, coincides with $f_1(z)$; however, in the third quadrant, $f_2(z)$ has to be different from $f_1(z)$. Thus, the Taylor expansion of $f_1(z)$ analytically continues the function into the third quadrant, and the analytic continuation is different from the original function.

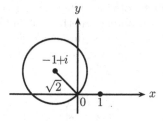

Figure 27.4

It can be shown that if analytic continuation of $f(z)$ is possible at all then it can be accomplished by following a chain of overlapping disks along some path. Furthermore, different paths may lead to different analytic continuations of $f(z)$. The function $F(z)$ obtained by means of an analytic continuation of $f(z)$ along all possible paths is called the *complete analytic function*. Its domain of definition S^* is called the *natural domain* of existence. The boundary of S^* is called the *natural boundary*.

Example 27.6. Consider the function

$$f(z) = 1 + z^2 + z^4 + z^8 + \cdots + z^{2^j} + \cdots.$$

Clearly, each root of any of the equations $z^2 = 1$, $z^4 = 1$, $z^8 = 1, \cdots$ is a singularity of $f(z)$. Hence, on any arc, however small, of the circle $|z| = 1$, the function $f(z)$ has an infinite number of singularities. Therefore, we cannot extend this function outside the circle $|z| = 1$. In conclusion, for the function $f(z)$, the circle $|z| = 1$ is the natural boundary.

Problems

27.1. Show that the functions

$$f_1(z) = -\frac{i\pi}{2} + \sum_{j=1}^{\infty} \frac{(-1)^{j-1}}{j}(z-1)^j, \quad |z-1| < 1$$

and

$$f_2(z) = \frac{i\pi}{2} + \sum_{j=1}^{\infty} \frac{1}{j}(z+1)^j, \quad |z+1| < 1$$

are analytic continuations of each other.

27.2. Suppose that $f(z) = \sum_{j=0}^{\infty} a_j z^j$ has radius of convergence R. Show that if $f\left(re^{i\theta}\right)$ tends to infinity as $r \to R$, then the point $Re^{i\theta}$ is a singular point of $f(z)$.

27.3. Suppose $f(z) = \sum_{j=0}^{\infty} a_j(z-z_0)^j$ has a finite radius of convergence $R > 0$. Show that $f(z)$ has at least one singular point on the boundary of the disk $B(z_0, R)$.

27.4. Show that for the function $f(z) = \sum_{j=1}^{\infty} z^{j!}$ the unit circle $|z| = 1$ is a natural boundary.

27.5 (Painlevé's Theorem). Assume that S_1 and S_2 are two regions with an adjacent piecewise smooth boundary arc L (see Figure 27.5(a)) and $f_1(z)$ and $f_2(z)$ are analytic functions on S_1 and S_2, respectively. Furthermore, assume that $f_1(z)$ and $f_2(z)$ are continuous on $S_1 \cup L$ and $S_2 \cup L$, respectively, and $f_1(z) = f_2(z)$, $z \in L$. Show that

$$F(z) = \begin{cases} f_1(z) & \text{if } z \in S_1 \cup L \\ f_2(z) & \text{if } z \in S_2 \end{cases}$$

is analytic in $S = S_1 \cup L \cup S_2$.

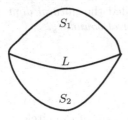

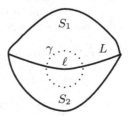

Figure 27.5(a)   Figure 27.5(b)

27.6 (Riemann's Theorem). Let $S = B(z_0, r) \backslash \{z_0\}$, and let $f(z)$ be an analytic and bounded function on S. Show that $f(z)$ can be analytically continued to $B(z_0, r)$.

Answers and Hints

27.1. The functions $f_1(z)$ and $f_2(z)$ are the restrictions of $-i\pi/2 + \log z = -i\pi/2 + \text{Log}|z| + i \arg z$, where $-\pi/2 < \arg z < 3\pi/2$ to the disks $|z - 1| < 1$ and $|z + 1| < 1$.

27.2. If $Re^{i\theta}$ is not a singular point, then there is a function $F(z)$ that is analytic in a disk with center $Re^{i\theta}$ and agrees with $f(z)$ in $|z| < R$. But then $\lim_{r \to R} f\left(re^{i\theta}\right) = \lim_{r \to R} F\left(re^{i\theta}\right) = F\left(Re^{i\theta}\right)$. But this is a contradiction.

27.3. Suppose that at all points on the boundary $\{z : |z - z_0| = R\}$ the function $f(z)$ can be continued analytically. Then, at each point ξ of the boundary, the function $f(z)$ can be expanded in terms of powers of $(z - \xi)$, say $f_\xi(z)$ in the disk $B(\xi)$ with radius of convergence $r(\xi)$. Since $\bigcup_\xi B(\xi)$ covers the compact set $\{z : |z - z_0| = R\}$, by Theorem 4.4 there also exists a finite subcover $\{B(\xi_1), B(\xi_2), \cdots, B(\xi_m)\}$. Now the function $F(z)$ defined by $F(z) = f(z)$, $z \in B(z_0, R)$ and $F(z) = f_{\xi_j}(z)$, $z \in B(\xi_j)$, $1 \le j \le m$ is analytic in $S = B(z_0, R) \cup B(\xi_1) \cup \cdots \cup B(\xi_m)$. Since S contains the closed disk $\overline{B}(z_0, R)$, it must also contain the disk $\{z : |z - z_0| \le R + \epsilon\}$. But, then the power series representation $F(z) = \sum_{j=0}^{\infty} a_j (z - z_0)^j$ is valid in the disk $\{z : |z - z_0| < R + \epsilon\}$, which contradicts the fact that $f(z)$ has the radius of convergence R.

27.4. $|f(Re^{ir\pi})| \to \infty$ as $R \to 1^-$ for any rational number r.

27.5. Clearly, F is continuous in S. Draw an arbitrary piecewise smooth closed contour γ so that its interior lies in S. If it lies entirely in S_1 or S_2, then, by Theorem 15.2, $\int_\gamma F(z)dz = 0$. If γ is contained partly in S_1 and partly in S_2 (see Figure 27.5(b)), denoted by γ_1 and γ_2, respectively, and the arc $L \cap \gamma$ is denoted by ℓ, then again by Theorem 15.2, $\int_{\gamma_1 + \ell} F(z)dz = 0$ and $\int_{\gamma_2 - \ell} F(z)dz = 0$. Hence, $\int_\gamma F(z)dz = 0$. Now use Theorem 18.4.

27.6. We assume that $z_0 = 0$. Define $g(z) = z^2 f(z)$, $z \in S$; $g(z) = 0$, $z = 0$. Clearly, $g'(0) = 0$ and since $g'(z) = z^2 f'(z) + 2z f(z)$, $z \neq 0$ we find $g'(z) \to 0$ as $z \to 0$. Thus, $g(z)$ is continuous and differentiable on $B(0, r)$ and satisfies the Cauchy-Riemann conditions, and hence is analytic on $B(0, r)$. Therefore, Taylor's series $g(z) = 0 + 0z + a_2 z^2 + a_3 z^3 + \cdots$ converges on $B(0, r)$. Now define $F(z) = f(z)/z^2 = a_2 + a_3 z + \cdots$ which has the same radius of convergence. Thus, $F(z)$ is analytic on $B(0, r)$, and $F(z) = f(z)$, $z \in S$.

Lecture 28
Symmetry and Reflection

The cross ratio defined in Problem 11.9 will be used here to introduce the concept of symmetry of two points with respect to a line or a circle. We shall also prove *Schwarz's Reflection Principle*, which is of great practical importance for analytic continuation.

Recall that in the real plane two points P and Q are said to be symmetric across the line L if L is the perpendicular bisector of the segment joining P and Q. In the complex plane, if L is the real axis, points z', z'' are said to be symmetric (reflections of each other) across L, if $z'' = \overline{z'}$. Thus, from the definition of the cross ratio, if three points z_1, z_2, z_3 are on the real line L, then the points z', z'' are symmetric if and only if

$$(z_1, z_2, z_3, z'') = \overline{(z_1, z_2, z_3, z')}. \tag{28.1}$$

Now, since linear fractional transformations map lines and circles to lines or circles and preserve the value of the cross ratio (see Problem 11.9(a)), it is natural to use (28.1) to define symmetry of points across any line or circle γ as follows: If the points z_1, z_2, z_3 are on a line or circle γ, the points z', z'' are said to be *symmetric with respect to* γ if (28.1) holds. Once again, from Problem 11.9(a) it follows that the points z', z'' are symmetric if and only if w', w'' are symmetric with respect to γ_1, where $w' = f(z')$, $w'' = f(z'')$ and γ_1 is the image of γ under the linear fractional transformation f. In the following theorems, we shall show that the definition of symmetry above is independent of the points z_1, z_2, z_3. For this, we need the following elementary lemma.

Lemma 28.1. Let z_1, z_2, z_3 be distinct complex numbers. Then, $(z_1, z_2, z_3, z'') = (z_1, z_2, z_3, z')$ if and only if $z'' = z'$.

Theorem 28.1. The points z' and z'' are symmetric with respect to a line γ in the complex plane if and only if γ is the perpendicular bisector of the segment joining z' and z''.

Proof. We can map γ onto the real axis in the w-plane by a linear transformation $w = az + b$, where $|a| = 1$. Then, for the real points w_j, $j = 1, 2, 3$ and the images $w' = az' + b$, $w'' = az'' + b$, we have $(w_1, w_2, w_3, w'') = \overline{(w_1, w_2, w_3, w')} = (w_1, w_2, w_3, \overline{w'})$, and hence, in view of Lemma 28.1, $w'' = \overline{w'}$; i.e., the real axis in the w-plane is the perpendicular bisector of the segment joining w' and w''. Now, since the inverse transformation $z =$

$(1/a)w + (-b/a)$ is just a translation and rotation, the segment joining w' and w'' is rotated and translated along with the real axis onto the segment joining z' and z'' and the line γ. Thus, γ is the perpendicular bisector of the segment joining z' and z''. ∎

Theorem 28.2. The points z' and z'' are symmetric with respect to a circle $\gamma : |z - z_0| = r$ if and only if z' and z'' lie on a ray from the center O of the circle, and $|z' - z_0||z'' - z_0| = r^2$.

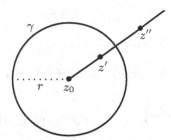

Figure 28.1

Proof. Without any loss of generality, we can assume that $z_0 = 0$. Since $\overline{z}_j = r^2/z_j$, $j = 1, 2, 3$, the points z', z'' are symmetric with respect to the circle $|z| = r$ if and only if

$$(z_1, z_2, z_3, z'') = (\overline{z}_1, \overline{z}_2, \overline{z}_3, \overline{z'}) = (r^2/z_1, r^2/z_2, r^2/z_3, \overline{z'}). \quad (28.2)$$

Now, since the function $f(z) = r^2/z$ is a linear fractional transformation, from (28.2) and Problem 11.9(a) it follows that

$$(z_1, z_2, z_3, z'') = (z_1, z_2, z_3, r^2/\overline{z'}).$$

But then, from Lemma 28.1, we have $z'' = r^2/\overline{z'}$; i.e., $z''\overline{z'} = r^2$. Hence, if $z' = \mu e^{i\theta}$ and $z'' = \nu e^{i\phi}$, then $z''\overline{z'} = \nu\mu e^{i(\phi-\theta)} = r^2$. Therefore, $\phi = \theta$ and $|z'||z''| = r^2$. ∎

Remark 28.1. From Theorem 28.2, it is clear that the center of the circle γ is symmetric to ∞ with respect to γ. Thus, the image of this center will be symmetric to the image of ∞ with respect to γ_1 under a linear fractional transformation.

We shall now prove a theorem that shows that some analytic functions possess the property that $\overline{f(z)} = f(\overline{z})$ at all points z in certain domains while others do not. For example, the functions $z + 1$ and z^2 have this property in the entire plane, but $z + i$ and iz^2 do not. The following result, which is known as *Schwarz's Reflection Principle*, provides sufficient conditions when the reflection of $f(z)$ in the real axis corresponds to the reflection of z.

Theorem 28.3. Suppose that a function f is analytic in some domain S that contains a segment of the x-axis and is symmetric to that axis. Then,

$$\overline{f(z)} = f(\bar{z}) \qquad (28.3)$$

for each point z in the domain if and only if $f(x)$ is real for each point x on the segment.

Proof. Let $f(x)$ be real at each point x on the segment. We shall show that the function

$$F(z) = \overline{f(\bar{z})} \qquad (28.4)$$

is analytic in S. For this, we write

$$f(z) = u(x,y) + iv(x,y) \quad \text{and} \quad F(z) = U(x,y) + iV(x,y).$$

Now, since

$$\overline{f(\bar{z})} = u(x,-y) - iv(x,-y), \qquad (28.5)$$

from (28.4) it follows that

$$U(x,y) = u(x,t) \quad \text{and} \quad V(x,y) = -v(x,t), \qquad (28.6)$$

where $t = -y$.

Now, since $f(x+it)$ is an analytic function of $x+it$, the first-order partial derivatives of the functions $u(x,t)$ and $v(x,t)$ are continuous throughout S and satisfy the Cauchy-Riemann equations

$$u_x = v_t, \quad u_t = -v_x. \qquad (28.7)$$

Next, in view of equations (28.6), we have

$$U_x = u_x \quad \text{and} \quad V_y = -v_t \frac{dt}{dy} = v_t \quad (t = -y),$$

and hence, from (28.7), we get $U_x = V_y$. Similarly,

$$U_y = u_t \frac{dt}{dy} = -u_t, \quad V_x = -v_x$$

and therefore from (28.7) we find $U_y = -V_x$. Thus, the partial derivatives of $U(x,y)$ and $V(x,y)$ are continuous and satisfy the Cauchy-Riemann equations, and hence the function $F(z)$ is analytic in S.

Now, since $f(x)$ is real on the segment of the real axis lying in S, $v(x,0) = 0$ on that segment. Thus, in view of (28.6) it follows that

$$F(x) = U(x,0) + iV(x,0) = u(x,0) - iv(x,0) = u(x,0);$$

i.e.,

$$F(z) = f(z) \tag{28.8}$$

at each point $z = x$ on the segment. But now, in view of Corollary 26.5, we can conclude that (28.8) holds throughout S. Hence, (28.4) gives

$$\overline{f(\overline{z})} = f(z), \quad z \in S, \tag{28.9}$$

which is the same as (28.3).

Conversely, assume that (28.3) holds. Then, (28.9) in view of (28.5) is the same as

$$u(x, -y) - iv(x, -y) = u(x, y) + iv(x, y).$$

Thus, if $(x, 0)$ is a point on the segment of the real axis that lies in S, then

$$u(x, 0) - iv(x, 0) = u(x, 0) + iv(x, 0),$$

which implies that $v(x, 0) = 0$; i.e., $f(x)$ is real on the segment of the real axis lying in S. ∎

Note that the functions $z + 1$ and z^2 satisfy the conditions of Theorem 28.3; in particular, for $z = x$, both the functions are real; i.e., $x + 1$ and x^2. However, $z + i$ and iz^2 are not real when $z = x$.

The following corollary of Theorem 28.3 provides a means of describing an analytic continuation across the interval on the real axis that is a part of the boundary of the domain.

Corollary 28.1.   Let S be a domain in the upper half-plane whose boundary includes an interval $J = (a, b)$ of the real axis. Let S' be the reflection of S across the real axis; i.e., $S' = \{\overline{z} : z \in S\}$. Let the function $f(z)$ be analytic on S and continuous on $S \cup J$. Furthermore, let f be real-valued on I. Then, the function $F(z)$ defined by

$$F(z) = \begin{cases} f(z), & z \in S \cup J \\ \overline{f(\overline{z})}, & z \in S' \end{cases}$$

is the unique analytic continuation of $f(z)$ to $S \cup J \cup S'$.

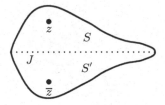

Figure 28.2

Corollary 28.2. Let S be the upper half-plane and let $f(z)$ be an analytic function on S that is continuous and real on the real axis. Then, $f(z)$ can be extended analytically to the entire complex plane.

Corollary 28.3. Let $f(z)$ be an analytic function on the right half-plane that is continuous and purely imaginary on the imaginary axis. Then, the function $F(z)$ defined by

$$F(z) = \begin{cases} f(z), & \text{Re}(z) \geq 0 \\ -\overline{f(-\overline{z})}, & \text{Re}(z) \leq 0 \end{cases}$$

is the unique analytic continuation of $f(z)$ to the entire complex plane.

Proof. The transformation $w = iz$ maps the right half-plane $\text{Re}(z) \geq 0$ onto the upper half-plane $\text{Im}(w) \geq 0$. The function $g(w) = if(-iw)$ satisfies the conditions of Corollary 28.2 and hence has an analytic continuation $G(w)$ to the entire w-plane. Now, from Corollary 28.1, we have

$$G(w) = \begin{cases} if(-iw), & \text{Im}(w) \geq 0 \\ -i\overline{f(-i\overline{w})}, & \text{Im}(w) \leq 0, \end{cases}$$

which is the same as $G(w) = iF(-iw)$. Thus, $F(z)$ is the desired extension. ∎

Corollary 28.4. Let $f(z)$ be an entire function taking real values on the real axis and imaginary values on the imaginary axis. Then, $f(z) + f(-z) = 0$ for all z.

Proof. From Corollaries 28.2 and 28.3, we have $f(z) = \overline{f(\overline{z})}$ and $f(z) = -\overline{f(-\overline{z})}$. On replacing $\overline{z}$ by z in these relations, we get the desired result. ∎

Lecture 29
Singularities and Poles I

In this lecture, we shall define, classify, and characterize singular points of complex functions. We shall also study the behavior of complex functions in the neighborhoods of singularities.

A point z_0 is called a *singular point* of the function $f(z)$ if $f(z)$ is not analytic at z_0 but is analytic at some point in $B(z_0, \epsilon)$ for all $\epsilon > 0$. A singular point z_0 of $f(z)$ is called *isolated* if there exists $R > 0$ such that $f(z)$ is analytic on some *punctured* open disk $0 < |z - z_0| < R$.

Example 29.1. For the function $f(z) = (z + 1)/[z^4(z^2 + 1)]$, the singular points are $z = 0$, i, $-i$ and all are isolated.

Example 29.2. The function $\text{Log}\, z$ is analytic in $\mathbf{C} \backslash (-\infty, 0]$. Each point in $(-\infty, 0]$ is a singular point of $\text{Log}\, z$ but not an isolated singularity.

Example 29.3. For the function $f(z) = 1/\sin(\pi/z)$, singularities occur where $\sin \pi/z = 0$; i.e., $\pi/z = n\pi$ or $z = 1/n$. Hence, the singularities of $f(z)$ are $\{0\} \cup \{1/n : n \in \mathbf{Z}\}$.

Now, let z_0 be an isolated singularity of $f(z)$. Then, $f(z)$ has a Laurent series expansion around z_0; i.e., $f(z) = \sum_{j=-\infty}^{\infty} a_j (z - z_0)^j$ on some punctured open disk, say, $0 < |z - z_0| < R$.

(i). If $a_j = 0$ for all $j < 0$, we say that z_0 is a *removable singularity* or a *regular point* of $f(z)$.

(ii). If $a_{-m} \neq 0$ for some positive integer m but $a_j = 0$ for all $j < -m$, we say that z_0 is a *pole of order* m for $f(z)$. A pole of order 1 is called a *simple pole*.

(iii). If $a_j \neq 0$ for an infinite number of negative values of j, we say that z_0 is an *essential singularity* of f.

When $f(z)$ has a removable singularity at z_0, its Laurent series takes the form

$$f(z) = a_0 + a_1(z - z_0) + a_2(z - z_0)^2 + \cdots, \quad 0 < |z - z_0| < R, \quad (29.1)$$

and hence $\lim_{z \to z_0} f(z) = a_0$. Thus, if $f(z)$ is not defined at z_0 (the specified value of $f(z_0)$ is not the same as a_0), we can define (redefine) the function $f(z)$ at z_0 as $f(z_0) = a_0$. The function $f(z)$ thus obtained will be analytic on the open disk $B(z_0, R)$.

Example 29.4. The following functions have removable singularities:

$$f_1(z) = \frac{\sin z}{z} = \frac{1}{z}\left(z - \frac{z^3}{3!} + \frac{z^5}{5!} - \cdots\right) = 1 - \frac{z^2}{3!} + \frac{z^4}{5!} - \cdots, \quad z_0 = 0,$$

$$f_2(z) = \frac{\cos z - 1}{z} = \frac{1}{z}\left(1 - \frac{z^2}{2!} + \frac{z^4}{4!} - \cdots - 1\right) = -\frac{z}{2!} + \frac{z^3}{4!} - \cdots, \quad z_0 = 0,$$

$$f_3(z) = \frac{z^3 - 1}{z - 1} = z^2 + z + 1 = 3 + 3(z-1) + (z-1)^2 + 0 + 0 + \cdots, \quad z_0 = 1.$$

In fact, we can remove the singularities of these functions by defining $f_1(0) = 1$, $f_2(0) = 0$, $f_3(1) = 3$.

Theorem 29.1. The function $f(z)$ has a removable singularity at z_0 if and only if any one of the following conditions holds:

(i). $f(z)$ has a (finite) limit as z approaches z_0,

(ii). $f(z)$ can be redefined at z_0 so that the new function is analytic at z_0,

(iii). $\lim_{z \to z_0}(z - z_0)f(z) = 0$,

(iv). $f(z)$ is bounded in some punctured neighborhood of z_0.

Proof. Parts (i)-(iii) follow immediately from (29.1). To prove part (iv), let z_0 be a removable singularity of $f(z)$. Then, from (29.1) it is clear that the function

$$g(z) = \begin{cases} f(z), & 0 < |z - z_0| < R \\ a_0, & z = z_0 \end{cases}$$

is analytic on $|z - z_0| < R$, and hence bounded in every closed neighborhood of z_0. This in turn implies that $f(z)$ is bounded in some punctured neighborhood of z_0. Conversely, suppose there is a punctured neighborhood $0 < |z - z_0| < r < R$ and a finite M such that on this neighborhood $|f(z)| \leq M$. Then, in Problem 25.10, taking $R_1 \to 0$, it follows that $b_j = 0$, $j = 1, 2, \cdots, \infty$. Hence, the Laurent expansion of $f(z)$ reduces to (29.1), and hence z_0 is a removable singularity of $f(z)$. ∎

Now, for the function $f(z)$, let the point z_0 be a pole of order m. Then, its Laurent series reduces to

$$f(z) = \frac{a_{-m}}{(z - z_0)^m} + \frac{a_{-(m-1)}}{(z - z_0)^{m-1}} + \cdots + \frac{a_{-1}}{(z - z_0)} \quad (29.2)$$
$$+ a_0 + a_1(z - z_0) + a_2(z - z_0)^2 + \cdots, \quad a_{-m} \neq 0,$$

which is valid in some punctured neighborhood of z_0.

Example 29.5. The functions

$$f_1(z) = \frac{\sin z}{z^2} = \frac{1}{z} - \frac{z}{3!} + \frac{z^3}{5!} - \frac{z^5}{7!} + \cdots,$$

$$f_2(z) = \frac{\sin z}{z^4} = \frac{1}{z^3} - \frac{1}{3!z} + \frac{z}{5!} - \frac{z^3}{7!} + \cdots,$$

$$f_3(z) \; = \; \frac{e^z}{z^3} \; = \; \frac{1}{z^3} + \frac{1}{z^2} + \frac{1}{2!z} + \frac{1}{3!} + \frac{z}{4!} + \cdots,$$

respectively, have poles of order 1, 3, and 3 at $z_0 = 0$.

Theorem 29.2. A function $f(z)$ has a pole of order m at z_0 if and only if, in some punctured neighborhood of z_0, $f(z) = g(z)/(z - z_0)^m$, where $g(z)$ is analytic at z_0 and $g(z_0) \neq 0$.

Proof. If $f(z)$ has a pole of order m at z_0, then from (29.2) it follows that, in some punctured neighborhood of z_0,

$$(z - z_0)^m f(z) \; = \; g(z) \; =: \; a_{-m} + a_{-m+1}(z - z_0) + \cdots.$$

Setting $g(z_0) = a_{-m} \neq 0$, we find that $g(z)$ is analytic and $f(z) = g(z)/(z - z_0)^m$. Conversely, if $f(z) = g(z)/(z - z_0)^m$, where $g(z)$ is analytic at z_0 and $g(z_0) \neq 0$, then from Taylor's series we have

$$g(z) \; = \; c_0 + c_1(z - z_0) + c_2(z - z_0)^2 + \cdots, \quad c_0 = g(z_0) \neq 0,$$

and hence the Laurent series for $f(z)$ around z_0 is

$$f(z) \; = \; \frac{g(z)}{(z - z_0)^m} \; = \; \frac{c_0}{(z - z_0)^m} + \frac{c_1}{(z - z_0)^{m-1}} + \cdots.$$

Since $c_0 \neq 0$, z_0 is a pole of order m for $f(z)$. ∎

Example 29.6. Since

$$f(z) \; = \; \frac{\cos z}{(z - 1)^3 (z^2 - 1)^2} \; = \; \frac{\cos z/(z + 1)^2}{(z - 1)^5}$$

and the numerator $\cos z/(z+1)^2$ is analytic and nonzero at $z = 1$, Theorem 29.2 implies that the function $f(z)$ has a pole of order 5. Similarly, in view of Example 24.3, the function $1/[(z^2 - 1)^2 \mathrm{Log}\, z]$ has a pole of order 3 at $z = 1$.

Corollary 29.1. If $f(z)$ and $g(z)$ have poles of orders m and n, respectively, at z_0, then $f(z)g(z)$ has a pole of order $m + n$ at z_0.

Corollary 29.2. If $f(z)$ has a zero of order m at z_0, then $1/f(z)$ has a pole of order m at z_0. Conversely, if $f(z)$ has a pole of order m at z_0, and if we define $1/f(z_0) = 0$, then $1/f(z)$ has a zero of order m at z_0.

Example 29.7. Classify the singularity at $z = 0$ of the function $f(z) = 1/[z(e^z - 1)]$. Clearly, $z = 0$ is a zero of the function $z(e^z - 1)$. Since

$$\frac{d}{dz}[z(e^z - 1)]\bigg|_{z=0} \; = \; (e^z - 1 + ze^z)\bigg|_{z=0} \; = \; 0$$

and
$$\frac{d^2}{dz^2}[z(e^z - 1)]\Big|_{z=0} = e^z(z+2)\Big|_{z=0} \neq 0,$$

$z = 0$ is a zero of order 2 of $z(e^z - 1)$. Hence, $z = 0$ is a pole of order 2 of the function $f(z)$.

Example 29.8. Find and classify the singularities of the function
$$f(z) = \frac{5z + 3}{(1 - z)^3 \sin^2 z}.$$

The singularities of $f(z)$ are the zeros of its denominator; i.e., $z = 1$ and $z = n\pi$, $n = 0, \pm 1, \pm 2, \cdots$. For $z = 1$, since
$$\frac{5z + 3}{(1 - z)^3 \sin^2 z} = \frac{(5z + 3)/\sin^2 z}{(1 - z)^3} \quad \text{and} \quad \frac{5z + 3}{\sin^2 z}\Big|_{z=1} \neq 0,$$

$z = 1$ is a pole of order 3. Since $z = n\pi$ are the zeros of order 2 for the function $\sin^2 z$ and the function $(1 - z)^3/(5z + 3)$ does not vanish at these points, $z = n\pi$ are zeros of order 2 for the function $1/f(z)$. Hence, $z = n\pi$ are poles of order 2 for $f(z)$.

Example 29.9. Classify the zeros and poles of the function $f(z) = \tan z/z$. Since $(\tan z)/z = (\sin z)/(z \cos z)$, the only possible zeros are those of $\sin z$; i.e., $z = n\pi$ $(n = 0, \pm 1, \pm 2, \cdots)$. However, $z = 0$ is a singularity. The points $z = (n + 1/2)\pi$, which are zeros of $\cos z$, are also singularities. If n is a nonzero integer, $z = n\pi$ is a simple zero for $f(z)$ since $f'(n\pi) = 1/(n\pi) \neq 0$. At $z = 0$, we have $\lim_{z\to 0}(z \tan z)/z = 0$. Hence, the origin is a removable singularity. Finally, since $\cos z$ has simple zeros at $z = (n + 1/2)\pi$, $n = 0, \pm 1, \pm 2, \cdots$ and $z/\sin z$ does not vanish at these points, $1/f(z)$ also has simple zeros at these points. Hence, $f(z)$ has simple poles at $z = (n + 1/2)\pi$, $n = 0, \pm 1, \pm 2, \cdots$.

Corollary 29.3. If $f(z)$ and $g(z)$ have zeros of orders m and n, respectively, at z_0, then the quotient function $h(z) = f(z)/g(z)$ has the following behavior:

(i).   If $m > n$, then $h(z)$ has a removable singularity at z_0. If we define $h(z_0) = 0$, then $h(z)$ has a zero of order $m - n$ at z_0.

(ii).   If $m < n$, then $h(z)$ has a pole of order $n - m$ at z_0.

(iii). If $m = n$, then $h(z)$ has a removable singularity at z_0.

Corollary 29.4. The only singularities of the rational function (5.2) are removable singularities or poles.

Theorem 29.3. If the function $f(z)$ has a pole of order m at z_0, then $|(z - z_0)^\ell f(z)| \to \infty$ as $z \to z_0$ for all $\ell = 0, 1, \cdots, m - 1$, while $(z - z_0)^m f(z)$ has a removable singularity at z_0.

Proof. From (29.2) in some punctured neighborhood of z_0, we have

$$(z - z_0)^m f(z) = a_{-m} + a_{-m+1}(z - z_0) + \cdots.$$

Thus, the singularity of $(z - z_0)^m f(z)$ at z_0 is removable, and $(z - z_0)^m f(z) \to a_{-m} \neq 0$ as $z \to z_0$. Hence, for any integer $\ell < m$, it follows that

$$|(z - z_0)^\ell f(z)| = \left| \frac{1}{(z - z_0)^{m-\ell}} (z - z_0)^m f(z) \right| \to \infty \quad \text{as} \quad z \to z_0. \quad \blacksquare$$

Theorem 29.4. The point z_0 is a pole of $f(z)$ if and only if $|f(z)| \to \infty$ as $z \to z_0$.

Proof. From Theorem 29.3, if z_0 is a pole of $f(z)$, then $\lim_{z \to z_0} |f(z)| = \infty$. Conversely, assume that $\lim_{z \to z_0} |f(z)| = \infty$; i.e., for any number $A > 0$ there is a punctured neighborhood $N_\rho(z_0) = \{z : 0 < |z - z_0| < \rho\}$ of z_0 such that on it $|f(z)| > A$. Then, the function $1/f(z)$ on $N_\rho(z_0)$ is analytic, bounded, and nonzero. But then, from Theorem 29.1, for the function $1/f(z)$, the point z_0 is a removable singularity. Consider the function

$$\phi(z) = \begin{cases} 1/f(z), & z \in N_\rho(z_0) \\ 0, & z = z_0. \end{cases}$$

Clearly, $\phi(z)$ is nonzero on $N_\rho(z_0)$, and since $\lim_{z \to z_0} \phi(z) = 0 = \phi(z_0)$, it is analytic and has a zero of order $m \geq 1$ at z_0. Therefore, in view of Theorem 26.1, $\phi(z) = (z - z_0)^m \psi(z)$, where $\psi(z)$ is analytic and nonzero on $N_\rho(z_0)$. Consequently, if we let $g(z) = 1/\psi(z)$, then $g(z)$ is analytic and nonzero on $N_\rho(z_0)$, and

$$f(z) = \frac{1}{\phi(z)} = \frac{1}{(z - z_0)^m \psi(z)} = \frac{1}{(z - z_0)^m} g(z), \quad 0 < |z - z_0| < \rho.$$

Hence, in view of Theorem 29.2, $f(z)$ has a pole of order m at z_0. $\quad \blacksquare$

A function $f(z)$ that is analytic in a region S, except for poles, is said to be *meromorphic* in S. It is clear that the class of meromorphic functions includes analytic and rational functions.

Lecture 30
Singularities and Poles II

Expanding a function in a Laurent series is often difficult. Therefore, in this lecture we shall find the behavior of an analytic function in the neighborhood of an essential singularity. We shall also discuss zeros and singularities of analytic functions at infinity.

Theorem 30.1 (Casorati-Weierstrass Theorem). Suppose that $f(z)$ is analytic on the punctured disk $0 < |z - z_0| < R$. Then, z_0 is an essential singularity of $f(z)$ if and only if the following conditions hold:

(i). There exists a sequence $\{\alpha_n\}$ such that $\alpha_n \to z_0$ and $\lim_{n \to \infty} |f(\alpha_n)| = \infty$.

(ii). For any complex number w, there exists a sequence $\{\alpha_n\}$ (depending on w) such that $\alpha_n \to z_0$ and $\lim_{n \to \infty} f(\alpha_n) = w$.

Proof. If (i) holds, then, in view of Theorem 29.1 (iv), z_0 cannot be a removable singularity. If (ii) holds, then, from Theorem 29.4, z_0 is not a pole. Hence, if (i) and (ii) hold, then z_0 is an essential singularity. Conversely, suppose that z_0 is an essential singularity. Clearly, $f(z)$ cannot be bounded near z_0 (otherwise, z_0 is a removable singularity), and hence (i) holds. If (ii) does not hold, then $|f(z) - w| \geq \epsilon > 0$ for some w and all z in a punctured neighborhood $N_r(z_0)$. Consider the function $g(z) = 1/(f(z) - w)$. It follows that $g(z)$ is analytic on $N_r(z_0)$, and since $|g(z)| \leq 1/\epsilon$ for all $z \in N_r(z_0)$, z_0 is a removable singularity for $g(z)$. In view of Theorem 29.4, we also note that $\lim_{z \to z_0} |f(z)| \neq \infty$, and hence $g(z_0) \neq 0$. But this in turn implies that the function $f(z) = w + 1/g(z)$ is analytic in a neighborhood of z_0. This contradiction ensures that (ii) holds. ∎

Corollary 30.1. Let z_0 be an essential singular point of $f(z)$, and let E_δ be the set of values taken by $f(z)$ in the punctured neighborhood $N_\delta(z_0)$. Then, $\overline{E}_\delta = \mathbf{C} \cup \{\infty\}$.

Theorem 30.1 has a generalization, which we state without proof.

Theorem 30.2 (Picard's Theorem). A function with an essential singularity assumes every complex value, except possibly one, infinitely often in every neighborhood of this singularity.

Example 30.1. Consider the function $f(z) = e^{1/z}$. In view of the Laurent expansion of $e^{1/z}$ in Example 25.1, this function has an essential singularity at $z = 0$. If we choose the sequence $\{z_n\} = \{1/n\}$, then $z_n \to 0$

and $f(z_n) = e^n \to \infty$ as $n \to \infty$, and hence condition (i) of Theorem 30.1 is satisfied. If we choose the sequence $\{z_n\} = \{-1/n\}$, then $z_n \to 0$ and $f(z_n) = e^{-n} \to 0$ as $n \to \infty$. If we choose the sequence $\{z_n\} = \{1/(\text{Log}\,|w|+ i\text{Arg}\,w + 2\pi ni)\}$, where $w \neq 0$, then $z_n \to 0$ and $f(z_n) = e^{\text{Log}\,|w|+i\text{Arg}\,w} = w$, and hence condition (ii) of Theorem 30.1 is also satisfied. Here, since, for each n, $f(z_n) = w$ and $e^{1/z}$ is never zero, this function verifies Theorem 30.2 also.

Example 30.2. Since

$$f(z) = \sin\frac{1}{z} = \frac{1}{z} - \frac{1}{3!z^3} + \frac{1}{5!z^5} - \cdots,$$

this function has an essential singularity at $z = 0$. If we choose the sequence $\{z_n\} = \{i/n\}$, then $z_n \to 0$ and $f(z_n) = \sin(n/i) = \sin(-in) = -i\sinh n \to \infty$ as $n \to \infty$, and hence condition (i) of Theorem 30.1 is satisfied. If we choose the sequence $\{z_n\} = \{i/[\text{Log}(iw + \sqrt{1 - w^2}) + 2n\pi i]\}$, where $w \neq \infty$, then $z_n \to 0$ and $f(z_n) = \sin(1/z_n) = \sin(-i\log[iw + \sqrt{1 - w^2}]) = \sin(\sin^{-1} w) = w$, and hence condition (ii) of Theorem 30.1 is also satisfied. Clearly, this function also confirms Theorem 30.2.

Example 30.3. For the function $f(z) = e^{z/\sin z}$, the only possible singularities are $z = k\pi$, $k = 0, \pm 1, \pm 2, \cdots$. Since $\lim_{z \to 0}(z/\sin z) = 1$, $\lim_{z \to 0} f(z) = e$, and hence $f(z)$ has a removable singularity at $z = 0$. We claim that $z = k\pi$, $k \neq 0$ are essential singularities of $f(z)$. For this, let k be even. Then, since $\lim_{z \to k\pi+}(z/\sin z) = \infty$ and $\lim_{z \to k\pi-}(z/\sin z) = -\infty$, it follows that $\lim_{z \to k\pi+} f(z) = \infty$ and $\lim_{k \to k\pi-} f(z) = 0$. From the first limit, we find that $z = k\pi$ cannot be a removable singularity, whereas from the second limit it is clear that $z = k\pi$ cannot be a pole. Thus, the only possibility left is that $z = k\pi$ is an essential singularity. A similar argument holds when k is odd.

Now we shall investigate the behavior of an analytic function in the neighborhood of infinity. For this, suppose that $f(z)$ is analytic for all $|z| > R$, except possibly at the point $z = \infty$. We note that the transformation $w = 1/z$ carries the point $z = \infty$ into the point $w = 0$, and every sequence $\{z_n\}$ converging to ∞ into a sequence $\{w_n\} = \{1/z_n\}$ converging to 0. Thus, the study of $f(z)$ in the neighborhood of infinity can be reduced to the study of $F(w) = f(1/w)$, which is analytic on the annulus $0 < |w| < 1/R$. We say $z = \infty$ is a *removable singularity*, a *pole of order m*, or an *essential singularity* of $f(z)$ if $w = 0$ is a removable singularity, a pole of order m, or an essential singularity of $F(w)$.

Thus, if $z = \infty$ is a removable singularity of $f(z)$, then

$$F(w) = a_0 + a_{-1}w + a_{-2}w^2 + \cdots + a_{-j}w^j + \cdots,$$

and hence

$$f(z) = a_0 + a_{-1}z^{-1} + a_{-2}z^{-2} + \cdots + a_{-j}z^{-j} + \cdots.$$

Therefore, the expansion of $f(z)$ does not have terms involving positive powers of z. Similarly, if $z = \infty$ is a pole of order m of $f(z)$, then

$$F(w) = a_m w^{-m} + \cdots + a_1 w^{-1} + a_0 + a_{-1}w + \cdots + a_{-j}w^j + \cdots, \quad a_m \neq 0,$$

and hence

$$f(z) = a_m z^m + \cdots + a_1 z + a_0 + a_{-1}z^{-1} + \cdots + a_{-j}z^{-j} + \cdots, \quad a_m \neq 0.$$

Therefore, the expansion of $f(z)$ contains a finite number of terms involving positive powers of z. Finally, if $z = \infty$ is an essential singularity of $f(z)$, then

$$F(w) = \sum_{j=-\infty}^{\infty} a_j w^{-j},$$

and hence

$$f(z) = \sum_{j=-\infty}^{\infty} a_j z^j.$$

Therefore, the expansion of $f(z)$ contains an infinite number of terms involving positive powers of z.

From the considerations above it is clear that the situation at $z = \infty$ is the same as at any finite point, except the roles of positive and negative powers are now interchanged. Furthermore, in view of Theorems 29.1, 29.2, and 30.1, we can state the following theorem.

Theorem 30.3.   The function $f(z)$ has a removable singularity at $z = \infty$ if $f(z)$ is bounded in a deleted neighborhood of ∞, a pole at $z = \infty$ if $\lim_{z \to \infty} f(z) = \infty$, and an essential singularity at $z = \infty$ if $\lim_{z \to \infty} f(z)$ does not exist.

When $f(z)$ has a removable singularity at ∞, $\lim_{z \to \infty} f(z)$ exists. If $\lim_{z \to \infty} f(z) = 0$, we say that $f(z)$ has a *zero* at ∞.

Example 30.4.   The function $z/(z^2 + 1)$ has a zero at ∞, whereas $z^2/(z^2 + 1)$ does not have a zero at ∞ but has a removable singularity. The function $f(z) = z^5$ has a pole of order 5 at $z = \infty$. The function $f(z) = ze^{-z}$ has an essential singularity at $z = \infty$.

In the following results, we shall characterize entire functions at $z = \infty$.

Theorem 30.4.   An entire function $f(z)$ has a removable singularity at $z = \infty$ if and only if it is a constant.

Proof. If $f(z)$ is a constant, then clearly $z = \infty$ is its removable singularity. Conversely, if $f(z)$ has a removable singularity at $z = \infty$, then $\lim_{z\to\infty} f(z) = w_0$ exists, and hence there is an $R > 0$ such that $|z| > R$ implies $|f(z) - w_0| < 1$; i.e., $|f(z)| < |w_0| + 1$. This in turn implies that there exists a finite constant $M > 0$ such that $|f(z)| < M$ on $\mathbf{C}$. But then, in view of Theorem 18.6, the function $f(z)$ has to be a constant. ∎

Theorem 30.5. An entire function $f(z)$ has a pole of order m at $z = \infty$ if and only if it is a polynomial of degree m.

Proof. If $f(z) = a_m z^m + a_{m-1} z^{m-1} + \cdots + a_1 z + a_0$, $a_m \neq 0$, then

$$F(w) = f(1/w) = a_m \frac{1}{w^m} + a_{m-1}\frac{1}{w^{m-1}} + \cdots + a_1 \frac{1}{w} + a_0$$
$$= \frac{1}{w^m}\left(a_m + a_{m-1} w + \cdots + a_1 w^{m-1} + a_0\right),$$

and hence $\lim_{w\to 0} w^m F(w) = a_m \neq 0$, while $\lim_{w\to 0} w^n F(w) = 0$ for all $n > m$. Hence, $F(w)$ has a pole of order m at 0; i.e., $f(z)$ has a pole of order m at ∞. Conversely, suppose that $f(z)$ has a pole of order m at ∞. Then, $F(w)$ has a pole of order m at 0; i.e., m is the least positive integer such that $\lim_{w\to 0} w^m F(w) = c$, a constant, which is the same as $\lim_{z\to\infty} f(z)/z^m = c$. This implies that there is $R > 0$ such that $|f(z)| \leq (|c|+1)|z|^m$ for $|z| > R$. But then, in view of Problem 24.8, $f(z)$ is a polynomial of degree at most m. However, since m is the least such integer, the degree of the polynomial is exactly m. ∎

From Theorems 30.4 and 30.5, it follows that $f(z)$ has an essential singularity at $z = \infty$ if and only if $f(z)$ is not a polynomial of finite degree; i.e., it is an entire *transcendental* function, such as e^z, $\sin z$.

We conclude this lecture by characterizing rational functions.

Theorem 30.6. An entire function $f(z)$ is rational (see (5.2)) if and only if it has at most poles on the extended complex plane $\mathbf{C} \cup \{\infty\}$.

Proof. Since a limiting point of poles is an essential singularity, the function $f(z)$ can have only a finite number of poles on the extended complex plane. Let the poles be at $a_1, a_2, \cdots, a_k, \infty$ of orders $n_1, n_2, \cdots, n_k, m$, respectively. Consider the function

$$g(z) = (z - a_1)^{n_1}(z - a_2)^{n_2} \cdots (z - a_k)^{n_k} f(z).$$

Clearly, $g(z)$ is analytic on every bounded domain. Since $h(z) = (z - a_1)^{n_1}(z - a_2)^{n_2} \cdots (z - a_k)^{n_k}$ has a pole of order $N = n_1 + n_2 + \cdots + n_k$ at infinity, the function $g(z)$ must have a pole of order $N + m$ at infinity. But then, from Theorem 30.5, it follows that $g(z)$ is a polynomial of degree $N + m$. This, however, implies that $f(z) = g(z)/h(z)$ is a rational function. The converse is immediate. ∎

Problems

30.1. Show that $z = 0$ is a removable singularity of the following functions. Furthermore, define $f(0)$ such that these functions are analytic at $z = 0$.

(a). $f(z) = \dfrac{e^{z^2} - 1}{z}$,   (b). $f(z) = \dfrac{\sin z - z}{z^2}$,   (c). $f(z) = \dfrac{1 - \frac{1}{2}z^2 - \cos z}{\sin z^2}$.

30.2. Show that the following functions are entire

(a). $\begin{cases} \dfrac{\cos z}{z^2 - \pi^2/4} & \text{when } z \neq \pm\dfrac{\pi}{2} \\ -\dfrac{1}{\pi} & \text{when } z \neq \pm\dfrac{\pi}{2}, \end{cases}$ (b). $\begin{cases} \dfrac{e^z - e^{-z}}{2z} & \text{when } z \neq 0 \\ 1 & \text{when } z = 0. \end{cases}$

30.3. Find and classify the isolated singularities of the following functions:

(a). $\dfrac{z^3 + 1}{z^2(z - 1)}$,   (b). $z^2 e^{1/z}$,   (c). $\dfrac{z}{\cos z}$,   (d). $\dfrac{\sin 3z}{z^2}$,

(e). $\cos\dfrac{1}{z}\dfrac{\sin(z - 1)}{z^2 + 1}$,   (f). $\dfrac{(z^2 - 1)\cos \pi z}{(z + 2)(2z - 1)(z^2 + 1)^2 \sin^2 \pi z}$.

30.4. Suppose that $f(z)$ has a pole of order m at z_0. Show that $f'(z)$ has a pole of order $m + 1$ at z_0.

30.5. Show that the function $f(z)$ whose Laurent series is

$$\sum_{j=-\infty}^{-1} z^j + \sum_{j=1}^{\infty} \frac{z^j}{3^j}$$

does not have an essential singularity at $z = 0$, although it contains infinitely many negative powers of z.

30.6. Classify the zeros and singularities of the functions

(a). $f(z) = \sin(1 - z^{-1})$,   (b). $f(z) = \dfrac{\tan z}{z}$,   (c). $f(z) = \tanh z$.

30.7. Verify Theorem 30.2 for the functions $\cos(1/z)$ and $\cosh(1/z)$.

30.8. Suppose that $f(z)$ is analytic on $|z| < R$, $R > 1$ except at z_0, $|z_0| = 1$, where it has a simple pole. Furthermore, suppose that the expansion $f(z) = \sum_{j=0}^{\infty} a_j z^j$ holds on $|z| < 1$. Show that $\lim_{j \to \infty} a_j/a_{j+1} = z_0$.

30.9. Investigate the behavior of the following functions at infinity:

(a). $\dfrac{1}{z^2(z^3+1)^2}$,   (b). $\dfrac{1-e^z}{1+e^z}$,   (c). ze^z,   (d). $e^{\cot z}$.

30.10. Suppose that $f(z)$ and $g(z)$ are entire functions such that $(f \circ g)(z)$ is a polynomial. Show that both $f(z)$ and $g(z)$ are polynomials.

30.11. Find all functions that are analytic everywhere in the extended complex plane $\mathbf{C} \cup \{\infty\}$.

Answers or Hints

30.1. (a). $f(z) = (z + \frac{z^3}{2!} + \frac{z^5}{3!} + \cdots)$, $f(0) = 0$, (b). $f(z) = (-\frac{z}{3!} + \frac{z^3}{5!} - \cdots)$, $f(0) = 0$, (c). $f(z) = z^2(-\frac{1}{4!} + \frac{z^2}{6!} - \cdots)/(1 - \frac{z^4}{3!} + \frac{z^8}{5!} - \cdots)$, $f(0) = 0$.

30.2. (a). First note that $\frac{\cos z}{z^2 - \pi^2/4}$ is analytic everywhere except $\pm\pi/2$. Hence, $\pm\pi/2$ are isolated singularities of the function. Next, we have $\lim_{z\to\pi/2} \frac{\cos z}{z^2-\pi^2/4} = \lim_{z\to\pi/2} \frac{-\sin z}{2z} = -\frac{1}{\pi}$ and $\lim_{z\to-\pi/2} \frac{\cos z}{z^2-\pi^2/4} = \lim_{z\to-\pi/2} \frac{-\sin z}{2z} = -\frac{1}{\pi}$.

Thus, $z = \pm\pi/2$ are removable singularities, and hence f is an entire function.

(b). Clearly, $\frac{e^z - e^{-z}}{2z}$ is analytic everywhere except $z = 0$. But $\lim_{z\to 0} \frac{e^z - e^{-z}}{2z} = \lim_{z\to 0} \frac{e^z + e^{-z}}{2} = 1$, and hence $z = 0$ is a removable singularity of $\frac{e^z - e^{-z}}{2z}$, and so g is an entire function. Alternatively, we note that $g(z) = \frac{1}{2z}\left(1 + z + \frac{z^2}{2!} + \cdots\right) - \frac{1}{2z}\left(1 - z + \frac{z^2}{2!} - \frac{z^3}{3!} + \cdots\right) = 1 + \frac{z^2}{3!} + \frac{z^4}{4!} + \cdots$, which is an entire function.

30.3. (a). $z = 0$ pole of order 2, $z = 1$ simple pole, (b). $z = 0$ essential singularity, (c). $z = (\pi/2) + k\pi$, $k \in \mathbf{Z}$ simple pole, (d). $z = 0$ simple pole, (e). $z = 0$ essential singularity, $z = i$ simple pole, $z = -i$ simple pole, (f). $z = 1/2$ removable singularity, $z = \pm 1$ simple pole, $z \in \mathbf{Z}\backslash\{-2, -1, 1\}$ and $z = \pm i$ pole of order 2, $z = -2$ pole of order 3.

30.4. In some punctured neighborhood of z_0, $f(z) = g(z)/(z - z_0)^m$, where g is analytic at z_0 and $g(z_0) \neq 0$ by Theorem 29.2. Then $f'(z) = h(z)/(z - z_0)^{m+1}$, where $h(z) = (z - z_0)g'(z) - mg(z)$ is analytic at z_0 and $h(z_0) = -mg(z_0) \neq 0$, so f' has a pole of order $m + 1$ by Theorem 29.2 again.

30.5. The series does not converge in a deleted neighborhood of $z = 0$.

30.6. (a). Simple zeros at $z = 1/(1 - n\pi)$, $n = 0, \pm 1, \pm 2, \cdots$. If we let $z \to 0$ through positive values, then $\sin(1 - z^{-1})$ oscillates between ± 1, and hence at $z = 0$ is an essential singularity, (b). $z = 0$ is a removable singularity, $z = n\pi$, $n = \pm 1, \pm 2, \cdots$ are simple zeros, and $z = (2n+1)/2$, $n = 0, \pm 1, \pm 2, \cdots$ are simple poles.

30.7. $\cos(1/z) = \sum_{j=0}^{\infty} (-1)^j z^{-2j}/(2j)!$, $0 < |z| < \infty$ has an essential singularity at 0. We want to show that for every complex value w, except possibly one, there is a sequence $z_n \to 0$, $z_n \neq 0$ such that $\cos(1/z_n) = w$.

$\cos^{-1} w = -i \log[w + i(1 - w^2)^{1/2}] = -i\text{Log}[w + i(1 - w^2)^{1/2}] + 2n\pi, \ n \in \mathbf{Z}$
is defined for all $w \in \mathbf{C}$ since the quantity inside the brackets is never zero.
Fix w and set $\zeta_n = -i\text{Log}[w + i(1 - w^2)^{1/2}] + 2n\pi$. Then $|\zeta_n| \to \infty$. So
we can take $z_n = 1/\zeta_n$. The result for $\cosh(1/z) = \cos(1/(iz))$ also follows
from this.

30.8. $f(z) = \frac{a_{-1}}{z - z_0} + h(z)$, where $h(z)$ is analytic in $|z| < R$. Thus, $h(z) = \sum_{j=0}^{\infty} b_j z^j$ in $|z| < R$. Clearly, $b_j \to 0$. For $|z| < 1$, we have $\sum_{j=0}^{\infty} a_j z^j = -(a_{-1}/z_0) \sum_{j=0}^{\infty} (z/z_0)^j + \sum_{j=0}^{\infty} b_j z^j$. Thus, $\lim_{j \to \infty} \frac{a_j}{a_{j+1}}$

$= \lim_{j \to \infty} \left[-\frac{a_{-1}}{z_0^{j+1}} + b_j \right] / \left[-\frac{a_{-1}}{z_0^{j+2}} + b_{j+1} \right] = z_0$.

30.9. (a). Zero of order 8, (b). simple poles at $(2k+1)\pi i$, $k = 0, \pm 1, \pm 2, \cdots$
and ∞ is a limit point of poles, (c). simple pole, (d). essential singularity.

30.10. If $f(z)$ is not a polynomial, then it has an essential singularity at ∞.
But, then from Theorem 30.1, $\lim_{z \to \infty} |(f \circ g)(z)| \neq \infty$, which contradicts
that $(f \circ g)(z)$ is a polynomial. A similar argument holds if $g(z)$ is not a
polynomial.

30.11. Let $f(z)$ be such a function. Since it is analytic at ∞, it is bounded
for $|z| > M$. But then, by continuity, $f(z)$ is also bounded for $|z| \leq M$.
Hence, $f(z)$ is a bounded entire function. By Theorem 18.6, $f(z)$ must be
a constant.

Lecture 31
Cauchy's Residue Theorem

In this lecture, we shall use Laurent's expansion to establish Cauchy's Residue Theorem, which has far-reaching applications. In particular, it generalizes Cauchy's integral formula for derivatives (18.5), so that integrals that have a finite number of isolated singularities inside a contour can be integrated rather easily.

Suppose that the function $f(z)$ is analytic in a punctured neighborhood $N(z_0)$ of an isolated singular point z_0, so that it can be expanded in a Laurent series $f(z) = \sum_{j=-\infty}^{\infty} a_j(z - z_0)^j$. The coefficient a_{-1} is called the *residue* of $f(z)$ at z_0 and is denoted by $R[f, z_0]$.

Example 31.1. From Examples 25.1-25.3, we have

$$R\left[e^{1/z}, 0\right] = 1, \quad R\left[\frac{z^2 - 2z + 7}{z - 2}, 1\right] = 7, \quad R\left[\frac{e^{2z}}{(z-1)^3}, 1\right] = 2e^2.$$

Example 31.2. Clearly,

$$R\left[e^{z+1/z}, 0\right] = (\text{coefficient of } 1/z)\left(\sum_{j=0}^{\infty} \frac{z^j}{j!}\right)\left(\sum_{j=0}^{\infty} \frac{1}{j!z^j}\right) = \sum_{j=0}^{\infty} \frac{1}{j!(j+1)!}.$$

If $f(z)$ has a removable singularity at z_0, all the coefficients of the negative powers of $(z-z_0)$ in its Laurent expansion are zero, and so, in particular, the residue at z_0 is zero.

Example 31.3. From Example 29.4, it is clear that

$$R\left[\frac{\sin z}{z}, 0\right] = 0, \quad R\left[\frac{\cos z - 1}{z}, 0\right] = 0, \quad R\left[\frac{z^3 - 1}{z - 1}, 1\right] = 0.$$

If $f(z)$ has a pole of order $m \geq 1$ at z_0, then the residue $R[f, z_0]$ can be calculated with the help of the following theorem.

Theorem 31.1. If $f(z)$ has a pole of order m at z_0, then

$$R[f, z_0] = \lim_{z \to z_0} \frac{1}{(m-1)!} \frac{d^{m-1}}{dz^{m-1}}\left[(z - z_0)^m f(z)\right]. \qquad (31.1)$$

Proof. The Laurent series for $f(z)$ around z_0 is given by

$$f(z) = \frac{a_{-m}}{(z-z_0)^m} + \cdots + \frac{a_{-2}}{(z-z_0)^2} + \frac{a_{-1}}{(z-z_0)} + a_0 + a_1(z-z_0) + \cdots.$$

Multiplying the series by $(z-z_0)^m$, we get

$$(z-z_0)^m f(z) = a_{-m} + a_{-m+1}(z-z_0) + \cdots + a_{-2}(z-z_0)^{m-2}$$
$$+a_{-1}(z-z_0)^{m-1} + a_0(z-z_0)^m + a_1(z-z_0)^{m+1} + \cdots.$$

Differentiating $m-1$ times, we obtain

$$\frac{d^{m-1}}{dz^{m-1}}[(z-z_0)^m f(z)] = (m-1)!a_{-1} + m!a_0(z-z_0)$$
$$+\frac{(m+1)!}{2}a_1(z-z_0)^2 + \cdots.$$

Hence, it follows that

$$\lim_{z\to z_0} \frac{d^{m-1}}{dz^{m-1}}[(z-z_0)^m f(z)] = (m-1)!a_{-1},$$

which is the same as (31.1). ∎

Corollary 31.1. If $f(z)$ has a simple pole at z_0, then

$$R[f, z_0] = \lim_{z\to z_0}(z-z_0)f(z). \tag{31.2}$$

Corollary 31.2. If $f(z) = P(z)/Q(z)$, where the functions $P(z)$ and $Q(z)$ are both analytic at z_0 and $Q(z)$ has a simple zero at z_0, while $P(z_0) \neq 0$, then

$$R[f, z_0] = \frac{P(z_0)}{Q'(z_0)}. \tag{31.3}$$

Proof. Since $Q(z_0) = 0$ and $f(z)$ has a simple pole at z_0, it follows that

$$R[f, z_0] = \lim_{z\to z_0}(z-z_0)\frac{P(z)}{Q(z)} = \lim_{z\to z_0}\frac{P(z)}{\left[\frac{Q(z)-Q(z_0)}{z-z_0}\right]} = \frac{P(z_0)}{Q'(z_0)}.$$

Corollary 31.3. If $f(z)$ has a simple pole at z_0 and $g(z)$ is analytic at z_0, then

$$R[fg, z_0] = g(z_0)R[f, z_0]. \tag{31.4}$$

Example 31.4. Find the residue at each of the singularities of $f(z) = e^{z^3}/[z(z+1)]$. The function $f(z)$ has simple poles at $z = 0$ and $z = -1$. Therefore, from (31.2), we have

$$R[f, 0] = \lim_{z\to 0} z f(z) = \lim_{z\to 0}\frac{e^{z^3}}{z+1} = 1.$$

and

$$R[f, -1] = \lim_{z \to -1} (z+1)f(z) = \lim_{z \to -1} \frac{e^{z^3}}{z} = -e^{-1}.$$

Example 31.5. Find the residue at each of the singularities of $f(z) = z/(z^n - 1)$. The function $f(z)$ has simple poles at $z_k = e^{2i\pi k/n}$, $k = 0, 1, \cdots, n-1$. Therefore, from (31.3) and the fact that $z_k^n = 1$, we have

$$R\left[\frac{z}{z^n - 1}, z_k\right] = \frac{z_k}{nz_k^{n-1}} = \frac{1}{n}z_k^2 = \frac{1}{n}e^{4i\pi k/n}.$$

Example 31.6. Find the residue at each of the singularities of $f(z) = \cot z$. The function $\cot z = \cos z / \sin z$ has simple poles at $z = k\pi$, $k = 0, \pm 1, \pm 2, \cdots$. Using (31.3), we get

$$R[\cot z, k\pi] = \left.\frac{\cos z}{(\sin z)'}\right|_{z=k\pi} = \frac{\cos k\pi}{\cos k\pi} = 1.$$

Example 31.7. Find the residue of $f(z) = (\cot z)/z^2$ at $z = 0$. Since $f(z)$ has a pole of order 3 at $z = 0$, from (31.1) and L'Hôpital's Rule (Theorem 6.3), we have

$$R\left[\frac{\cot z}{z^2}, 0\right] = \frac{1}{2!}\lim_{z \to 0}\frac{d^2}{dz^2}\left[z^3\frac{\cot z}{z^2}\right] = \frac{1}{2!}\lim_{z \to 0}\frac{d^2}{dz^2}[z\cot z]$$

$$= \frac{1}{2}\lim_{z \to 0}\frac{d}{dz}[\cot z - z\mathrm{cosec}^2 z] = \frac{1}{2}\lim_{z \to 0}[-2\mathrm{cosec}^2 z + 2z\mathrm{cosec}^2 z \cot z]$$

$$= \lim_{z \to 0}\frac{z\cos z - \sin z}{\sin^3 z} = \lim_{z \to 0}\frac{-z\sin z}{3\sin^2 z \cos z}$$

$$= -\frac{1}{3}\lim_{z \to 0}\frac{z}{\sin z}\lim_{z \to 0}\frac{1}{\cos z} = -\frac{1}{3}.$$

Now we shall state and prove the main theorem of this lecture.

Theorem 31.2 (Cauchy's Residue Theorem). If γ is a positively oriented simple closed contour and f is analytic inside and on γ except at the points $z_1, z_2, \cdots, z_n$ inside γ, then

$$\int_\gamma f(z)dz = 2\pi i \sum_{j=1}^n R[f, z_j]. \tag{31.5}$$

Proof. Let γ_j, $j = 1, 2, \cdots, n$ be positively oriented circles centered at z_j, $j = 1, 2, \cdots, n$ respectively. Furthermore, suppose that each γ_j has a radius r_j so small that γ_j, $j = 1, 2, \cdots, n$ are mutually disjoint and lie interior to γ (see Figure 16.4). Then, from Theorem 16.2, the relation (16.1) holds. At z_j the function $f(z)$ has a Laurent series expansion

$$f(z) = \sum_{k=-\infty}^{\infty} a_k(z - z_j)^k,$$

which is valid for all z on a small, positively oriented circle γ_j centered at z_j. But then, it follows that

$$\int_{\gamma_j} f(z)dz = \int_{\gamma_j} \sum_{k=-\infty}^{\infty} a_k(z - z_j)^k dz$$

$$= \sum_{k=-\infty}^{\infty} a_k \int_{\gamma_j} (z - z_j)^k dz = 2\pi i a_{-1} = 2\pi i R[f, z_j].$$

Substituting this in (16.1), the relation (31.5) follows. ∎

Example 31.8. Let γ be any positively oriented, simple, closed contour containing only the isolated singularity at $z = 0$ of the functions considered below. Then, from Examples 31.1-31.4, 31.6 and 31.7, it follows that

$$\int_{\gamma} e^{1/z}dz = 2\pi i, \quad \int_{\gamma} e^{z+1/z}dz = 2\pi i \sum_{j=0}^{\infty} \frac{1}{j!(j+1)!}, \quad \int_{\gamma} \frac{\sin z}{z} = 0,$$

$$\int_{\gamma} \frac{e^{z^3}}{z(z+1)}dz = 2\pi i, \quad \int_{\gamma} \cot z\, dz = 2\pi i, \quad \int_{\gamma} \frac{\cot z}{z^2}dz = -\frac{2}{3}\pi i.$$

Example 31.9. Evaluate $\displaystyle\int_{\gamma} dz/(z^4 + 1)$, where γ is the following contour.

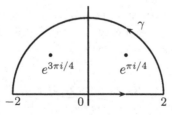

Figure 31.1

The singularities of the integrand occur at the fourth roots of -1; i.e., $e^{\pi i/4}$, $e^{3\pi i/4}$, $e^{5\pi i/4}$, and $e^{7\pi i/4}$. However, only $e^{\pi i/4}$ and $e^{3\pi i/4}$ are inside the contour. Hence, by Theorem 31.2, we have

$$\int_\gamma \frac{dz}{z^4+1} = 2\pi i \left(R\left[\tfrac{1}{z^4+1}, e^{\pi i/4}\right] + R\left[\tfrac{1}{z^4+1}, e^{3\pi i/4}\right] \right)$$

$$= 2\pi i \left(\frac{1}{4(e^{\pi i/4})^3} + \frac{1}{4(e^{3\pi i/4})^3} \right)$$

$$= 2\pi i \left(-\frac{e^{\pi i/4}}{4} + \frac{e^{-\pi i/4}}{4} \right) = \frac{\pi}{\sqrt{2}}.$$

Now suppose that $f(z)$ is analytic in a deleted neighborhood of $z = \infty$, so that $f(z) = \sum_{j=-\infty}^{\infty} a_j z^j$ holds for $R < |z| < \infty$. Let $\gamma_\rho : |z| = \rho > R$. From the definition of a curve positively oriented surrounding infinity (Lecture 12), it follows that

$$\int_{\gamma_\rho} f(z)dz = \int_{\gamma_\rho} \frac{a_{-1}}{z} dz = ia_{-1} \int_{2\pi}^{0} d\theta = -2\pi i a_{-1}.$$

This suggests that $R[f, \infty] = -a_{-1}$. It is interesting to note that at ∞ the residue of a function can be nonzero even when ∞ is a regular point. For example, for the function $f(z) = 1/z$, ∞ is a regular point, in fact, a simple zero, but $R[1/z, \infty] = -1$.

Theorem 31.3. Let the function $f(z)$ be analytic in the extended complex plane, except at isolated singular points. Then, the sum of all residues of $f(z)$ is equal to zero.

Proof. The function $f(z)$ can have only a finite number of singular points; otherwise, the singular points will have a limit point α (maybe at infinity), which will be a nonisolated singular point of $f(z)$. Thus, there exists a positively oriented circle γ_ρ such that all finite singular points $z_1, \cdots, z_n$ of $f(z)$ lie inside γ_ρ. But then, from Theorem 31.2, we have

$$\int_{\gamma_\rho} f(z)dz = 2\pi i \sum_{j=1}^{n} R[f, z_j]. \tag{31.6}$$

Since relative to the point $z = \infty$ the direction of γ_ρ is negative, we also have

$$\int_{\gamma_\rho} f(z)dz = -2\pi i R[f, \infty]. \tag{31.7}$$

Subtracting (31.7) from (31.6), we find

$$\sum_{j=1}^{n} R[f, z_j] + R[f, \infty] = 0. \qquad \blacksquare \tag{31.8}$$

Since rational functions have only isolated singular points (poles), Theorem 31.3 is directly applicable to rational functions.

Example 31.10. Consider $I = \int_\gamma \dfrac{dz}{(z-7)(z^{23}-1)}$, where $\gamma : z = 3e^{i\theta}$, $0 \le \theta \le 2\pi$. Clearly, $I = 2\pi i \sum_{j=1}^{23} R[f, \omega_j]$, where ω_j are the 23 roots of unity. Obviously, this sum is not easy to compute. However, since $R[f, \infty] = 0$ and $R[f, 7] = 1/(7^{23} - 1)$, from (31.8) we get $I = -2\pi i(R[z, \infty] + R[f, 7]) = -2\pi i/(7^{23} - 1)$.

Problems

31.1. For each of the functions in Problem 30.3 (a)-(d), compute $R[f, z_0]$, where z_0 is an isolated singularity of $f(z)$.

31.2. Show that

(a). $R\left[\dfrac{z^{2n}}{(z+1)^n}, -1\right] = (-1)^{n+1}\dfrac{(2n)!}{(n-1)!(n+1)!}$,

(b). $R\left[\cos\dfrac{1}{z}\dfrac{\sin(z-1)}{z^2+1}, i\right] = \dfrac{1}{2}\cosh 1 \sinh(1+i)$,

(c). $R\left[\dfrac{\cos z}{z^2(z-\pi)^3}, 0\right] = -\dfrac{3}{\pi^4}$,

(d). $R\left[\dfrac{\cos z}{z^2(z-\pi)^3}, \pi\right] = -\dfrac{(6-\pi^2)}{2\pi^4}$,

(e). $R\left[z^3\cos\dfrac{1}{z-2}, 2\right] = -\dfrac{143}{24}$,

(f). $R\left[\dfrac{e^{1/z}}{z^2+1}, 0\right] = \sin 1$.

31.3. Show that

(a). $R\left[\dfrac{z^{2n}}{(z+1)^n}, \infty\right] = (-1)^n\dfrac{(2n)!}{(n-1)!(n+1)!}$,

(b). $R\left[z^3\cos\dfrac{1}{z-2}, \infty\right] = \dfrac{143}{24}$.

31.4. Evaluate the following integrals:

(a). $\int_\gamma \dfrac{z^2}{(z-1)^2(z+2)}dz$, where $\gamma : |z| = 3$,

(b). $\int_\gamma \dfrac{z^3}{(z-2)(z-1-i)}dz$, where $\gamma : |z| = 1.5$, (c). $\displaystyle\int_\gamma \frac{dz}{z^4 - 1}$, where γ is the rectangle defined by $x = -0.5$, $x = 2$, $y = -2$, $y = 2$,

(d). $\displaystyle\int_\gamma \frac{z}{(z^2 - 1)^2(z^2 + 1)}dz$, where $\gamma : |z - 1| = \sqrt{3}$,

(e). $\displaystyle\int_\gamma \cos(1/z^2)e^{1/z}dz$, where $\gamma : |z| = 1$,

(f). $\displaystyle\int_\gamma \tan z$, where $\gamma : |z| = 2$.

31.5. Let γ be the circle $|z + 1| = 2$ traversed twice in the clockwise direction. Evaluate $\displaystyle\int_\gamma \frac{dz}{(z^2 + z)^2}$.

31.6. Suppose that $f(z)$ is an even function with an isolated singularity at 0. Show that $R[f, 0] = 0$.

31.7. Suppose that $f(z) = \sum_{j=-\infty}^{\infty} a_j(z-z_0)^j$ and $g(z) = \sum_{j=-\infty}^{\infty} b_j(z-z_0)^j$ are valid in an annulus around z_0. Show that

$$R[fg, z_0] = \sum_{j=-\infty}^{\infty} a_j b_{-j-1}.$$

31.8. Suppose that $f(z)$ and $g(z)$ are analytic functions with zeros of respective orders k and $k + 1$ at z_0. Show that

$$R\left[\frac{f}{g}, z_0\right] = (k + 1)\frac{f^{(k)}(z_0)}{g^{(k+1)}(z_0)}.$$

31.9. Let α be a nonzero fixed complex number such that $|\alpha| = a < 1$. Show that

$$\int_\gamma \frac{|dz|}{|z - \alpha|^2} = \frac{2\pi}{1 - a^2},$$

where γ is the unit circle $|z| = 1$.

31.10. Find all functions $f(z)$ having the following properties: $f(z)$ has a double pole with residue 3 at $z = 0$; $f(z)$ has a simple pole with residue 7 at $z = 1$; and $f(z)$ is bounded in a neighborhood of $z = \infty$.

Answers or Hints

31.1. (a). $R[f, 0] = \frac{d}{dz}\frac{z^3+1}{z-1} = -1$, $R[f, 1] = 2$, (b). $R[f, 0] = 1/3!$,
(c). $R[f, (2k + 1)\pi/2] = \lim_{z\to(2k+1)\pi/2} \frac{z-(2k+1)\pi/2}{\cos z} = \frac{-(2k+1)\pi/2}{\sin(2k+1)\pi/2}$,

(d). $R[f, 0] = 3$.

31.2. For (a)-(d), use Theorem 31.2; for (e) and (f), expand the functions in Laurent's series.

31.3. Use the definition.

31.4. (a). $2\pi i$, (b). $4\pi i$, (c). $\pi i/2$, (d). $\pi i/4$, (e). $2\pi i$, (f). $-4\pi i$.

31.5. $\int_\gamma \frac{dz}{(z^2+z)^2} = -2 \int_\Gamma \frac{dz}{z^2(z+1)^2}$ where Γ is the circle $|z+1| = 2$ anticlockwise, so $-4\pi i(\text{Res}(0) + \text{Res}(-1)) = 0$.

31.6. Use Problem 25.7.

31.7. Use the Cauchy product.

31.8. $f(z) = (z - z_0)^k \phi(z)$, $\phi(z_0) \neq 0$, $g(z) = (z - z_0)^{k+1} \psi(z)$, $\psi(z_0) \neq 0$. Thus, $R[f/g, z_0] = \lim_{z \to z_0} (z - z_0)[(z - z_0)^k \phi(z)/(z - z_0)^{k+1} \psi(z)] = \phi(z_0)/\psi(z_0)$. Now use $\phi(z_0) = f^{(k)}(z_0)/k!$ and $\psi(z_0) = g^{(k+1)}(z_0)/(k + 1)!$.

31.9. $|dz| = |ie^{i\theta}|d\theta = d\theta = dz/(iz)$ and $|z - \alpha|^2 = |z|^2 - \alpha\bar{z} - \bar{\alpha}z + |\alpha|^2 = 1 - \alpha\bar{z} - \bar{\alpha}z + a^2$. Hence, $\int_\gamma \frac{|dz|}{|z-\alpha|^2} = -\frac{1}{i} \int_\gamma \frac{dz}{\bar{\alpha}z^2 - (1+a^2)z + \alpha}$. Now, $\bar{\alpha}z^2 - (1 + a^2)z + \alpha = 0$ gives $z_1 = 1/\bar{\alpha}$, $z_2 = a^2/\bar{\alpha}$. Since $|\alpha| = a < 1$, z_1 lies outside γ, whereas z_2 is inside γ. Finally, note that $R\left[\frac{1}{\bar{\alpha}z^2 - (1+a^2)z + \alpha}, z_2\right] = \frac{1}{a^2-1}$.

31.10. In view of Theorem 18.6, the function $g(z) = f(z) - \frac{3}{z} - \frac{A}{z^2} - \frac{7}{z-1}$ is a constant. Thus, $f(z) = \frac{3}{z} + \frac{A}{z^2} + \frac{7}{z-1} + B$.

Lecture 32
Evaluation of Real Integrals by Contour Integration I

In this lecture and the next, we shall show that the theory of residues can be applied to compute certain types of definite as well as improper real integrals. Some of these integrals occur in physical and engineering applications, and often cannot be evaluated by using the methods of calculus.

First we shall apply Cauchy's Residue Theorem to evaluate definite real integrals of the form

$$\int_0^{2\pi} R(\cos\theta, \sin\theta)d\theta, \qquad (32.1)$$

where $R(\cos\theta, \sin\theta)$ is a rational function with real coefficients of $\cos\theta$ and $\sin\theta$ and whose denominator does not vanish on $[0, 2\pi]$. We shall transform (32.1) into a simple closed contour integral. For this, we let γ be the positively oriented unit circle $|z| = 1$ parameterized by $z = e^{i\theta}$. Thus, $1/z = e^{-i\theta}$. Since $\cos\theta = (e^{i\theta} + e^{-i\theta})/2$ and $\sin\theta = (e^{i\theta} - e^{-i\theta})/(2i)$, we have

$$\cos\theta = \frac{1}{2}\left(z + \frac{1}{z}\right) \quad \text{and} \quad \sin\theta = \frac{1}{2i}\left(z - \frac{1}{z}\right). \qquad (32.2)$$

Also, differentiating $z = e^{i\theta}$ along γ, we find

$$\frac{dz}{d\theta} = ie^{i\theta} = iz,$$

and hence

$$d\theta = \frac{dz}{iz} = -i\frac{dz}{z}.$$

Therefore, the integral (32.1) can be written as the contour integral

$$\int_0^{2\pi} R(\cos\theta, \sin\theta)d\theta = \int_\gamma R\left(\frac{z + z^{-1}}{2}, \frac{z - z^{-1}}{2i}\right)\frac{dz}{iz}. \qquad (32.3)$$

If in (32.1) the function R involves $\cos n\theta$ and $\sin n\theta$, then we can directly use

$$\cos n\theta = \frac{1}{2}\left(z^n + \frac{1}{z^n}\right) \quad \text{and} \quad \sin n\theta = \frac{1}{2i}\left(z^n - \frac{1}{z^n}\right), \qquad (32.4)$$

which follow from De Moivre's formula (3.4).

Example 32.1. Evaluate $I = \int_0^{2\pi} \dfrac{1}{1 + a\sin\theta} d\theta$, $0 < |a| < 1$. This integral can be transformed into a contour integral:

$$I = \int_\gamma \frac{1}{1 + a(z - z^{-1})/2i} \frac{dz}{iz} = \int_\gamma \frac{2}{az^2 + 2iz - a} dz$$

$$= \frac{2}{a} \int_\gamma \frac{1}{[z + i(1 + \sqrt{1 - a^2})/a][z + i(1 - \sqrt{1 - a^2})/a]} dz.$$

Since $|-i(1 + \sqrt{1 - a^2})/a| > 1$ and $|-i(1 - \sqrt{1 - a^2})/a| < 1$, it follows that

$$I = \frac{4\pi i}{a} R\left[\frac{1}{[z+i(1+\sqrt{1-a^2})/a][z+i(1-\sqrt{1-a^2})/a]}, -i(1-\sqrt{1-a^2})/a\right]$$

$$= \frac{4\pi i}{a} \frac{1}{[-i(1 - \sqrt{1 - a^2})/a + i(1 + \sqrt{1 - a^2})/a]} = \frac{2\pi}{\sqrt{1 - a^2}}.$$

Similarly, we have

$$I = \int_0^{2\pi} \frac{1}{1 + a\cos\theta} d\theta = \frac{2\pi}{\sqrt{1 - a^2}}, \quad 0 < |a| < 1.$$

Example 32.2. Evaluate $I = \int_0^\pi \dfrac{d\theta}{(a + \cos\theta)^2}$, $a > 1$. Since the integrand is symmetric in $[0, 2\pi]$ about $\theta = \pi$, it is clear that

$$I = \frac{1}{2} \int_0^{2\pi} \frac{d\theta}{(a + \cos\theta)^2}.$$

We transform this integral into a contour integral to obtain

$$I = \frac{1}{2} \int_\gamma \frac{4z^2}{(z^2 + 2az + 1)^2} \frac{dz}{iz} = \frac{2}{i} \int_\gamma \frac{z\,dz}{(z - \alpha)^2(z - \beta)^2},$$

where $\alpha = -a + \sqrt{a^2 - 1}$ and $\beta = -a - \sqrt{a^2 - 1}$. Thus, the integrand has two double poles at α and β, of which only α is inside γ. Hence, it follows that

$$I = \frac{4\pi i}{i} R\left[\frac{z}{(z - \alpha)^2(z - \beta)^2}, \alpha\right] = 4\pi \lim_{z \to \alpha} \frac{d}{dz}\left(\frac{z}{(z - \beta)^2}\right)$$

$$= -4\pi \frac{\alpha + \beta}{(\alpha - \beta)^3} = 4\pi \frac{a}{4(a^2 - 1)^{3/2}} = \frac{\pi a}{(a^2 - 1)^{3/2}}.$$

Next, we shall evaluate integrals of certain functions over infinite intervals. If $f(x)$ is a function continuous on the nonnegative real axis

$0 \leq x < \infty$, then the improper integral of $f(x)$ over $[0, \infty)$ is defined by

$$I_1 = \int_0^\infty f(x)dx = \lim_{b \to \infty} \int_0^b f(x)dx. \tag{32.5}$$

If the limit exists, the integral I_1 is said to be *convergent*; otherwise it is *divergent*. Similarly, the improper integral of $f(x)$ over $(-\infty, 0]$ is defined as

$$I_2 = \int_{-\infty}^0 f(x)dx = \lim_{c \to \infty} \int_{-c}^0 f(x)dx. \tag{32.6}$$

If $f(x)$ is continuous for all x, the improper integral of $f(x)$ over $(-\infty, \infty)$ is defined by

$$I = \int_{-\infty}^\infty f(x)dx = \lim_{c \to \infty} \int_{-c}^0 f(x)dx + \lim_{b \to \infty} \int_0^b f(x)dx = I_2 + I_1 \tag{32.7}$$

provided both the limits in (32.7) exist.

The *Cauchy principal value* (p.v.) of I is defined to be the number

$$\text{p.v.} \int_{-\infty}^\infty f(x)dx = \lim_{\rho \to \infty} \int_{-\rho}^\rho f(x)dx \tag{32.8}$$

provided this limit exists. Cauchy's principal value of an integral may exist even though the integral itself is not convergent. For example, $\int_{-\rho}^\rho x dx = 0$ for all ρ, and hence p.v. $\int_{-\infty}^\infty x dx = 0$, but $\int_0^\infty x dx = \infty$. However, if the improper integral I exists, then clearly it must be equal to its principal value (p.v.). We also note that if $f(x)$ is an even function; i.e., $f(-x) = f(x)$ for all real numbers x, and if the Cauchy principal value of I exists, then I exists. Moreover, we have

$$\frac{1}{2} \text{ p.v.} \int_{-\infty}^\infty f(x)dx = \int_0^\infty f(x)dx. \tag{32.9}$$

In the following theorem, we use residue theory to compute the Cauchy principal value of I for a certain class of functions $f(x)$.

Theorem 32.1. Let $f(z)$ be a rational function having no real poles, and there exists μ such that for any $|z| > \mu$, $|f(z)| \leq M/|z|^2$ for some $M > 0$. Then,

$$\text{p.v.} \int_{-\infty}^\infty f(x)dx = 2\pi i \sum R[f, z_j], \tag{32.10}$$

holds, where the sum is taken over all the poles z_j of $f(z)$ that fall in the upper half-plane.

Proof. Suppose $\rho > \mu$ is so large that the semicircle γ_ρ^+ contains all the poles of $f(z)$ that are in the upper half-plane. Then, on $\Gamma_\rho = [-\rho, \rho] \cup \{\gamma_\rho^+\}$, by Cauchy's Residue Theorem, we have

$$\int_{\Gamma_\rho} f(z)dz = \int_{-\rho}^{\rho} f(x)dx + \int_{\gamma_\rho^+} f(z)dz = 2\pi i \sum R[f, z_j]. \qquad (32.11)$$

Clearly, the right-hand side of (32.11) is independent of ρ, while for the left-hand side we find

$$\lim_{\rho \to \infty} \int_{-\rho}^{\rho} f(x)dx = \int_{-\infty}^{\infty} f(x)dx$$

and

$$\left| \int_{\gamma_\rho^+} f(z)dz \right| = \left| \int_0^\pi f(\rho e^{i\theta}) \rho i e^{i\theta} d\theta \right| \leq \int_0^\pi |f(\rho e^{i\theta})| \rho d\theta \leq \frac{M}{\rho^2} \pi \rho = \frac{M\pi}{\rho}$$

so that $\lim_{\rho \to \infty} \int_{\gamma_\rho^+} f(z)dz = 0$. Hence, (32.10) follows. ∎

Corollary 32.1. If $f(z) = P(z)/Q(z)$ is a rational function such that $P(z)$ and $Q(z)$ do not have a common zero, $Q(z)$ has no zeros on the real line, and degree $Q \geq 2 +$ degree P, then $\lim_{\rho \to \infty} \int_{\gamma_\rho^+} f(z)dz = 0$, where γ_ρ^+ is the upper half-circle of radius ρ.

Remark 32.1. In the proof of Theorem 32.1, the upper semicircle is used only for convenience. One may as well use the lower semicircle or any other suitable contour having the real interval $[-\rho, \rho]$ as one of its parts.

Example 32.3. Evaluate

$$I = \text{p.v.} \int_{-\infty}^{\infty} \frac{dx}{x^4 + a^4}, \quad a > 0.$$

Clearly, the function $f(z) = 1/(z^4 + a^4)$ satisfies the conditions of Corollary 32.1. Its singular points in the upper half-plane are the simple poles at $z_1 = ae^{\pi i/4}$ and $z_2 = ae^{3\pi i/4}$. Thus, from (32.10), it follows that

$$\begin{aligned} I &= 2\pi i \left(R\left[\frac{1}{z^4 + a^4}, z_1 \right] + R\left[\frac{1}{z^4 + a^4}, z_2 \right] \right) \\ &= 2\pi i \left[\frac{1}{4z_1^3} + \frac{1}{4z_2^3} \right] = \frac{2\pi i}{4a^3} \left(e^{-3\pi i/4} + e^{-9\pi i/4} \right) \\ &= \frac{2\pi i}{4a^3} \left(-e^{\pi i/4} + e^{-\pi i/4} \right) = \frac{\pi}{\sqrt{2}a^3}. \end{aligned}$$

Example 32.4. Evaluate

$$I = \int_0^\infty \frac{dx}{(x^2 + a^2)^2}, \quad a > 0.$$

Note that the integral is even and that the function $1/(z^2+a^2)^2$ has a double pole at ai and $-ai$, and hence, from Corollary 32.1, (32.9) and (32.10), it follows that

$$I = \frac{1}{2} \text{ p.v. } \int_{-\infty}^{\infty} \frac{dx}{(x^2 + a^2)^2} = \frac{1}{2}(2\pi i)R[f, ai]$$

$$= \pi i \lim_{z \to ai} \frac{d}{dz} \frac{1}{(z + ai)^2} = \pi i \frac{-2}{(2ai)^3} = \frac{\pi}{4a^3}.$$

Lecture 33
Evaluation of Real Integrals by Contour Integration II

We begin this lecture with two examples where we need to use appropriate contours to evaluate certain proper and improper real integrals. We shall also prove Jordan's Lemma, which plays a fundamental role in the computation of integrals involving rational functions multiplied by trigonometric functions.

Example 33.1. We shall derive the *Fresnel integrals*

$$\int_0^\infty \cos x^2 dx = \int_0^\infty \sin x^2 dx = \frac{1}{2}\sqrt{\frac{\pi}{2}}, \qquad (33.1)$$

which occur in the theory of diffraction. For this, we need the inequality

$$\frac{2}{\pi}\theta < \sin\theta, \quad 0 < \theta < \frac{\pi}{2}, \qquad (33.2)$$

and the integral

$$L = \int_0^\infty e^{-x^2} dx = \frac{\sqrt{\pi}}{2}. \qquad (33.3)$$

To show (33.2), we consider the function $f(\theta) = \sin\theta - 2\theta/\pi$, $\theta \in [0, \pi/2]$. Since $f''(\theta) = -\sin\theta$, $\min_{0\leq\theta\leq\pi/2} f(\theta) = f(0) = f(\pi/2) = 0$. To establish (33.3), we note first that

$$L^2 = \left(\int_0^\infty e^{-x^2} dx\right)\left(\int_0^\infty e^{-y^2} dy\right) = \int_0^\infty \int_0^\infty e^{-(x^2+y^2)} dx dy$$

and then use the polar coordinates $x = r\cos\phi$, $y = r\sin\phi$, to obtain

$$L^2 = \int_0^{\pi/2}\int_0^\infty e^{-r^2} r dr d\theta = \left[-\frac{1}{2}e^{-r^2}\right]_0^\infty \times \frac{\pi}{2} = \frac{\pi}{4}.$$

Now, to establish the integrals in (33.1), we consider the entire function $f(z) = e^{iz^2}$ and the contour $\gamma = OA + AB + BO$ as in Figure 33.1, so that, by Theorem 15.2, we have

$$\int_\gamma e^{iz^2} dz = \int_{OA} e^{iz^2} dz + \int_{AB} e^{iz^2} dz + \int_{BO} e^{iz^2} dz = 0.$$

Clearly, $z = x$, $0 \leq x \leq \rho$ on the line OA; $z = \rho e^{i\theta}$, $0 \leq \theta \leq \pi/4$ on the arc AB; and $z = re^{\pi i/4}$, $0 \leq r \leq \rho$ on the line BO. Hence it follows that

$$\int_0^\rho e^{ix^2} dx + \int_0^{\pi/4} \exp\left(i\rho^2 e^{2i\theta}\right) \rho i e^{i\theta} d\theta + \int_\rho^0 \exp\left(ir^2 e^{\pi i/2}\right) e^{\pi i/4} dr = 0.$$

$$(33.4)$$

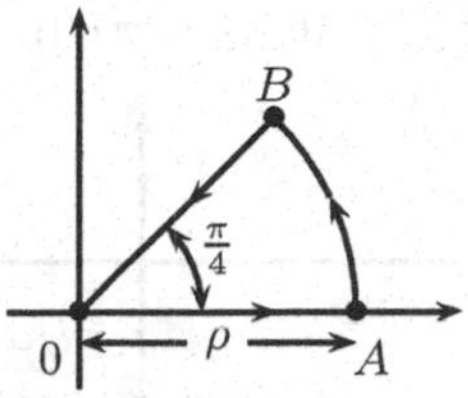

Figure 33.1

Now, in view of (33.2), we have

$$\left| \int_0^{\pi/4} \exp\left(i\rho^2 e^{2i\theta}\right) \rho i e^{i\theta} d\theta \right| \leq \rho \int_0^{\pi/4} \left| \exp[i\rho^2(\cos 2\theta + i \sin 2\theta)] \right| d\theta$$

$$\leq \rho \int_0^{\pi/4} e^{-\rho^2 \sin 2\theta} d\theta \leq \rho \int_0^{\pi/4} e^{-\rho^2(4\theta/\pi)} d\theta$$

$$= \frac{\pi}{4} \frac{1 - e^{-\rho^2}}{\rho} \to 0 \quad \text{as} \quad \rho \to \infty,$$

$$(33.5)$$

whereas (33.3) gives

$$\lim_{\rho \to \infty} \int_\rho^0 \exp\left(ir^2 e^{\pi i/2}\right) e^{\pi i/4} dr = -e^{\pi i/4} \int_0^\infty e^{-r^2} dr = -\frac{1+i}{\sqrt{2}} \frac{\sqrt{\pi}}{2}.$$

$$(33.6)$$

Combining (33.4)-(33.6), we get

$$\lim_{\rho \to \infty} \int_0^\rho e^{ix^2} dx = \int_0^\infty (\cos x^2 + i \sin x^2) dx = \frac{1+i}{\sqrt{2}} \frac{\sqrt{\pi}}{2},$$

which on comparing the real and imaginary parts gives (33.1).

Example 33.2. We shall show that

$$I = \int_0^\infty \frac{1}{t^\alpha + 1} dt = \frac{\pi}{\alpha \sin(\pi/\alpha)}, \quad \alpha > 1. \qquad (33.7)$$

For this, we use the substitution $t = e^x$, so that

$$I = \text{p.v.} \int_{-\infty}^\infty \frac{e^x}{e^{\alpha x} + 1} dx. \qquad (33.8)$$

The function $f(z) = e^z/(e^{\alpha z} + 1)$ has an infinite number of singularities on both the upper and lower half-planes at the points $z_k = (2k+1)\pi i/\alpha$, $k = 0, \pm 1, \pm 2, \cdots$. Thus, to compute the integral in (33.8), as in Examples 32.3 and 32.4, semicircles cannot be used. Therefore, we use a rectangular contour $\Gamma_\rho = \gamma_1 + \gamma_2 + \gamma_3 + \gamma_4$ (see Figure 33.2), which as $\rho \to \infty$ expands in the horizontal direction with the fixed vertical length $2\pi/\alpha$, and contains only one singularity of $f(z)$ at $z_0 = \pi i/\alpha$.

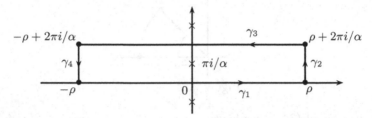

Figure 33.2

By Cauchy's Residue Theorem, we have

$$\int_{\Gamma_\rho} f(z)dz = 2\pi i R[f, \pi i/\alpha] = 2\pi i \left[\frac{e^z}{\frac{d}{dz}(e^{\alpha z} + 1)} \right]\Bigg|_{z = \pi i/\alpha} = -2\pi i \frac{e^{\pi i/\alpha}}{\alpha}.$$

$$(33.9)$$

We also have

$$\int_{\Gamma_\rho} f(z)dz = \int_{\gamma_1} f(z)dz + \int_{\gamma_2} f(z)dz + \int_{\gamma_3} f(z)dz + \int_{\gamma_4} f(z)dz$$

$$= \int_{-\rho}^{\rho} f(x)dx + \int_0^{2\pi/\alpha} f(\rho + iy)i dy$$

$$+ \int_\rho^{-\rho} f(x + 2\pi i/\alpha)dx + \int_{2\pi/\alpha}^0 f(-\rho + iy)i dy.$$

$$(33.10)$$

Now, we note that

$$\int_\rho^{-\rho} f(x + 2\pi i/\alpha)dx = \int_\rho^{-\rho} \frac{e^{x + 2\pi i/\alpha}}{e^{\alpha x + 2\pi i} + 1}dx = -e^{2\pi i/\alpha} \int_{-\rho}^{\rho} \frac{e^x}{e^{\alpha x} + 1}dx,$$

$$(33.11)$$

$$\left| \int_0^{2\pi/\alpha} f(\rho + iy)i dy \right| = \left| \int_0^{2\pi/\alpha} \frac{e^{\rho + iy}}{e^{\alpha(\rho + iy)} + 1}i dy \right| \le \frac{2\pi}{\alpha} \frac{e^\rho}{e^{\alpha\rho} - 1},$$

$$(33.12)$$

and

$$\left| \int_{2\pi/\alpha}^0 f(-\rho + iy)i dy \right| = \left| \int_{2\pi/\alpha}^0 \frac{e^{-\rho + iy}}{e^{\alpha(-\rho + iy)} + 1}i dy \right| \le \frac{2\pi}{\alpha} \frac{e^{-\rho}}{1 - e^{-\alpha\rho}}.$$

$$(33.13)$$

Clearly, the right-hand sides of (33.12) and (33.13) go to 0 as $\rho \to \infty$. Hence, from (33.9)-(33.13), it follows that

$$
\begin{aligned}
-2\pi i \frac{e^{\pi i/\alpha}}{\alpha} &= \lim_{\rho \to \infty} \int_{\Gamma_\rho} f(z)dz \\
&= \int_{-\infty}^{\infty} \frac{e^x}{e^{\alpha x}+1}dx + 0 - e^{2\pi i/\alpha}\int_{-\infty}^{\infty}\frac{e^x}{e^{\alpha x}+1}dx + 0 \\
&= (1 - e^{2\pi i/\alpha})I,
\end{aligned}
$$

which gives

$$
I = -\frac{2\pi i}{\alpha}\frac{e^{\pi i}}{1 - e^{2\pi i/\alpha}} = \frac{\pi}{\alpha}\frac{-2i}{e^{-\pi i/\alpha} - e^{\pi i/\alpha}} = \frac{\pi}{\alpha \sin(\pi/\alpha)}.
$$

We shall now prove the following result.

Lemma 33.1 (Jordan's Lemma). Let the function $f(z)$ be analytic in the upper half-plane with the exception of a finite number of isolated singularities. Furthermore, let $\lim_{\rho \to \infty} \max_{\gamma_\rho} |f(z)| = 0$, where γ_ρ is the semicircle in the upper half-plane. Then, the following holds

$$
\lim_{\rho \to \infty} \int_{\gamma_\rho} e^{iaz}f(z)dz = 0, \quad a > 0. \tag{33.14}
$$

Proof. Let $z = \rho e^{i\theta}$, $0 \le \theta \le \pi$, and $M_\rho = \max_{\gamma_\rho} |f(z)|$. Then, from the inequality (33.2), we have

$$
\begin{aligned}
\left|\int_{\gamma_\rho} e^{iaz}f(z)dz\right| &\le \int_0^\pi \left|e^{(ia\rho[\cos\theta + i\sin\theta])}\right||f(z)|\rho d\theta \le \rho M_\rho \int_0^\pi e^{-a\rho\sin\theta}d\theta \\
&= 2\rho M_\rho \int_0^{\pi/2} e^{-a\rho\sin\theta}d\theta \le 2\rho M_\rho \int_0^{\pi/2} e^{-2a\rho\theta/\pi}d\theta \\
&= \frac{\pi}{a}M_\rho\left(1 - e^{-a\rho}\right) \to 0 \quad \text{as} \quad \rho \to \infty. \quad \blacksquare
\end{aligned}
$$

Remark 33.1. If $a < 0$ and the function $f(z)$ satisfies the conditions of Lemma 33.1 in the lower half-plane and γ_ρ is the semicircle in the lower half-plane, then (33.14) holds. For $a = \pm i\alpha$, $(\alpha > 0)$, (33.14) also holds; in fact, respectively, we have

$$
\lim_{\rho_1 \to \infty} \int_{\gamma_{\rho_1}} e^{-\alpha z}f(z)dz = 0 \quad \text{and} \quad \lim_{\rho_2 \to \infty} \int_{\gamma_{\rho_2}} e^{\alpha z}f(z)dz = 0, \quad \alpha > 0,
$$

$$
\tag{33.15}
$$

where γ_{ρ_1} is the semicircle in the right half-plane and γ_{ρ_2} is the semicircle in the left half-plane.

Remark 33.2. If the function $f(z)$ satisfies the conditions of Lemma 33.1 in the half-plane $\text{Im } z \geq y_0$, where y_0 is a fixed positive or negative real number, and γ_ρ is the circular arc $|z - iy_0| = \rho$ in the half-plane $\text{Im } z \geq y_0$, then (33.14) also holds.

Theorem 33.1. Let the function $f(z)$ be as in Lemma 33.1 and have no singularities on the real axis. Then, for $a > 0$,

$$\text{p.v.} \int_{-\infty}^{\infty} e^{iax} f(x)dx = 2\pi i \sum R[e^{iaz} f, z_j], \qquad (33.16)$$

holds, where the sum is taken over all the poles z_j of $f(z)$ that fall in the upper half-plane.

Corollary 33.1. If $f(z) = P(z)/Q(z)$ is a rational function such that $P(z)$ and $Q(z)$ do not have a common zero, $Q(z)$ has no zeros on the real line, and degree $Q \geq 1+$ degree P, then (33.16) holds.

Example 33.3. Clearly, the function $f(z) = 1/(z^2 + b^2)$, $b > 0$ satisfies the conditions of Corollary 33.1, and hence, for $a > 0$, we have

$$\text{p.v.} \int_{-\infty}^{\infty} \frac{e^{iax}}{x^2 + b^2} dx = 2\pi i R\left[\frac{e^{iaz}}{z^2 + b^2}, ib\right] = \frac{\pi}{b} e^{-ab}.$$

Therefore, on comparing the real and imaginary parts, it follows that

$$\text{p.v.} \int_{-\infty}^{\infty} \frac{\cos ax}{x^2 + b^2} dx = \frac{\pi}{b} e^{-ab} \quad \text{and} \quad \text{p.v.} \int_{-\infty}^{\infty} \frac{\sin ax}{x^2 + b^2} dx = 0.$$

Similarly, we find

$$\text{p.v.} \int_{-\infty}^{\infty} \frac{x \cos ax}{x^2 + b^2} dx = 0 \quad \text{and} \quad \text{p.v.} \int_{-\infty}^{\infty} \frac{x \sin ax}{x^2 + b^2} dx = \pi e^{-ab}.$$

Problems

33.1. Show that:

(a). $\displaystyle\int_0^{2\pi} \frac{\sin^2 \theta}{a + b \cos \theta} d\theta = \frac{2\pi}{b^2}(a - \sqrt{a^2 - b^2}), \quad a > b > 0.$

(b). $\displaystyle\int_0^{\pi} \frac{\cos 2\theta}{1 + a^2 - 2a \cos \theta} d\theta = \frac{\pi a^2}{1 - a^2}, \quad a^2 < 1.$

(c). $\displaystyle\int_0^{2\pi} \frac{1}{a + b \cos^2 \theta} d\theta = \int_0^{2\pi} \frac{1}{a + b \sin^2 \theta} d\theta = \frac{2\pi}{\sqrt{a}\sqrt{a + b}}, \quad 0 < b < a.$

(d). $\displaystyle\int_0^{2\pi} \frac{1}{a\cos\theta + b\sin\theta + c}\,d\theta = \frac{2\pi}{\sqrt{c^2 - a^2 - b^2}}, \quad a^2 + b^2 < c^2.$

(e). $\displaystyle\int_0^{2\pi} \frac{1}{a\cos^2\theta + b\sin^2\theta + c}\,d\theta = \frac{2\pi}{\sqrt{(a+c)(b+c)}}, \quad 0 \le c < a, \ c < b.$

(f). $\displaystyle\int_0^{2\pi} \frac{1}{(1 + 2a\cos\theta + a^2)^2}\,d\theta = \frac{2\pi(1 + a^2)}{(1 - a^2)^3}, \quad |a| < 1.$

(g). $\displaystyle\int_0^{\pi} \frac{\sin^2\theta}{(1 - 2a\cos\theta + a^2)(1 - 2b\cos\theta + b^2)}\,d\theta = \frac{\pi}{2(1 - ab)}, \quad 0 < a < b < 1.$

(h). $\displaystyle\int_0^{2\pi} \sin^{2n}\theta\,d\theta = \int_0^{2\pi} \cos^{2n}\theta\,d\theta = \frac{2\pi}{4^n}\binom{2n}{n}, \quad n = 1, 2, \cdots.$

33.2. Show that:

(a). p.v. $\displaystyle\int_{-\infty}^{\infty} \frac{dx}{x^2 + x + 1} = \frac{2\pi}{\sqrt{3}}.$

(b). $\displaystyle\int_0^{\infty} \frac{dx}{(x^2 + a^2)(x^2 + b^2)} = \frac{\pi}{2ab(a + b)}, \quad a > 0, \ b > 0.$

(c). $\displaystyle\int_0^{\infty} \frac{2x^2 - 1}{x^4 + 5x^2 + 4}\,dx = \frac{\pi}{4}.$

(d). p.v. $\displaystyle\int_{-\infty}^{\infty} \frac{x^2}{(x^2 + a^2)^2} = \frac{\pi}{2a}, \quad a > 0.$

(e). $\displaystyle\int_0^{\infty} \frac{dx}{(x^2 + 1)^n} = \frac{1 \cdot 3 \cdot 5 \cdots (2n - 3)}{2 \cdot 4 \cdot 6 \cdots (2n - 2)}\frac{\pi}{2}, \quad n > 1.$

33.3. Let $f(z)$ be a meromorphic function in the upper half-plane having no real poles, and $zf(z) \to 0$ uniformly as $|z| \to \infty$ for $0 \le \arg z \le \pi$. Show that (32.10) holds.

33.4. Derive the *Poisson integral*

$$\text{p.v.} \int_{-\infty}^{\infty} e^{-ax^2} \cos(2abx)\,dx = \sqrt{\frac{\pi}{a}}\, e^{-ab^2}, \quad a > 0, \ b > 0.$$

33.5. Show that

$$\text{p.v.} \int_{-\infty}^{\infty} \frac{e^{\alpha x}}{e^x + 1}\,dx = \frac{\pi}{\sin\alpha\pi}, \quad 0 < \alpha < 1.$$

33.6. Show that

$$\int_0^{\infty} \frac{x}{x^k + 1}\,dx = \frac{\pi}{k\sin(2\pi/k)}, \quad k > 2.$$

33.7. Show that

$$\int_0^\infty \frac{x^{k-1}}{x+1}dx = \int_0^\infty \frac{x^{-k}}{x+1}dx = \frac{\pi}{\sin k\pi}, \quad 0 < k < 1.$$

33.8. With the help of the contour γ_ρ as shown, show that

$$\int_0^\infty \frac{dx}{x^3+1} = \frac{2\pi}{3\sqrt{3}}.$$

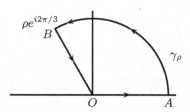

33.9. Let $a > 0$, $b > 0$, $c > 0$ and k be real. Show that:

(a). p.v. $\displaystyle\int_{-\infty}^\infty \frac{\cos ax}{x^4+b^4}dx = \frac{\pi}{\sqrt{2}b^3}e^{-ab/\sqrt{2}}\left(\cos\frac{ab}{\sqrt{2}} + \sin\frac{ab}{\sqrt{2}}\right).$

(b). $\displaystyle\int_0^\infty \frac{\cos ax}{(x^2+b^2)^2}dx = \frac{\pi}{4b^3}e^{-ab}(ab+1).$

(c). p.v. $\displaystyle\int_{-\infty}^\infty \frac{\cos x}{(x^2+b^2)(x^2+c^2)}dx = \frac{\pi}{(b^2-c^2)}\left(\frac{e^{-c}}{c} - \frac{e^{-b}}{b}\right), \quad b > c.$

(d). p.v. $\displaystyle\int_{-\infty}^\infty \frac{\cos ax}{(x+k)^2+b^2}dx = \frac{\pi}{b}e^{-ab}\cos ka.$

33.10. Show that:

(a). p.v. $\displaystyle\int_{-\infty}^\infty \frac{x^3\sin x}{(x^2+1)^2}dx = \frac{\pi}{2e}.$

(b). p.v. $\displaystyle\int_{-\infty}^\infty \frac{x\sin \pi x}{x^2+2x+5}dx = -\frac{\pi}{e^{2\pi}}.$

(c). p.v. $\displaystyle\int_{-\infty}^\infty \frac{\sin x}{x+i}dx = \frac{\pi}{e}.$

(d). p.v. $\displaystyle\int_{-\infty}^\infty \frac{\cos x + x\sin x}{x^2+1}dx = \frac{2\pi}{e}.$

33.11. The *gamma function* is denoted and defined as

$$\Gamma(z) =: \int_0^\infty e^{-t}t^{z-1}dt, \tag{33.17}$$

where $\operatorname{Re} z > 0$ and $t^{z-1} = e^{(z-1)\operatorname{Log} t}$ is the principal branch of t^{z-1}. This integral is improper at both ends, and hence it is interpreted as $\lim_{b \to 0, c \to \infty} \int_b^c e^{-t} t^{z-1} dt$.

(a). Show that $|e^{-t} t^{z-1}| \le e^{-t} t^{\operatorname{Re} z - 1}$, and hence (33.17) converges absolutely for all z such that $\operatorname{Re} z > 0$.

(b). Write $\Gamma(z) = \sum_{j=0}^{\infty} f_j(z)$, where $f_j(z) = \int_j^{j+1} e^{-t} t^{z-1} dt$. Show that each $f_j(z)$ is analytic for all z such that $\operatorname{Re} z > 0$.

(c). For all z such that $|z| \le A$, $0 < \operatorname{Re} z \le B$, show that $|f_j(z)| \le M_j = (j+1)^{B-1}(e^{-j} - e^{-(j+1)})$. Hence, deduce that $\Gamma(z)$ is analytic for all z such that $\operatorname{Re} z > 0$.

(d). Use integration by parts to obtain the recurrence relation $\Gamma(z+1) = z\Gamma(z)$.

(e). Show that $\Gamma(1) = 1$ and, for an integer $n > 1$, $\Gamma(n+1) = n!$. Thus, the gamma function is a generalization of the factorial function.

(f). Show that

$$\Gamma(\alpha)\Gamma(1-\alpha) = \int_0^\infty \int_0^\infty e^{-(s+t)} s^{-\alpha} t^{\alpha-1} ds\, dt = \frac{\pi}{\sin \pi \alpha}, \quad 0 < \alpha < 1.$$

(g). The analytic continuation of $\Gamma(z)$ is

$$\Gamma_1(z) = e^{-\gamma z} z^{-1} \prod_{n=1}^{\infty} \left(1 + \frac{z}{n}\right)^{-1} e^{z/n}, \quad z \ne 0, -1, -2, \cdots;$$

here γ is the *Euler constant* given by

$$\gamma = \lim_{m \to \infty} \left[1 + \frac{1}{2} + \frac{1}{3} + \cdots + \frac{1}{m} - \ln m\right] \simeq 0.5772.$$

Answers or Hints

33.1. Similar to Examples 32.1 and 32.2.
33.2. Similar to Examples 32.3 and 32.4.
33.3. Similar to Theorem 32.1.
33.4. Take a rectangle as a contour and follow Example 33.1.
33.5. Use the substitution $\alpha x = t$ in (33.8).
33.6. In (33.7), use the substitutions $t = x^2$, $2\alpha = k$.
33.7. In (33.7), use the substitutions $t^\alpha = x$, $\alpha = 1/(1-k)$.
33.8. Since $z^3 + 1 = 0$ implies $z = e^{i\pi/3}$, $e^{i\pi}$, $e^{i5\pi/3}$, for $\rho > 1$ only $e^{i\pi/3}$ is enclosed by γ_ρ. Hence, $\int_{\gamma_\rho} \frac{dz}{z^3+1} dz = 2\pi i R[1/(z^3+1), e^{i\pi/3}] =$

$2\pi i \frac{1}{3e^{i2\pi/3}}$. Next, note that $\left| \int_{AB} \frac{dz}{z^3+1} \right| \leq \frac{2\pi\rho}{3} \frac{1}{\rho^3-1} \to 0$ as $\rho \to \infty$. Furthermore, $\int_{BO} \frac{dz}{z^3+1} = -\int_0^\rho \frac{1}{t^3+1} e^{i2\pi/3} dt$. Thus, we have $\frac{2\pi i}{3e^{i2\pi/3}} = (1 - e^{i2\pi/3}) \int_0^\infty \frac{1}{x^3+1} dx$, or $\int_0^\infty \frac{dx}{x^3+1} = \frac{2\pi i}{3(e^{i2\pi/3} - e^{i4\pi/3})} = \frac{2\sqrt{3}\pi}{9}$.

33.9. Similar to Example 33.3.

33.10. (a). and (b). similar to Example 33.3, (c). use $\sin x = (e^{ix} - e^{-ix})/(2i)$, (d). use $e^{iz}/(z-i)$.

33.11. (a). Compute directly, (b). integral of an analytic function, (c). use Theorem 22.4 and note that A and B are arbitrary, (d). verify directly, (e). compute directly, (f). use $x = s + t$, $y = t/s$ and Problem 33.7, (g). use Weierstrass's Factorization Theorem (Theorem 43.3).

Lecture 34
Indented Contour Integrals

In previous lectures, when evaluating the real improper integrals we assumed that the integrand has no singularity over the whole interval of integration. In this lecture, we shall show that by using indented contours some functions which have simple poles at certain points on the interval of integration can be computed.

Recall from calculus that if $f(x)$ is continuous on the interval $[a, b)$ and discontinuous at b, then the improper integral of $f(x)$ over $[a, b]$ is defined by

$$\int_a^b f(x)dx = \lim_{r \to b^-} \int_a^r f(x)dx,$$

provided the limit exists. Similarly, if $f(x)$ is continuous on the interval $(b, c]$ and discontinuous at b, then the improper integral of $f(x)$ over $[b, c]$ is defined by

$$\int_b^c f(x)dx = \lim_{s \to b^+} \int_s^c f(x)dx,$$

provided the limit exists. If $f(x)$ is continuous on the interval $[a, c]$ and discontinuous at $b \in (a, c)$, then the improper integral of $f(x)$ over $[a, c]$ and its Cauchy principal value are defined by

$$\int_a^c f(x)dx := \lim_{r \to b^-} \int_a^r f(x)dx + \lim_{s \to b^+} \int_s^c f(x)dx$$

and

$$\text{p.v.} \int_a^c f(x)dx := \lim_{t \to 0^+} \left(\int_a^{b-t} f(x)dx + \int_{b+t}^c f(x)dx \right),$$

provided the appropriate limits exist. If the function $f(x)$ is continuous on the interval $(-\infty, \infty)$ and discontinuous at b, then the Cauchy principal value of its integral is defined by

$$\text{p.v.} \int_{-\infty}^{\infty} f(x)dx := \lim_{\rho \to \infty} \lim_{t \to 0^+} \left(\int_{-\rho}^{b-t} f(x)dx + \int_{b+t}^{\rho} f(x)dx \right),$$

provided the limits as $t \to 0^+$ and $\rho \to \infty$ exist independently. If $f(x)$ has discontinuities at several points, then the definitions above can be extended rather easily.

Example 34.1. Clearly,

$$\int_0^1 \frac{1}{x^{1/5}}\,dx = \lim_{t\to 0+}\int_t^1 \frac{1}{x^{1/5}}\,dx = \lim_{t\to 0+}\frac{5}{4}(1 - t^{4/5}) = \frac{5}{4},$$

$$\int_3^9 \frac{1}{x-3}\,dx = \lim_{t\to 0+}\int_{3+t}^9 \frac{1}{x-3}\,dx = \lim_{t\to 0+}\ln|x-3|\Big|_{3+t}^9$$
$$= \lim_{t\to 0+}[\ln 6 - \ln t] = \infty,$$

and

$$\int_1^3 \frac{1}{x-3}\,dx = \lim_{t\to 0+}\int_1^{3-t} \frac{1}{x-3}\,dx = \lim_{t\to 0+}\ln|x-3|\Big|_1^{3-t}$$
$$= \lim_{t\to 0+}[\ln t - \ln 2] = -\infty;$$

however,

$$\text{p.v.}\int_1^9 \frac{1}{x-3}\,dx = \lim_{t\to 0+}\left(\int_1^{3-t} \frac{1}{x-3}\,dx + \int_{3+t}^9 \frac{1}{x-3}\,dx\right)$$
$$= \lim_{t\to 0+}[\ln t - \ln 2 + \ln 6 - \ln t] = \ln 3.$$

To prove the main results of this lecture, we shall need the following lemma.

Lemma 34.1. Suppose that $f(z)$ has a simple pole at the point b on the x-axis. If γ_r is the contour $\gamma_r : z = b + re^{i\theta}$, $\theta_1 \le \theta \le \theta_2$ (see Figure 34.1), then

$$\lim_{r\to 0+}\int_{\gamma_r} f(z)\,dz = i(\theta_2 - \theta_1)R[f,b]. \tag{34.1}$$

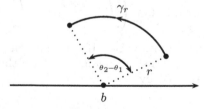

Figure 34.1

Proof. The Laurent series for $f(z)$ at $z = b$ can be written as

$$f(z) = \frac{R[f,b]}{z-b} + g(z), \tag{34.2}$$

where $g(z)$ is analytic at b. Integrating (34.2) and using the parametrization of γ_r, we obtain

$$\int_{\gamma_r} f(z)dz = R[f, b] \int_{\theta_1}^{\theta_2} \frac{ire^{i\theta}}{re^{i\theta}} d\theta + ir \int_{\theta_1}^{\theta_2} g(b + re^{i\theta})e^{i\theta} d\theta$$

$$= i(\theta_2 - \theta_1)R[f, b] + +ir \int_{\theta_1}^{\theta_2} g(b + re^{i\theta})e^{i\theta} d\theta.$$

Now, since $g(z)$ is analytic at b, there exists an $M > 0$ such that $|g(b + re^{i\theta})| \leq M$ in some neighborhood of b. Thus, it follows that

$$\lim_{r \to 0^+} \left| ir \int_{\theta_1}^{\theta_2} g(b + re^{i\theta})e^{i\theta} d\theta \right| \leq \lim_{r \to 0^+} r \int_{\theta_1}^{\theta_2} M d\theta$$

$$= \lim_{r \to 0^+} r(\theta_2 - \theta_1)M = 0. \quad \blacksquare$$

The following theorems extend the results we presented in earlier lectures.

Theorem 34.1. If $f(z) = P(z)/Q(z)$ is a rational function such that $P(z)$ and $Q(z)$ do not have a common zero, $Q(z)$ has simple zeros at the points $b_1, \cdots, b_\ell$ on the real line, and degree $Q \geq 2+$ degree P, then

$$\text{p.v.} \int_{-\infty}^{\infty} \frac{P(x)}{Q(x)} dx = 2\pi i \sum R[f, z_j] + \pi i \sum_{k=1}^{\ell} R[f, b_k] \qquad (34.3)$$

holds, where the first sum is taken over all the poles z_j of $f(z)$ that fall in the upper half-plane.

Theorem 34.2. If $f(z) = P(z)/Q(z)$ is a rational function such that $P(z)$ and $Q(z)$ do not have a common zero, $Q(z)$ has simple zeros at the points $b_1, \cdots, b_\ell$ on the real line, and degree $Q \geq 1+$ degree P, then for $a > 0$,

$$\text{p.v.} \int_{-\infty}^{\infty} \frac{P(x)}{Q(x)} \cos ax\, dx = -2\pi \sum \text{Im } R[e^{iaz} f, z_j] - \pi \sum_{k=1}^{\ell} \text{Im } R[e^{iaz} f, b_k]$$

$$(34.4)$$

and

$$\text{p.v.} \int_{-\infty}^{\infty} \frac{P(x)}{Q(x)} \sin ax\, dx = 2\pi \sum \text{Re } R[e^{iaz} f, z_j] + \pi \sum_{k=1}^{\ell} \text{Re } R[e^{iaz} f, b_k]$$

$$(34.5)$$

hold, where the first sum in (34.4) as well as in (34.5) is taken over all the poles z_j of $f(z)$ that fall in the upper half-plane.

Proof of Theorems 34.1 and 34.2. Let ρ be sufficiently large so that all poles of $f(z)$ lie under the semicircle $\gamma_\rho : z = \rho e^{i\theta}$, $0 \le \theta \le \pi$. Also, let r be so small that the semicircles $C_k : z = b_k + re^{i\theta}$, $0 \le \theta \le \pi$, $k = 1, \cdots, \ell$ are disjoint and the poles of $f(z)$ lie above them (see Figure 34.2). Furthermore, let Γ be the closed positively oriented contour consisting of γ_ρ, $-C_j$, $j = 1, \cdots, \ell$ and the segments between these small semicircles $I\rho = [-\rho, \rho] \setminus \cup_{k=1}^{\ell} (b_k - r, b_k + r)$. Then, from Theorem 31.2, it follows that

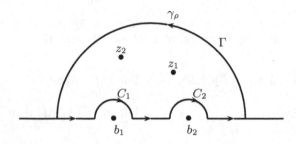

Figure 34.2

$$2\pi i \sum R[f, z_j] = \int_\Gamma f(z)dz = \int_{\gamma_\rho} f(z)dz + \int_{I_\rho} f(x)dx - \sum_{k=1}^{\ell} \int_{C_k} f(z)dz.$$
(34.6)

Now, in (34.6), we let $\rho \to \infty$ and $r \to 0^+$. If $f(z)$ satisfies the conditions of Theorem 34.1, then, in view of Corollary 32.1 and Lemma 34.1, (34.6) reduces to

$$2\pi i \sum R[f, z_j] = \int_{-\infty}^{\infty} f(x)dx - \pi i \sum_{k=1}^{\ell} R[f, b_k],$$
(34.7)

which is the same as (34.3). If $f(z)$ satisfies the conditions of Theorem 34.2, then also from Corollary 33.1 and Lemma 34.1 the relation (34.7) holds, but for the function $f(z)$ multiplied by e^{iaz}, $a > 0$; i.e.,

$$2\pi i \sum R[e^{iaz} f, z_j] = \int_{-\infty}^{\infty} e^{iax} f(x)dx - \pi i \sum_{k=1}^{\ell} R[e^{iaz} f, b_k].$$
(34.8)

Finally, equating the real and imaginary parts of (34.8), we obtain (34.4) and (34.5), respectively. ∎

Example 34.2. We shall show that

$$\text{p.v.} \int_{-\infty}^{\infty} \frac{x}{x^3 - a^3} dx = \frac{\pi}{\sqrt{3}\, a}, \quad a > 0.$$

Clearly, the function $f(z) = z/(z^3 - a^3)$ has simple poles at $z_1 = a$, $z_2 = (-1 - i\sqrt{3})a/2$, and $z_3 = (-1 + \sqrt{3})a/2$. Since the pole z_2 is in the lower half-plane, from Theorem 34.1 it follows that

$$\text{p.v.} \int_{-\infty}^{\infty} \frac{x}{x^3 - a^3}dx = 2\pi i R[f, z_3] + \pi i R[f, z_1]$$

$$= 2\pi i \times \frac{2}{3(-1 + \sqrt{3}\,i)a} + \pi i \times \frac{1}{3a} = \frac{\pi}{\sqrt{3}\,a}.$$

Example 34.3. Since $R[e^{iaz}/z, 0] = 1$, $a > 0$. From Theorem 34.2, it follows that

$$\text{p.v.} \int_{-\infty}^{\infty} \frac{\cos ax}{x}dx = 0 \quad \text{and} \quad \text{p.v.} \int_{-\infty}^{\infty} \frac{\sin ax}{x}dx = \pi, \quad a > 0.$$

We also note that the function $(\sin ax)/x$ is even, and hence

$$\int_0^{\infty} \frac{\sin ax}{x}dx = \frac{\pi}{2}, \quad a > 0.$$

Example 34.4. Since the function $f(z) = z/(z^2 - b^2)$, $b > 0$ has simple poles at b and $-b$ and, for $a > 0$,

$$R\left[\frac{ze^{iaz}}{z^2 - b^2}, b\right] = \frac{e^{iab}}{2}, \quad R\left[\frac{ze^{iaz}}{z^2 - b^2}, -b\right] = \frac{e^{-iab}}{2},$$

from Theorem 34.2, it follows that

$$\text{p.v.} \int_{-\infty}^{\infty} \frac{x \cos ax}{x^2 - b^2}dx = -\pi\text{Im}\left(\frac{e^{iab}}{2} + \frac{e^{-iab}}{2}\right) = -\pi\text{Im}(\cos ab) = 0$$

and

$$\text{p.v.} \int_{-\infty}^{\infty} \frac{x \sin ax}{x^2 - b^2}dx = \pi\text{Re}\left(\frac{e^{iab}}{2} + \frac{e^{-iab}}{2}\right) = \pi\text{Re}(\cos ab) = \pi\cos ab.$$

Problems

34.1. Show that:

(a). $\text{p.v.} \displaystyle\int_{-\infty}^{\infty} \frac{\cos x}{x - a}dx = -\pi\sin a$, $a > 0$.

(b). $\text{p.v.} \displaystyle\int_{-\infty}^{\infty} \frac{\sin x}{x - a}dx = \pi\cos a$, $a > 0$.

(c). $\displaystyle\int_0^\infty \frac{\sin^2 x}{x^2}\,dx \;=\; \frac{\pi}{2}.$

(d). $\displaystyle\int_0^\infty \frac{\cos ax - \cos bx}{x^2}\,dx \;=\; \frac{\pi}{2}(b-a), \quad a,b>0.$

(e). $\displaystyle\int_0^\infty \frac{\sin ax}{x(x^2+b^2)}\,dx \;=\; \frac{\pi}{2b^2}\left(1-e^{-ab}\right), \quad a,b>0.$

(f). $\displaystyle\int_0^\infty \frac{\sin ax}{x(x^2+b^2)^2}\,dx \;=\; \frac{\pi}{4b^4}\left[2-(2+ab)e^{-ab}\right], \quad a,b>0.$

Answers or Hints

34.1. Similar to Examples 34.1-34.3.

Lecture 35
Contour Integrals Involving Multi-valued Functions

In previous lectures, we successfully applied contour integration theory to evaluate integrals of real-valued functions. However, often it turns out that the extension of a real function to the complex plane is a multi-valued function. In this lecture, we shall show that by using contours cleverly some such functions can also be integrated.

We begin with the evaluation of the integral

$$I = \int_0^\infty x^{\alpha-1} f(x) dx, \quad 0 < \alpha < 1. \tag{35.1}$$

For this, we shall assume that: (i) $f(z)$ is a single-valued analytic function, except for a finite number of isolated singularities not on the positive real semiaxis, (ii) $f(z)$ has a removable singularity at $z = 0$, and (iii) $f(z)$ has a zero of order at least one at $z = \infty$.

Consider the domain $S : 0 < \arg z < 2\pi$, which is the z-plane cut along the positive real semiaxis. Clearly, the function $F(z) = z^{\alpha-1} f(z)$ in the domain S is single-valued, its singularities are the same as those of $f(z)$, and it coincides with $x^{\alpha-1} f(x)$ on the upper side of the cut; i.e., $\arg z = 0$. Let $\rho > 0$ be sufficiently large so that all singularities z_j of $f(z)$ lie inside the circle γ_ρ, and let $r > 0$ be sufficiently small so that all singularities of $f(z)$ lie outside the circle γ_r. In the domain S, we consider the closed contour Γ, which consists of open circles γ_ρ, γ_r and the segments of the real axis $[r, \rho]$ on the upper and lower sides of the cut (see Figure 35.1). Then, by Theorem 31.3, it follows that

$$\begin{aligned}
\int_\Gamma F(z) dz &= 2\pi i \sum R[z^{\alpha-1} f(z), z_j] \\
&= \int_r^\rho x^{\alpha-1} f(x) dx + \int_{\gamma_\rho} z^{\alpha-1} f(z) dz \\
&\quad + \int_\rho^r z^{\alpha-1} f(z) dz - \int_{\gamma_r} z^{\alpha-1} f(z) dz \\
&= I_1 + I_2 + I_3 - I_4,
\end{aligned} \tag{35.2}$$

where the sum is taken over all singularities of $f(z)$.

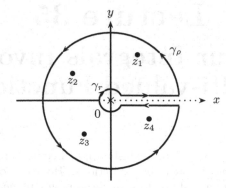

Figure 35.1

From assumption (iii) it is clear that $|f(z)| \leq M/|z|$ for all sufficiently large $|z|$. Thus, it follows that

$$|I_2| = \left| \int_{\gamma_\rho} z^{\alpha-1} f(z) dz \right| \leq \frac{M\rho^{\alpha-1}}{\rho} 2\pi\rho = 2\pi M \rho^{\alpha-1} \to 0 \quad \text{as} \quad \rho \to \infty.$$

For the integral I_3, we note that $\arg z = 2\pi$, and hence $z = xe^{2\pi i}$, $x > 0$. Thus, we have

$$I_3 = \int_\rho^r z^{\alpha-1} f(z) dz = -e^{2\pi i(\alpha-1)} \int_r^\rho x^{\alpha-1} f(x) dx = -e^{2\pi i(\alpha-1)} I_1.$$

For the integral I_4, in view of assumption (ii), we find

$$|I_4| = \left| \int_{\gamma_r} z^{\alpha-1} f(z) dz \right| \leq C r^{\alpha-1} 2\pi r \to 0 \quad \text{as} \quad r \to 0.$$

Finally, in (35.2), letting $r \to 0$, $\rho \to \infty$ and using the relations above, we obtain

$$\int_0^\infty x^{\alpha-1} f(x) dx = \frac{2\pi i}{1 - e^{2\pi\alpha i}} \sum R[z^{\alpha-1} f(z), z_j], \quad 0 < \alpha < 1. \quad (35.3)$$

Example 35.1. For $0 < \alpha < 1$, from (35.3), we have

$$\int_0^\infty \frac{x^{\alpha-1}}{x+1} dx = \frac{2\pi i}{1 - e^{2\pi\alpha i}} R\left[\frac{z^{\alpha-1}}{z+1}, -1 \right]$$

$$= \frac{2\pi i}{1 - e^{2\pi\alpha i}} (-1)^{\alpha-1} = \frac{2\pi i e^{\pi(\alpha-1)i}}{1 - e^{2\pi(\alpha-1)i}} = \frac{\pi}{\sin \alpha\pi},$$

which is the same as given in Problem 33.7.

Example 35.2. From (35.3), we have

$$\int_0^\infty \frac{\sqrt{x}}{x^3+1}dx = \int_0^\infty x^{1/2-1}\frac{x}{x^3+1}dx$$

$$= \frac{2\pi i}{1-e^{\pi i}}\left(R\left[\frac{z^{1/2}}{z^3+1},e^{\pi i/3}\right] + R\left[\frac{z^{1/2}}{z^3+1},e^{\pi i}\right] + R\left[\frac{z^{1/2}}{z^3+1},e^{5\pi i/3}\right]\right)$$

$$= \pi i\left[\frac{i}{3} + \frac{1}{3}e^{-\pi i/2} + \frac{1}{3}e^{-5\pi i/2}\right] = \frac{\pi i}{3}[i - i - i] = \frac{\pi}{3}.$$

Now, we shall consider the integral of the form

$$I = \int_0^1 x^{\alpha-1}(1-x)^{-\alpha}f(x)dx, \quad 0 < \alpha < 1. \tag{35.4}$$

We shall assume that: (i) $f(z)$ is a single-valued analytic function, except for a finite number of isolated singularities not on the interval $0 \le x \le 1$, and (ii) $f(z)$ has a removable singularity at $z = \infty$. The function $F(z) = z^{\alpha-1}(1-z)^{-\alpha}f(z)$ has two branch points, namely $z = 0$ and $z = 1$, and the point $z = \infty$ is a removable singularity. Now, as for the integral (35.1), from Figure 35.2 it is clear that

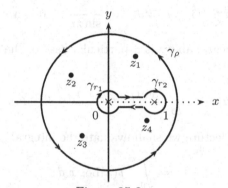

Figure 35.2

$$\int_\Gamma F(z)dz = 2\pi i\sum R[z^{\alpha-1}(1-z)^{-\alpha}f(z), z_j]$$

$$= \int_r^{1-r} x^{\alpha-1}(1-x)^{-\alpha}f(x)dx - \int_{\gamma_{r_2}} z^{\alpha-1}(1-z)^{-\alpha}f(z)dz$$

$$+ \int_{1-r}^r z^{\alpha-1}(1-z)^{-\alpha}f(z)dz - \int_{\gamma_{r_1}} z^{\alpha-1}(1-z)^{-\alpha}f(z)dz$$

$$+ \int_{\gamma_\rho} z^{\alpha-1}(1-z)^{-\alpha}f(z)dz$$

$$= I_1 - I_2 + I_3 - I_4 + I_5, \tag{35.5}$$

where the sum is taken over all singularities of $f(z)$.

Now, as for the integrals in (35.2), it follows that I_2 and I_4 tend to zero, and $I_3 = -e^{2\pi i\alpha}I_1$ as $r \to 0$. Next, since $f(z)$ has a removable singularity at $z = \infty$, it has the expansion $f(z) = a_0 + (a_{-1}/z) + \cdots$, where $a_0 = \lim_{z\to\infty} f(z)$. This leads to $I_5 = 2\pi i a_0 e^{\pi i\alpha}$. Thus, as $r \to 0$ and $\rho \to \infty$, we get

$$2\pi i \sum R[z^{\alpha-1}(1-z)^{-\alpha}f(z), z_j] = (1 - e^{2\pi i\alpha})I + 2\pi i a_0 e^{\pi i\alpha},$$

which gives

$$\int_0^1 x^{\alpha-1}(1-x)^{-\alpha}f(x)dx = \frac{\pi a_0}{\sin \pi\alpha} + \frac{2\pi i}{(1-e^{2\pi i\alpha})} \sum R[z^{\alpha-1}(1-z)^{-\alpha}f(z), z_j],$$

(35.6)

where $0 < \alpha < 1$ and $a_0 = \lim_{z\to\infty} f(z)$.

Example 35.3. From (35.6), it immediately follows that

$$\int_0^1 x^{\alpha-1}(1-x)^{-\alpha}dx = \frac{\pi}{\sin \pi\alpha}, \quad 0 < \alpha < 1.$$

Recall that the integral above is a particular case of the *beta function* denoted and defined as

$$B(p,q) := \int_0^1 x^{p-1}(1-x)^{q-1}dx, \quad 0 < p, q < 1.$$

Finally, in this lecture we shall evaluate the integral

$$I = \int_0^\infty f(x)\text{Log}\, x \, dx. \tag{35.7}$$

We shall assume that $f(x)$ is an even function and that $f(z)$ is analytic onto the upper half-plane $\text{Im}\, z > 0$ and satisfies the conditions of Theorem 32.1. From Figure 35.3 it follows that

$$\begin{aligned}
\int_\Gamma f(z)\text{Log}\, z\, dz &= 2\pi i \sum R[f(z)\text{Log}\, z, z_j] \\
&= \int_r^\rho f(x)\text{Log}\, x\, dx + \int_{\gamma_\rho} f(z)\text{Log}\, z\, dz \\
&\quad + \int_r^\rho f(x)[\text{Log}\, x + \pi i]dx - \int_{\gamma_r} f(z)\text{Log}\, z\, dz \\
&= I_1 + I_2 + I_3 - I_4,
\end{aligned}$$

(35.8)

where the sum is taken over all singularities of $f(z)$ on the upper half-plane.

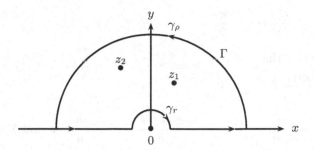

Figure 35.3

Now, following as earlier, it follows that

$$|I_2| \leq \frac{M}{\rho^2} \int_0^\pi |\mathrm{Log}\,\rho e^{i\theta}||\rho e^{i\theta}|d\theta$$

$$\leq \frac{M}{\rho} \int_0^\pi |\mathrm{Log}\,\rho + i\theta|d\theta \leq \frac{M\pi}{\rho}\sqrt{\mathrm{Log}^2\rho + \pi^2} \to 0 \ \text{ as } \ \rho \to \infty.$$

Furthermore, I_4 also tends to zero as $r \to 0$. From our earlier lectures, we also have

$$\int_0^\infty f(x)dx = \pi i \sum R[f(z), z_j].$$

Thus, from (35.8), we get

$$\int_0^\infty f(x)\mathrm{Log}\,x\,dx = \pi i \sum R\left[f(z)\left(\mathrm{Log}\,z - \frac{\pi i}{2}\right), z_j\right]. \tag{35.9}$$

Example 35.4. From (35.9), it is clear that

$$\int_0^\infty \frac{\mathrm{Log}\,x}{(x^2+1)^2}dx = \pi i R\left[\frac{1}{(z^2+1)^2}\left(\mathrm{Log}\,z - \frac{\pi i}{2}\right), i\right] = -\frac{\pi}{4}.$$

Problems

35.1. Show that:

(a). $\displaystyle\int_0^\infty \frac{1}{\sqrt{x}(x^2+1)}dx = \frac{\pi}{\sqrt{2}}.$

(b). $\displaystyle\int_0^\infty \frac{\sqrt{x}}{(x^2+1)^2}dx = \frac{\pi}{4\sqrt{2}}.$

(c). $\displaystyle\int_0^\infty \frac{x^{1/4}}{x^2+x+1}dx = \pi(1 - 1/\sqrt{3}).$

(d). $\displaystyle\int_0^\infty \frac{\sqrt{x}}{(x+1)^2}dx = \frac{\pi}{2}.$

35.2. Show that:

(a). $\displaystyle\int_0^\infty \frac{\text{Log}\, x}{x^2+1}dx = 0.$

(b). $\displaystyle\int_0^\infty \frac{\text{Log}\, x}{(x+a)^2+b^2}dx = \frac{\text{Log}\sqrt{a^2+b^2}}{b}\tan^{-1}\frac{b}{a},$ a and b are real.

(c). $\displaystyle\int_0^\infty \frac{\text{Log}\, x}{x^4+1}dx = \frac{\pi^2}{8\sqrt{2}}.$

(d). $\displaystyle\int_0^\infty \frac{(\text{Log}\, x)^2}{x^2+1}dx = \frac{\pi^3}{8}.$

(e). $\displaystyle\int_0^\infty \frac{\sqrt{x}\,\text{Log}\, x}{(x+1)^2}dx = \pi.$

(f). $\displaystyle\int_0^\infty \frac{(\text{Log}\, x)^2}{x^3+1}dx = \frac{3\sqrt{2}\,\pi^2}{64}.$

35.3. Show that the substitution $t = x/(1-x)$ reduces the integral (35.4) to the form (35.1).

35.4. Show that:

(a). $\displaystyle\int_0^\infty \frac{\cos ax}{\cosh \pi x}dx = \frac{1}{2}\text{sech}\frac{a}{2},$ a is real.

(b). $\displaystyle\int_0^\infty \frac{\sin ax}{\sinh \pi x}dx = \frac{1}{2}\tanh\frac{a}{2},$ a is real.

(c). $\displaystyle\int_{-\infty}^\infty \frac{e^{ax}}{\cosh x}dx = \pi \sec\left(\frac{\pi a}{2}\right),$ $|a| < 1.$

(d). $\displaystyle\int_0^\infty e^{-\pi x}\frac{\sin ax}{\sinh \pi x}dx = \frac{1}{2}\frac{1+e^{-a}}{1-e^{-a}} - \frac{1}{a},$ $a > 0.$

(e). $\displaystyle\int_0^\infty \frac{\cos x}{x^\alpha}dx = \Gamma(1-\alpha)\sin\frac{\pi\alpha}{2},$ $0 < \alpha < 1.$

(f). $\displaystyle\int_0^\infty \frac{\sin x}{x^\alpha}dx = \Gamma(1-\alpha)\cos\frac{\pi\alpha}{2},$ $0 < \alpha < 1.$

Answers or Hints

35.1. Similar to Examples 35.1 and 35.2.

35.2. Similar to Example 35.4.

35.3. Verify directly.

35.4. Use proper function and contour.

Lecture 36
Summation of Series

One of the remarkable applications of the Residue Theorem is that we can sum $\sum_{n \in \mathbf{Z}} f(n)$ for certain types of functions $f(z)$. For this, we shall prove the following result.

Theorem 36.1. Let the function $f(z)$ be analytic on $\mathbf{C}$ except at finitely many points $z_1, z_2, \cdots, z_k$, none of which is a real integer. Furthermore, let there exist $M > 0$ such that $|z^2 f(z)| \leq M$ for all $|z| > \rho$ for some $\rho > 0$. Consider the functions

$$g(z) = \pi \frac{\cos \pi z}{\sin \pi z} f(z) \quad \text{and} \quad h(z) = \frac{\pi}{\sin \pi z} f(z), \quad z \in \mathbf{C}.$$

Then, the following hold:

$$\sum_{n=-\infty}^{\infty} f(n) = -\sum_{j=1}^{k} R[g, z_j] \tag{36.1}$$

and

$$\sum_{n=-\infty}^{\infty} (-1)^n f(n) = -\sum_{j=1}^{k} R[h, z_j]. \tag{36.2}$$

Proof. Recall that the function $\sin \pi z$ vanishes at each $n \in \mathbf{Z}$. Thus, the functions $\pi \cos \pi z / \sin \pi z$ and $\pi / \sin \pi z$ have simple poles at each $n \in \mathbf{Z}$. Furthermore, by hypothesis, $f(z)$ is analytic at each n. Now, we consider the following two cases.

Case (i). Let $f(n) \neq 0$ for each n. Then, both $g(z)$ and $h(z)$ have simple poles at each $n \in \mathbf{Z}$ and singularities at $z_1, z_2, \cdots, z_k$. Let γ be a large rectangle containing all singularities $z_1, z_2, \cdots, z_k$ of $f(z)$, the integers $-n, \cdots, -1, 0, 1, \cdots, n$, and not passing through any integer. Then, by the Residue Theorem, it follows that

$$\int_\gamma g(z) dz = 2\pi i \left[\sum_{j=1}^{k} R[g, z_j] + \sum_{m=-n}^{n} R[g, m] \right].$$

Now since $g(z)$ has a simple pole at each m, we have

$$R[g, m] = \lim_{z \to m} (z - m) g(z) = \lim_{z \to m} f(z) \frac{\pi(z - m)}{\sin \pi z} \cos \pi z = f(m).$$

242

Hence, we obtain

$$\int_\gamma y(z)dz \;=\; 2\pi i \left[\sum_{j=1}^k R[g, z_j] + \sum_{m=-n}^n f(m) \right]. \tag{36.3}$$

Similarly, we find

$$\int_\gamma h(z)dz \;=\; 2\pi i \left[\sum_{j=1}^k R[h, z_j] + \sum_{m=-n}^n (-1)^m f(m) \right]. \tag{36.4}$$

Case (ii). If $f(m) = 0$ for some m, then $g(z)$ and $h(z)$ have removable singularities at m. Thus, the contribution of the point m to $\int_\gamma g(z)dz$ and to $\int_\gamma h(z)dz$ is zero, which is also $f(m)$. Thus, in this case also, (36.3) and (36.4) remain valid.

If we can show that, as the rectangle γ gets larger, enclosing all the integers m in the limit, $\lim \int_\gamma g(z)dz = 0 = \lim \int_\gamma h(z)dz$, then from (36.3) and (36.4) the relations (36.1) and (36.2), respectively, follow immediately. To show this, we consider the rectangular region S_n with boundary γ_n.

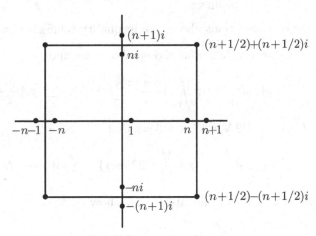

Figure 36.1

Let n be so large that $|z_j| < n$ for all j and $\rho < n$. Then, for any $z \in S_n$, $|z| > n > \rho$, we have $|f(z)| \le M/|z|^2 \le M/\rho^2 < M/n^2$.

Next, we need to show that

$$\left| \frac{\cos \pi z}{\sin \pi z} \right| \;\le\; A \quad \text{for some} \quad A > 0 \quad \text{and} \quad |z| > n. \tag{36.5}$$

For this, note that, for $z = x + iy$, $i\pi z = \pi i x - \pi y$ and

$$\left| \frac{\cos \pi z}{\sin \pi z} \right| = \left| \frac{e^{\pi i x - \pi y} + e^{-\pi i x + \pi y}}{e^{\pi i x - \pi y} - e^{-\pi i x + \pi y}} \right| \leq \frac{\left| e^{\pi i x - \pi y} \right| + \left| e^{-\pi i x + \pi y} \right|}{\left| |e^{-\pi i x + \pi y}| - |e^{\pi i x - \pi y}| \right|}.$$

Hence, it follows that

$$\left| \frac{\cos \pi z}{\sin \pi z} \right| \leq \frac{e^{-\pi y} + e^{\pi y}}{e^{\pi y} - e^{-\pi y}} = \frac{1 + e^{-2\pi y}}{1 - e^{-2\pi y}} \leq \frac{1 + e^{-\pi}}{1 - e^{-\pi}}$$

provided $y \geq 1/2$, and similarly,

$$\left| \frac{\cos \pi z}{\sin \pi z} \right| \leq \frac{e^{-\pi y} + e^{\pi y}}{e^{-\pi y} - e^{\pi y}} = \frac{1 + e^{2\pi y}}{1 - e^{2\pi y}} \leq \frac{1 + e^{-\pi}}{1 - e^{-\pi}}$$

provided $y \leq -1/2$. Now let $-1/2 < y < 1/2$, and $z = N + 1/2 + iy$ where $N > n$. Then, we have

$$\left| \frac{\cos \pi z}{\sin \pi z} \right| = \left| \cot \pi \left(N + \frac{1}{2} + iy \right) \right| = \left| \cot \left(\frac{\pi}{2} + \pi i y \right) \right| = |\tanh \pi y| \leq \tanh \frac{\pi}{2}.$$

Similarly, if $z = -(N + 1/2) + iy$, we get

$$\left| \frac{\cos \pi z}{\sin \pi z} \right| = |\tanh \pi y| \leq \tanh \frac{\pi}{2}.$$

From the considerations above, the inequality (36.5) is clear.

Finally, for any $N > n$ as chosen above, we find

$$\left| \int_{\gamma_N} g(z)dz \right| \leq \pi \int_{\gamma_N} \left| \frac{\cos \pi z}{\sin \pi z} \right| |f(z)||dz| \leq \pi A \frac{M}{N^2} L(\gamma_N).$$

Since, $L(\gamma_N) = 4(2N + 1)$, it follows that

$$\left| \int_{\gamma_N} g(z)dz \right| \leq \pi A \frac{M}{N^2} 4(2N + 1) \rightarrow 0 \quad \text{as} \quad N \rightarrow \infty.$$

Following the same arguments, we can show that

$$\left| \int_{\gamma_N} h(z)dz \right| \rightarrow 0 \quad \text{as} \quad N \rightarrow \infty.$$

This completes the proof of Theorem 36.1. ∎

Remark 36.1. If the function $f(z)$ has a singularity at some real integer K, then Theorem 36.1 also remains valid; however, then, from the sum $\sum f(m)$, we need to exclude the term $f(K)$ and include the residue $R[g, K]$ in the sum $\sum R[g, z_j]$.

Example 36.1. The function $f(z) = 1/z^2$ satisfies the hypothesis of Theorem 36.1 except that it has a pole of order 2 at 0. Hence, the function $g(z) = \dfrac{\pi}{z^2} \dfrac{\cos \pi z}{\sin \pi z}$ has a pole of order 3 at 0. Now, since

$$R\left[\frac{\pi}{z^2} \frac{\cos \pi z}{\sin \pi z}, 0\right] = -\frac{\pi^2}{3}$$

and

$$\sum_{n=1}^{\infty} \frac{1}{n^2} = \frac{1}{2} \sum_{n=-\infty,\, n\neq 0}^{\infty} \frac{1}{n^2},$$

from (36.1) and Remark 36.1 it follows that

$$\sum_{n=1}^{\infty} \frac{1}{n^2} = -\frac{1}{2} R\left[\frac{\pi}{z^2} \frac{\cos \pi z}{\sin \pi z}, 0\right] = \frac{\pi^2}{6}.$$

For the same function $f(z) = 1/z^2$, if we use the relation (36.2) and Remark 36.1, we get

$$1 - \frac{1}{2^2} + \frac{1}{3^2} - \cdots = \sum_{n=1}^{\infty} (-1)^{n+1} \frac{1}{n^2} = \frac{\pi^2}{12}.$$

Similarly, we can show that

$$\sum_{n=0}^{\infty} \frac{1}{(2n+1)^2} = \frac{\pi^2}{8}, \qquad \sum_{n=1}^{\infty} \frac{1}{n^4} = \frac{\pi^4}{90}, \qquad \sum_{n=0}^{\infty} \frac{1}{(2n+1)^4} = \frac{\pi^4}{96},$$

$$1 - \frac{1}{3^3} + \frac{1}{5^3} - \frac{1}{7^3} + \cdots = \frac{\pi^3}{32},$$

and

$$1 - \frac{1}{3^5} + \frac{1}{5^5} - \frac{1}{7^5} + \cdots = \frac{5\pi^5}{1536}.$$

Example 36.2. The function $f(z) = 1/(z^2 - a^2)$, where a is real and noninteger, satisfies the hypothesis of Theorem 36.1. Thus, from (36.1), it follows that

$$\sum_{n=-\infty}^{\infty} \frac{1}{n^2 - a^2} = -R\left[\frac{\pi}{z^2 - a^2} \frac{\cos \pi z}{\sin \pi z}, a\right] - R\left[\frac{\pi}{z^2 - a^2} \frac{\cos \pi z}{\sin \pi z}, -a\right]$$

$$= -\frac{\pi}{2a} \cot \pi a - \frac{\pi}{2a} \cot \pi a = -\frac{\pi}{a} \cot \pi a.$$

Furthermore, since

$$\sum_{n=-\infty}^{\infty} \frac{1}{n^2 - a^2} = \sum_{n=-\infty}^{-1} \frac{1}{n^2 - a^2} - \frac{1}{a^2} + \sum_{n=1}^{\infty} \frac{1}{n^2 - a^2}$$

and

$$\sum_{n=-\infty}^{-1} \frac{1}{n^2 - a^2} = \sum_{n=1}^{\infty} \frac{1}{n^2 - a^2},$$

we have

$$\sum_{n=1}^{\infty} \frac{1}{n^2 - a^2} = \frac{1}{2a^2} - \frac{\pi}{2a} \cot \pi a.$$

Problems

36.1. Let a be real and noninteger. Show that:

(a). $\displaystyle\sum_{n=-\infty}^{\infty} \frac{1}{n^2 + a^2} = \frac{\pi}{a} \coth \pi a.$

(b). $\displaystyle\sum_{n=-\infty}^{\infty} \frac{1}{(n-a)^2} = \pi^2 \mathrm{cosec}^2 \pi a.$

(c). $\displaystyle\sum_{n=-\infty}^{\infty} \frac{1}{(n^2+a^2)^2} = \frac{\pi}{2a^3} \coth \pi a + \frac{\pi^2}{2a^2} \mathrm{cosec}^2 \pi a.$

(d). $\displaystyle\sum_{n=1}^{\infty} \frac{(-1)^{n+1}}{n^2 + a^2} = \frac{1}{2a^2} - \frac{\pi}{2a} \mathrm{cosec}\, \pi a.$

(e). $\displaystyle\sum_{n=-\infty}^{\infty} \frac{(-1)^n}{(n+a)^2} = \pi^2 \cot \pi a\, \mathrm{cosec}\, \pi a.$

36.2. Show that, for a positive integer k,

$$\sum_{n=1}^{\infty} \frac{1}{n^{2k}} = (-1)^{k-1} \pi^{2k} \frac{2^{2k-1}}{(2k)!} B_{2k}$$

and

$$\sum_{n=1}^{\infty} \frac{(-1)^n}{n^{2k}} = (-1)^k \pi^{2k} \frac{(2^{2k-1} - 1)}{(2k)!} B_{2k},$$

where B_{2k} is the $2k$th *Bernoulli number* (see Problem 24.10).

Answers or Hints

36.1. Similar to Examples 36.1 and 36.2.
36.2. Consider the function $f(z) = 1/z^{2k}$ and see Problem 24.10.

Lecture 37
Argument Principle and Rouché and Hurwitz Theorems

We begin this lecture with an extension of Theorem 26.3 known as the Argument Principle. This result is then used to establish Rouché's Theorem, which provides locations of the zeros and poles of meromorphic functions. We shall also prove an interesting result due to Hurwitz.

Theorem 37.1 (Argument Principle). Let $f(z)$ be meromorphic inside and on a positively oriented contour γ. Furthermore, let $f(z) \neq 0$ on γ. Then,

$$\frac{1}{2\pi i} \int_\gamma \frac{f'(z)}{f(z)} dz = Z_f - P_f, \tag{37.1}$$

holds, where Z_f (P_f) is the number of zeros (poles), counting multiplicities of $f(z)$ that lie inside γ.

Proof. Let $a_1, \cdots, a_\ell$ be the zeros and $b_1, \cdots, b_p$ be the poles of $f(z)$ in γ with respective multiplicities $m_1, \cdots, m_\ell$ and $n_1, \cdots, n_p$ such that $Z_f = m_1 + \cdots + m_\ell$ and $P_f = n_1 + \cdots + n_p$. Then, from (26.3), we obtain

$$R\left[\frac{f'}{f}, a_k\right] = m_k. \tag{37.2}$$

Now, similar to (26.2) at the pole b_s, we have

$$f(z) = \frac{1}{(z - b_s)^{n_s}} h(z), \tag{37.3}$$

where $h(z)$ is analytic and nonzero in a neighborhood of b_s. From (37.3), it follows that

$$\frac{f'(z)}{f(z)} = -\frac{n_s}{z - b_s} + \frac{h'(z)}{h(z)}, \tag{37.4}$$

and hence

$$R\left[\frac{f'}{f}, b_s\right] = -n_s. \tag{37.5}$$

Finally, using Theorem 31.2 and the formulas (37.2) and (37.5), we find

$$\int_\gamma \frac{f'(z)}{f(z)} dz = 2\pi i \left[(m_1 + \cdots + m_\ell) - (n_1 + \cdots + n_p)\right] = 2\pi i [Z_f - P_f],$$

which is the same as (37.1).   ∎

Remark 37.1. The nomenclature argument principle for (37.1) comes from the fact that its left-hand side can be interpreted as the change in the argument as one runs around the image path $f(\gamma)$. More precisely, the relation (37.1) implies that $Z_f - P_f = (1/2\pi)$ [change in $\arg(f(z))$ as z traverses γ once in the positive direction].

Example 37.1. Consider the function

$$f(z) \;=\; \frac{(z-2)^3(z-1)^7 z^3}{(z-i)^4(z+3)^5(z-2i)^7}$$

and $\gamma : |z| = 2.5$. Clearly, $Z_f = 3 + 7 + 3 = 13$ and $P_f = 4 + 7 = 11$, and hence $Z_f - P_f = 2$. We also note that

$$\frac{f'(z)}{f(z)} \;=\; \frac{3}{z-2} + \frac{7}{z-1} + \frac{3}{z} - \frac{4}{z-i} - \frac{5}{z+3} - \frac{7}{z-2i},$$

and hence

$$\frac{1}{2\pi i}\int_\gamma \frac{f'(z)}{f(z)}dz \;=\; \frac{1}{2\pi i}[2\pi i(3+7+3) - 2\pi i(4+7)] \;=\; 2 \;=\; Z_f - P_f.$$

Lemma 37.1. Suppose that $f(z)$ is continuous and assumes only integer values on a domain S. Then, $f(z)$ is a constant on S.

Proof. For each integer n, let $S_n \subset S$ be such that, for $z \in S_n$, $f(z) = n$. Clearly, the sets S_n are disjoint and their union is S. We claim that each S_n is open. For this, let $z_0 \in S_n$, so that $f(z_0) = n$. Since $f(z)$ is continuous at z_0, there exists an open disk $B(z_0, r)$ such that $|f(z) - n| < 1/2$ for all $z \in B(z_0, r)$. However, since $f(z)$ is integer-valued, this implies that $f(z) = n$ for all $z \in B(z_0, r)$. Hence, S_n is open. Now, since a connected set cannot be the union of disjoint open sets, only one of the S_n can be nonempty, say S_{n_0}, and thus $f(z) = n_0$ for all z in S.   ∎

Now we shall use Theorem 37.1 and Lemma 37.1 to prove the following result.

Theorem 37.2 (Rouché's Theorem). Suppose $f(z)$ and $g(z)$ are meromorphic functions in a domain S. If $|f(z)| > |g(z)|$ for all z on γ, where γ is a simply closed positively oriented contour in S and $f(z)$ and $g(z)$ have no zeros or poles on γ, then

$$Z_f - P_f \;=\; Z_{f+g} - P_{f+g}. \tag{37.6}$$

Proof. We claim that the function $f(z) + g(z)$ has no zeros on γ. Indeed, if $f(z_0) + g(z_0) = 0$ for some z_0 on γ, then $|f(z_0)| = |g(z_0)|$, which contradicts the hypothesis. Now, since $|f(z)| - |g(z)|$ is continuous on the compact set γ, there exists a constant $m > 0$ such that $|f(z)| - |g(z)| > m > 0$ on γ. Thus, for any $\lambda \in [0, 1]$, it follows that

$$|f(z) + \lambda g(z)| \geq |f(z)| - |g(z)| \geq m > 0$$

for all z on γ. Therefore,

$$J(\lambda) = \frac{1}{2\pi i} \int_\gamma \frac{f'(z) + \lambda g'(z)}{f(z) + \lambda g(z)} dz$$

is a continuous function of λ. But, in view of Theorem 37.1, $J(\lambda) = Z_{f+\lambda g} - P_{f+\lambda g}$, which is an integer, and hence, from Lemma 37.1, we have $J(0) = J(1)$; i.e., (37.6) holds. ∎

Corollary 37.1. Suppose $f(z)$ and $g(z)$ are analytic functions in a domain S. If $|f(z)| > |g(z)|$ for all z on γ, where γ is a simply closed contour in S, then $f(z)$ and $f(z) + g(z)$ have the same number of zeros inside γ counting multiplicities.

Example 37.2. We shall show that the function $\phi(z) = 2z^5 - 6z^2 + z + 1$ has three zeros in the annulus $1 \leq |z| \leq 2$. For this, first we let $f(z) = -6z^2$ and $g(z) = 2z^5 + z + 1$ on $|z| = 1$, so that $|f(z)| = 6$ and $|g(z)| \leq 2 + 1 + 1 = 4$, and hence $|f(z)| > |g(z)|$ on $|z| = 1$. Thus, by Corollary 37.1, $\phi(z)$ has two zeros in $|z| < 1$. Next, we let $f(z) = 2z^5$ and $g(z) = -6z^2 + z + 1$ on $|z| = 2$, so that $|f(z)| = 64$ and $|g(z)| \leq 24 + 2 + 1 = 27$, and hence $|f(z)| > |g(z)|$ on $|z| = 2$. Thus, by Corollary 37.1, $\phi(z)$ has five zeros in $|z| < 2$. Therefore, $\phi(z)$ has three zeros in $1 \leq |z| \leq 2$.

Example 37.3. We shall show that the function $\phi(z) = 2 + z^2 + e^{iz}$ has exactly one zero in the open upper half-plane. For this, we let $f(z) = 2 + z^2$ and $g(z) = e^{iz}$ on $\gamma : [-R, R] \cup \{z : \operatorname{Im} z \geq 0, |z| = R\}$, $R > \sqrt{3}$. Clearly, for $z \in [-R, R]$, $|f(z)| \geq 2 > 1 = |g(z)|$, and for $z = Re^{i\theta}$, $0 \leq \theta \leq \pi$, $|f(z)| \geq R^2 - 2 > 1 \geq e^{-R \sin \theta} = |g(z)|$. Thus, from Corollary 37.1, $\phi(z)$ has the same number of roots in $\{z : \operatorname{Im} z > 0, |z| < R\}$ as $2 + z^2$, which is exactly one.

Example 37.4. As an application of Corollary 37.1, we can prove the Fundamental Theorem of Algebra (Theorem 19.1). For this, we let $f(z) = a_n z^n$ and $g(z) = a_{n-1} z^{n-1} + \cdots + a_1 z + a_0$ on $|z| = R$, so that $|f(z)| = |a_n| R^n$ and $|g(z)| \leq |a_{n-1}| R^{n-1} + \cdots + |a_1| R + |a_0|$. Hence, $|f(z)| > |g(z)|$ on $|z| = R$ if we choose $R > 1$ so large that

$$\frac{|a_{n-1}|}{|a_n|} + \cdots + \frac{|a_1|}{|a_n|} + \frac{|a_0|}{|a_n|} < R.$$

Since $f(z)$ has n zeros in $|z| = R$, $P_n(z) = f(z) + g(z)$ also has exactly n zeros in $|z| = R$.

Now, for each n, let $p_n(z)$ be the polynomial

$$p_n(z) = 1 + z + \frac{z^2}{2!} + \cdots + \frac{z^n}{n!}.$$

It is clear that it has n zeros and the sequence $\{p_n(z)\}$ converges to e^z uniformly on any closed disk $\overline{B}(0, R)$. Thus, as n increases, the number of zeros of $p_n(z)$ being n increases while the limit function e^z has no zeros at all. In the following result, we shall show that in the limit all zeros of $p_n(z)$ go into the point at infinity.

Theorem 37.3. For a given $R > 0$, there exists an n such that, for any $m \geq n$, $p_m(z)$ has no zero in the open disk $B(0, R)$.

Proof. Let $\epsilon = \inf\{|e^z| : |z| \leq R\}$. Since e^z is continuous and nonzero, and $\overline{B}(0, R)$ is compact, it is clear that $\epsilon > 0$. On $\overline{B}(0, R)$, the sequence $\{p_n(z)\}$ converges uniformly to e^z, so there is n such that, for all $m \geq n$, $|p_m(z) - e^z| < \epsilon < |e^z|$ for $z \in \overline{B}(0, R)$. Now let $f(z) = e^z$ and $g(z) = p_m(z) - e^z$. Then, from Corollary 37.1, it follows that e^z and $f(z) + g(z) = p_m(z)$ have the same number of zeros in $B(0, R)$. But e^z has no zeros in $B(0, R)$. ∎

We conclude this lecture by proving the following result, which generalizes Theorem 37.3.

Theorem 37.4 (Hurwitz's Theorem). Suppose that $\{f_n(z)\}$ is a sequence of analytic functions on a domain S converging uniformly on every compact subset of S to a function $f(z)$. Then, either

(i). $f(z)$ is identically zero on S or

(ii). if $B(z_0, r)$ is any open disk in S such that $f(z)$ does not have zeros on its boundary, then $f_n(z)$ and $f(z)$ have the same number of zeros in $B(z_0, r)$ for all sufficiently large n.

Proof. From Theorem 22.6, the function $f(z)$ is analytic on S. Now suppose that $f(z)$ is not identically zero on S, and let $z_0 \in S$ be a zero of $f(z)$. Let $\overline{B}(z_0, r)$ be a closed disk such that $f(z)$ does not vanish on its boundary. Let $0 < \epsilon = \inf|f(z)|$ on the boundary of $\overline{B}(z_0, r)$. Then, in view of uniform convergence, there exists an n such that, for all $m \geq n$, $|f_m(z) - f(z)| < \epsilon \leq |f(z)|$ on the boundary of $\overline{B}(z_0, r)$. Now let $g(z) = f_m(z) - f(z)$. Then, from Corollary 37.1, it follows that $f(z)$ and $f(z) + g(z) = f_m(z)$ have the same number of zeros in $B(z_0, r)$. ∎

Corollary 37.2. Let $\{f_n(z)\}$ be a sequence of one-to-one analytic functions defined on a domain S, and suppose $\{f_n(z)\}$ converges uniformly to $f(z)$ on every compact subset of S. Then, the function $f(z)$ is also one-to-one and analytic on S.

Example 37.5. The sequence $\{e^z/n\}$ converges uniformly to zero on every compact subset of $\mathbf{C}$. Note that functions e^z/n have no zeros in the complex plane.

Problems

37.1. Let $f(z)$ be meromorphic in a simply connected domain S, and let γ be a simply closed positively oriented contour in S such that, for all z on γ, $f(z) \neq 0$ and $f(z) \neq \infty$. Then,

$$w(f(\gamma), a) = \frac{1}{2\pi i} \int_\gamma \frac{f'(z)}{f(z) - a} dz$$

is called the *winding number* of $f(\gamma)$ about a. It represents the number of times the curve $f(\gamma)$ winds around the point a. Clearly, Theorem 37.1 gives $w(f(\gamma), 0) = Z_f - P_f$. If γ is not a simple closed curve, but crosses itself several times, then $w(f(\gamma), a)$ also gives the number of times the curve $f(\gamma)$ winds around the point a. Find $w(f(\gamma), a)$ for the function $f(z) = z^7 - z$ when

(a). $\gamma : |z| = 2$, $a = 0$,

(b). $\gamma : |z| = 0.5$, $a = 0$,

(c). $\gamma : |z - i| = 1.5$, $a = 0$.

37.2. If $a > e$, show that the equation $e^z = az^n$ has n roots inside the circle $|z| = 1$.

37.3. Show that the function $z^8 - 5z^5 - 2z + 1$ has five zeros inside the unit circle $|z| = 1$.

37.4. Show that all nine zeros of $z^9 - 8z^2 + 5$ lie in the annulus $1/2 \leq |z| \leq 3/2$.

37.5. Show that all five zeros of $z^5 - 2z + 16$ lie in the annulus $1 \leq |z| \leq 2$.

37.6. Show that the equation $z + e^{-z} = c$ $(c > 1)$ has a unique real root in the right half-plane.

Answers or Hints

37.1. (a). 7, (b). 1, (c). 5.
37.2. Let $f(z) = az^n$ and $g(z) = -e^z$ on $|z| = 1$, and apply Corollary 37.1.
37.3. Let $f(z) = -5z^5 + 1$ and $g(z) = z^8 - 2z$.

37.4. Follow Example 37.2.
37.5. Follow Example 37.2.
37.6. Follow Example 37.3.

Lecture 38
Behavior of Analytic Mappings

In this lecture, we shall use the Rouché Theorem to investigate the behavior of the mapping f generated by an analytic function $w = f(z)$. Then, we shall study some properties of the inverse mapping f^{-1}. We shall also discuss functions that map the boundaries of their domains to the boundaries of their ranges. Such results are of immense value for constructing solutions of the Laplace equation with boundary conditions.

Let the function $w = f(z)$ be analytic at z_0 and $w_0 = f(z_0)$. We say that f has *order* $m \geq 1$ at z_0 if $f(z) - w_0$ has a zero of order m at z_0. Clearly, f has order $m \geq 2$ at z_0 if and only if $f^{(k)}(z_0) = 0$, $k = 1, \cdots, m-1$ and $f^{(m)}(z_0) \neq 0$. In particular, f has order one at z_0 if and only if $f'(z_0) \neq 0$.

Theorem 38.1 (Local Mapping Theorem). Let $w = f(z)$ be a nonconstant analytic function in a neighborhood of z_0. Let $w_0 = f(z_0)$ and m be the order of f at z_0. Then, there exist $r > 0$ and $\rho > 0$ such that every $w \neq w_0$ in $B(w_0, \rho)$ is attained by f at exactly m distinct points in $B(z_0, r)$.

Proof. Since $f(z) - w_0$ has a zero of order $m \geq 1$, from Theorem 26.1 it follows that

$$f(z) - w_0 = a_m(z - z_0)^m + a_{m+1}(z - z_0)^{m+1} + \cdots, \qquad a_m \neq 0. \quad (38.1)$$

Thus, $\lim_{z \to z_0} (f(z) - w_0)/(z - z_0)^m = a_m \neq 0$, and therefore there exists an $r > 0$ such that

$$\frac{|f(z) - w_0|}{|z - z_0|^m} > \frac{|a_m|}{2}$$

for all $0 < |z - z_0| \leq r$. Hence, for $z \neq z_0$, it follows that $f(z) \neq w_0$, and for $|z - z_0| = r$, we have

$$|f(z) - w_0| > \frac{|a_m| r^m}{2}. \qquad (38.2)$$

Again, since $m \geq 1$, in view of (38.1), we can choose r sufficiently small so that $f'(z) \neq 0$ for all $0 < |z - z_0| < r$. Now let $\rho = |a_m| r^m / 2$ and, for a fixed w such that $|w - w_0| < \rho$, set $g(z)$ to be the constant $w_0 - w$. From (38.2), for $|z - z_0| = r$, we have

$$|g(z)| = |w_0 - w| < \frac{|a_m| r^m}{2} < |f(z) - w_0|.$$

Thus, the Rouché Theorem with $f = f - w_0$ and $g = w_0 - w$ is applicable, and we have $Z_{f-w_0} = Z_{f-w_0+g} = Z_{f-w}$. Since $Z_{f-w_0} = m$, it follows that $Z_{f-w} = m$ for all $|w - w_0| < \rho$. This implies that w has m inverse images under the mapping $w = f(z)$ in $|z - z_0| < r$. Finally, the fact that $f'(z) \neq 0$ for all $0 < |z - z_0| < r$ ensures that $f(z) - w$ does not have repeated roots, and hence the inverse images are distinct. ∎

The particular case $m = 1$ of Theorem 38.1 will be used later, and hence we state and prove it separately.

Theorem 38.2 (Inverse Function Theorem). Let $w = f(z)$ be an analytic function in a neighborhood of z_0. Then, there exists an $r > 0$ such that f is one-to-one on $B(z_0, r)$ if and only if $f'(z_0) \neq 0$.

Proof. In Problem 7.12 (Answers or Hints), we used elementary calculus to show that there exists an open disk at z_0 on which f is one-to-one provided $f'(z_0) \neq 0$. Here, we shall provide an alternative proof: Consider the disk $B(z_0, r_1)$ where $r_1 > 0$ is sufficiently small. On $B(z_0, r_1)$, we define the function

$$F(z) = f(z) - f(z_0) - (z - z_0)f'(z_0).$$

Clearly, F is continuous and, for $z_1, z_2 \in B(z_0, r_1)$ with $z_1 \neq z_2$,

$$F(z_2) - F(z_1) = (f(z_2) - f(z_1)) - (z_2 - z_1)f'(z_0) = \int_{z_1}^{z_2} (f'(z) - f'(z_0))dz.$$

Let $r > 0$ be such that $0 < r < r_1$ and, for $|z - z_0| < r$, $|f'(z) - f'(z_0)| < |f'(z_0)|/2$. Then, for $z_1, z_2 \in B(z_0, r)$, $|F(z_1) - F(z_2)| \leq |z_2 - z_1||f'(z_0)|/2$. Hence, it follows that

$$
\begin{aligned}
|f(z_2) - f(z_1)| &= |(z_2 - z_1)f'(z_0) + (F(z_2) - F(z_1))| \\
&\geq |z_2 - z_1||f'(z_0)| - |F(z_2) - F(z_1)| \\
&\geq |z_2 - z_1||f'(z_0)|/2 > 0,
\end{aligned}
$$

and therefore $f(z_1) \neq f(z_2)$. Conversely, if $f'(z_0) = 0$, we can apply Theorem 38.1 to find a neighborhood of z_0 on which f is not one-to-one. ∎

Remark 38.1. Theorem 38.2 says that an analytic function $f(z)$ is one-to-one in some neighborhood $B(z_0, r)$ of a given point z_0 where $f'(z_0) \neq 0$; i.e., one-to-one is only a *local* property. For example, the exponential function e^z is one-to-one only locally, although its derivative does not vanish anywhere. Similarly, the function z^2 is locally one-to-one at every point other than the origin. In the following result, we shall estimate the size of the neighborhood $B(z_0, r)$; i.e., provide an upper bound on r.

Theorem 38.3 (Landau's Estimate). Let $w = f(z)$ be an analytic function on a closed disk $\overline{B}(z_0, R)$ and $f'(z_0) \neq 0$. Then, f is one-to-one on $B(z_0, r)$, where $r = R^2|f'(z_0)|/(4M)$ and M is the maximum value of $|f(z)|$ on $|z - z_0| = R$.

Proof. Clearly, it suffices to choose r such that $f'(z) \neq 0$ for all $z \in B(z_0, r)$. We assume that both z_0 and $w_0 = f(z_0)$ are zero; otherwise, we can use the transformation $Z = z - z_0$, $W = w - w_0$. Since $f(z)$ is analytic on $\overline{B}(0, R)$, for $|z| < R$ it can be written as a convergent Taylor series $f(z) = a_1 z + a_2 z^2 + \cdots$, where $|f'(0)| = |a_1| > 0$. Let λ be so small that

$$\sum_{j=2}^{\infty} j |a_j| \lambda^{j-1} R^{j-1} < |a_1|. \tag{38.3}$$

Then, for any $z \in B(0, \lambda R)$, we have

$$|f'(z)| = \left| a_1 + \sum_{j=2}^{\infty} j a_j z^{j-1} \right| \geq |a_1| - \sum_{j=2}^{\infty} j |a_j| \lambda^{j-1} R^{j-1} > 0.$$

Thus, if we can find a value of λ such that (38.3) holds, then we can take $r = \lambda R$. For this, since M is the maximum value of $|f(z)|$ on $|z - z_0| = R$, from Theorem 18.5 it follows that $|a_j| \leq M/R^j$, $j = 1, 2, \cdots$, and therefore

$$\sum_{j=2}^{\infty} j |a_j| \lambda^{j-1} R^{j-1} \leq \sum_{j=2}^{\infty} \frac{M}{R} j \lambda^{j-1} = \frac{M}{R} \frac{\lambda(2-\lambda)}{(1-\lambda)^2} < \frac{M}{R} \frac{2\lambda}{(1-\lambda)^2}.$$

We now take $\lambda = R|a_1|/(4M)$. Clearly, then $\lambda \leq 1/4$, and it follows that

$$\sum_{j=2}^{\infty} j |a_j| \lambda^{j-1} R^{j-1} < \frac{8}{9} |a_1| < |a_1|. \quad \blacksquare$$

Now we shall deduce the following corollaries from Theorem 38.1.

Corollary 38.1 (Open Mapping Property). A nonconstant analytic function f maps open sets into open sets.

Proof. Let S be an open set. For $w_0 \in f(S)$, let $z_0 \in S$ be such that $f(z_0) = w_0$. Since S is open, there exists an $r > 0$ such that $B(z_0, r) \subset S$. In view of Theorem 38.1, there exists a ρ such that every $w \in B(w_0, \rho)$ is attained by f; i.e., $w = f(z)$, $z \in B(z_0, r)$. Thus, $B(w_0, \rho) \subset f(S)$, and hence $f(S)$ is open. $\blacksquare$

Remark 38.2. If a nonconstant function is just continuous, then the corollary above does not hold. Indeed, the function $f(z) = \operatorname{Re} z$ is continuous everywhere in $\mathbf{C}$, but $f(\mathbf{C}) = \mathbb{R}$ is not open in $\mathbf{C}$.

Corollary 38.2. Let $f(z)$ be analytic on a domain S and $z_0 \in S$. Then, there exist two distinct points z_1 and z_2 in S such that

$$f'(z_0) = \frac{f(z_1) - f(z_2)}{z_1 - z_2}.$$

Proof. Consider the function $g(z) = f(z) - f'(z_0)z$. Clearly, $g(z)$ is analytic in S and $g'(z_0) = 0$. Thus, by Theorem 31.1 there exist points z_1 and z_2 such that $g(z_1) = g(z_2)$. ∎

The next corollary shows that the analytic image of a domain is a domain.

Corollary 38.3 (Mapping of Domains).

Let S be a nonempty domain, and f be a nonconstant analytic function on S. Then, $f(S)$ is a domain.

Proof. From Corollary 38.1, $f(S)$ is open, and since f is continuous from Problem 5.13(ii) (Answer or Hints) it is clear that $f(S)$ is connected. Thus, $f(S)$ is a domain. ∎

Remark 38.3. The Maximum Modulus Principle (Theorem 20.1) follows from Corollary 38.3. For this, suppose that $f(z)$ is nonconstant. Let $z_0 \in S$ and $w_0 = f(z_0)$. The image of $B(z_0, r) \subset S$ under f is a domain containing w_0, and hence contains $B(w_0, \rho)$. Therefore, $B(z_0, r)$ contains a point z whose image $w = f(z)$ is farther from the origin of the w-plane than the point w_0; i.e., a point z such that $|f(z)| > |f(z_0)|$.

The following result proves the existence of the inverse function and provides an expression for its derivative.

Theorem 38.4 (Inverse Function Theorem).

Let $w = f(z)$ be an analytic and one-to-one function on an open set S. Then, the inverse function f^{-1} exists and is analytic on the open set $f(S)$. Moreover,

$$\frac{d}{dw} f^{-1}(w) = \frac{1}{f'(z)}. \tag{38.4}$$

Proof. In view of Corollary 38.1, $f(S)$ is an open set. Since f is one-to-one, there exists an inverse function $g = f^{-1} : f(G) \to G$. The continuity of g also follows from Corollary 38.1. Now let $z_0 = g(w_0) \in S$. By Theorem 38.2, $f'(g(w_0)) \neq 0$. Thus, we have

$$\lim_{w \to w_0} \frac{g(w) - g(w_0)}{w - w_0} = \lim_{g(w) \to g(w_0)} \frac{g(w) - g(w_0)}{f(g(w)) - f(g(w_0))} = \frac{1}{f'(g(w_0))},$$

i.e., (38.4) for $z = z_0$, $w = w_0$ holds. ∎

The next result we state gives a formula for the inverse function of f in terms of an integral involving f and its derivative.

Theorem 38.5.

Let $w = f(z)$ be an analytic function on an open set containing $\overline{B}(z_0, r)$, and suppose that f is one-to-one on $B(z_0, r)$. If

$S' = f(B(z_0, r))$ and $\gamma : |z - z_0| = r$, then $f^{-1}(w)$ is defined for each w in S' and given by

$$f^{-1}(w) = \frac{1}{2\pi i} \int_\gamma \frac{z f'(z)}{f(z) - w} dz.$$

Now let S be a domain. We say that f maps ∂S onto $\partial f(S)$ if, for every point $w_0 \in \partial f(S)$, there exists a sequence $\{z_n\} \subseteq S$ converging to $z_0 \in \partial S$, and $f(z_n) \to w_0$.

Theorem 38.6. Let $w = f(z)$ be an analytic and one-to-one function on a domain S. Then, f maps ∂S onto $\partial f(S)$.

Proof. We shall prove only that f maps ∂S to $\partial f(S)$. Let $\{z_n\} \subseteq S$ be a sequence such that $f(z_n) \to w_0$, where w_0 is an interior point of $f(S)$. Since in view of Theorem 38.4 the function f^{-1} exists and is continuous, it follows that

$$\lim_{n\to\infty} z_n = \lim_{n\to\infty} f^{-1}(f(z_n)) = f^{-1}\left(\lim_{n\to\infty} f(z_n)\right) = f^{-1}(w_0);$$

i.e., $\{z_n\}$ converges to an interior point of S. ∎

The following corollary of Theorem 38.6 is immediate.

Corollary 38.4. Let $w = f(z)$ be an analytic and one-to-one function on a domain S. Let $z_0 \in \partial S$ and let f be continuous at z_0. Then, $f(z_0) \in \partial f(S)$.

We conclude this lecture by stating the following interesting result.

Theorem 38.7. Let $w = f(z)$ be an analytic function on a simply connected domain S, and let γ be a simple, closed, smooth curve in S. If f is one-to-one on γ, then it is also one-to-one in the interior of γ.

Lecture 39
Conformal Mappings

In Lecture 10, we saw that the nonconstant linear mapping (10.1) is an expansion or contraction and a rotation, followed by a translation. Thus, under a linear mapping, the angle between any two intersecting arcs in the z-plane is equal to the angle between the images in the w-plane. Mappings that have this *angle-preserving* property are called *conformal mappings*. These mappings are of immense importance in solving boundary value problems involving Laplace's equation. We begin with the following definitions.

Let z_0 be a fixed point in the z-plane and γ_1 and γ_2 be smooth paths that intersect at z_0, and let ℓ_1 and ℓ_2 denote the tangent lines on γ_1 and γ_2 at z_0. The paths γ_1 and γ_2 are said to intersect at an angle α at z_0 if the tangent lines ℓ_1 and ℓ_2 intersect at an angle α at z_0 (see Figure 39.1(a)). To determine this angle precisely in magnitude and sense, we assume that the paths γ_1 and γ_2 are parameterized by $z_1(t)$ and $z_2(t)$, $t \in [\alpha, \beta]$, which intersect at $z_1(t_0) = z_2(t_0) = z_0$. Now, since γ_1 is smooth, $z_1' = z_1'(t_0)$ is nonzero, and hence the angle between the tangent line ℓ_1 and the positive x-axis, denoted as $\arg(z_1')$, is well-defined. The angle $\arg(z_2')$ is defined similarly. The angle α (see Figure 39.1(b)) between γ_1 and γ_2 at z_0 is then

$$\alpha = \arg(z_2') - \arg(z_1'). \tag{39.1}$$

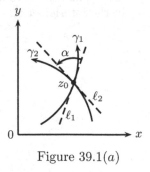

Figure 39.1(a)

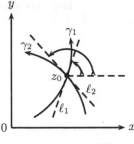

Figure 39.1(b)

Now, let $w = f(z)$ be a mapping defined on a domain $S \subseteq \mathbf{C}$ and let $z_0 \in S$. This mapping is said to be *conformal* at z_0 if it is one-to-one in a neighborhood of the point z_0 and for every pair of smooth paths γ_1 and γ_2 in S intersecting at z_0 the angle between γ_1 and γ_2 at z_0 is equal to the angle between the image paths γ_1' and γ_2' at $f(z_0)$ in magnitude and sense.

Furthermore, if $w = f(z)$ maps S onto a domain S' and is conformal at every point in S, then $w = f(z)$ is called a *conformal mapping* of S onto S'. Moreover, S' is called a *conformal image* of S.

The following result provides an easier test for the conformality of the mapping $w = f(z)$ at the point z_0.

Theorem 39.1. If $f(z)$ is analytic at z_0 and $f'(z_0) \neq 0$, then the mapping $w = f(z)$ is conformal at z_0.

Proof. Since $f'(z_0) \neq 0$ from Theorem 38.2, it follows that there exists an open disk at z_0 on which f is one-to-one. Let γ be a smooth path passing through z_0. We parameterize γ by $z(t)$ and assume that $z(t_0) = z_0$ and $z'(t_0) \neq 0$. The image of γ by f is a path parameterized by $f(z(t))$ that passes through $w_0 = f(z(t_0)) = f(z_0)$ in the w-plane. Since $z'(t_0) \neq 0$, the angle between the tangent line ℓ on γ at z_0 and the positive x-axis is $\arg(z'(t_0))$. Now, since

$$\frac{d}{dt} f(z(t)) \bigg|_{t_0} = f'(z(t_0)) z'(t_0) \neq 0,$$

the angle between the tangent line ℓ' on $f(z(t))$ at $w_0 = f(z_0)$ and the positive u-axis is

$$\arg(f'(z(t_0)) z'(t_0)) = \arg(f'(z(t_0))) + \arg(z'(t_0)).$$

Thus, $f(z)$ rotates the tangent line at z_0 by a fixed angle $\arg(f'(z_0))$. The result now follows from (39.1) and the fact that f rotates any two tangent lines intersecting at z_0 by the same angle $\arg(f'(z_0))$. ∎

Remark 39.1. In view of Remark 38.1, it is clear that Theorem 39.1 ensures conformality of the mapping $w = f(z)$ only *locally* at points where its derivative does not vanish.

Remark 39.2. If a one-to-one analytic function f maps an open set S onto the open set $f(S)$, then in view of Theorem 38.4 the inverse function $f^{-1} : f(S) \to S$ exists and is analytic. Now, since f^{-1} is one-to-one on $f(S)$, from Theorems 38.2 and 39.1 it follows that f^{-1} is also a conformal map.

Example 39.1. The exponential map $f(z) = e^z$ is conformal at all $z \in \mathbf{C}$. The map $f(z) = z^2$ is conformal everywhere except at the origin. The Möbius transformation $f(z) = (az+b)/(cz+d)$ is conformal everywhere except at $z = -d/c$. The mapping $w = P_n(z)$, where $P_n(z)$ is a polynomial of degree n, is conformal everywhere except possibly at ∞ and the points $z_1, z_2, \cdots, z_r$ $(1 \leq r \leq n-1)$, where the derivative $P'_n(z)$ vanishes.

Let $f(z)$ be a nonconstant analytic function at z_0. If $f'(z_0) = 0$, then z_0 is called a *critical point* of $f(z)$. Our next result shows that at critical points magnification of angles occurs, which in turn implies that at critical points analytic functions are not conformal.

Theorem 39.2. Let $f(z)$ be analytic at z_0. If $f'(z_0) = f''(z_0) = \cdots = f^{(n-1)}(z_0) = 0$ and $f^{(n)}(z_0) \neq 0$, then the mapping $w = f(z)$ magnifies angles at z_0 by a factor n.

Proof. From Taylor's series expansion of $f(z)$ at z_0, it follows that

$$f(z) = f(z_0) + (z - z_0)^n g(z), \tag{39.2}$$

where the function

$$g(z) = a_n + a_{n+1}(z - z_0) + a_{n+2}(z - z_0)^2 + \cdots$$

is analytic at z_0 and $g(z_0) = a_n = f^{(n)}(z_0)/n! \neq 0$. From (39.2), we obtain

$$\arg(w - w_0) = \arg(f(z) - f(z_0)) = n \arg(z - z_0) + \arg(g(z)). \tag{39.3}$$

Now, if γ is a smooth curve that passes through z_0 and $z \to z_0$ along γ, then $w \to w_0$ along the image curve γ'. Thus, the angle of inclination of the tangents ℓ on γ and ℓ' on γ', respectively, are $\theta = \lim_{z \to z_0} \arg(z - z_0)$ and $\phi = \lim_{w \to w_0} \arg(w - w_0)$. Hence, from (39.3), it follows that

$$\phi = \lim_{z \to z_0} (n \arg(z - z_0) + \arg(g(z))) = n\theta + \psi, \tag{39.4}$$

where $\psi = \lim_{z \to z_0} \arg(g(z))$. The result now follows from (39.1) and (39.4). ∎

Example 39.2. For the mapping $f(z) = \cos z$ critical points are $z = n\pi$, $n = 0, 1, 2, \cdots$. Since $f''(z) = -\cos z = \pm 1$ at the critical points, Theorem 39.2 indicates that angles at these critical points are magnified by a factor of 2.

Now we shall show that, given any two finite simply connected domains S and S', there exists a conformal mapping of S onto S'. For this, we need the following existence theorem of Riemann.

Theorem 39.3 (Riemann Mapping Theorem). If S is any simply connected domain in the complex plane (other than the entire plane itself), then there exists a one-to-one conformal mapping $w = f(z)$ that maps S onto the unit disk $|w| < 1$.

Corollary 39.1. Let S and S' be any two simply connected domains (none of them is the entire plane). Then, S and S' are conformal images of each other.

Proof. We use Theorem 39.3 to find a conformal mapping f from S onto the open unit disk $|w| < 1$. We apply Theorem 39.3 again to find a conformal mapping g from S' onto the open unit disk $|w| < 1$. Since g is one-to-one, g^{-1} exists and maps the open disk $|w| < 1$ onto S'. Then, $w = g^{-1} \circ f(z)$ is the desired mapping. ∎

Remark 39.3. The function $w = f(z)$ that maps S conformally onto S' is not unique. For this, first we map S conformally onto the unit disk, then map the unit disk conformally onto itself, and then map the unit disk conformally onto S'. The resulting composite mapping maps S conformally onto S'. However, in view of Problem 39.8, the unit disk can be mapped conformally onto itself in an infinite number of ways, and hence S can be mapped onto S' conformally in an infinite number of ways. In our next result we shall provide sufficient conditions so that the conformal mapping of S onto S' is unique. For this, we need the following lemma, which is of independent interest.

Lemma 39.1 (Schwarz's Lemma). Let $f(z)$ be analytic on the open disk $B(0, R)$, and $f(0) = 0$, $|f(z)| \leq M < \infty$. Then, the inequality

$$|f(z)| \leq \frac{M}{R}|z| \qquad (39.5)$$

holds for all $z \in B(0, R)$, and

$$|f'(0)| \leq \frac{M}{R}. \qquad (39.6)$$

In (39.5) equality occurs at some point $0 \neq z \in B(0, R)$ if and only if $f(z) = (M/R)\alpha z$, where α is a complex number of absolute value 1.

Proof. Consider the function $f_1(z) = f(Rz)/M$. Obviously, $f_1(z)$ is analytic for $|z| < 1$ and $f_1(0) = 0$. Also, for $|z| < 1$, we have $|f_1(z)| = |f(Rz)/M| \leq M/M = 1$ (since $|Rz| = R|z| < R$). Let $f_1(z) = a_1 z + a_2 z^2 + \cdots$ be the power series for $f_1(z)$. The constant term is 0 because $f_1(0) = 0$. Then, $f_1(z)/z$ is analytic, and $|f_1(z)/z| \leq 1/r$ for $|z| = r < 1$. This inequality also holds for $|z| \leq r$ by the Maximum Modulus Principle. Letting $r \to 1$, we find $|f_1(z)/z| \leq 1$ for all $z \in B(0, 1)$, which is the same as (39.5). For the inequality (39.6), it suffices to note that

$$\lim_{z \to 0} \frac{f_1(z)}{z} = \lim_{z \to 0} \frac{f_1(z) - f_1(0)}{z - 0} = f_1'(0) = \frac{R}{M} f'(0).$$

Finally, if $|f_1(z_0)/z_0| = 1$ for some z_0 in the unit disk, then again by the Maximum Modulus Principle $f_1(z)/z$ cannot have a maximum unless it is a constant, and therefore there is a constant α with $|\alpha| = 1$ such that $f_1(z)/z = \alpha$. ∎

Theorem 39.4 (Uniqueness Theorem). Let S be as in Theorem 39.3, and let $z_0 \in S$ be an arbitrary point. Then, there exists a unique function $w = f(z)$ that maps S conformally onto the unit disk $B(0,1)$ and satisfies $f(z_0) = 0$, $f'(z_0) > 0$.

Proof. Let $w = g(z)$ be another such function. Then, the function $\phi(w) = f(g^{-1}(w))$ is analytic on $B(0,1)$, satisfies $\phi(0) = f(g^{-1}(0)) = f(z_0) = 0$, $\phi'(0) = f'(z_0)/g'(z_0) > 0$, and maps $B(0,1)$ conformally onto itself. Thus, from Lemma 39.1, it follows that $|\phi(w)| \leq |w|$; i.e., $|f(z)| \leq |g(z)|$ for all $z \in S$. Now, interchanging the roles of f and g, we also have $|g(z)| \leq |f(z)|$ for all $z \in S$. This implies that $|f(z)| = |g(z)|$ for all $z \in S$; or equivalently, $|\phi(w)| = |w|$. But then, from Lemma 39.1, we find $\phi(w) = \alpha w$, where $|\alpha| = 1$. However, since $\phi'(0) > 0$, we must have $\alpha = 1$. Therefore, $\phi(w) = w$, and hence $f(z) = g(z)$, $z \in S$. ∎

Finally, we state a result that shows that the knowledge of $f(z)$ on the boundary of a domain G can be used to conclude that f maps S onto $f(S)$ conformally.

Theorem 39.5. Let γ be a closed rectifiable Jordan curve. Suppose $f(z)$ is analytic on $\overline{I(\gamma)}$ and one-to-one on γ. Then, f maps $I(\gamma)$ conformally onto $I(\Gamma)$, where $\Gamma = f(\gamma)$.

Problems

39.1. If $f(z)$ is analytic at z_0 and $f'(z_0) \neq 0$, show that the function $F(z) = \overline{f(z)}$ preserves the magnitude but reverses the sense of the angles at z_0.

39.2 (Noshiro-Warschawski Theorem). Let $f(z)$ be analytic in a convex domain S. If $\operatorname{Re} f'(z) > 0$ for all $z \in S$, show that $f(z)$ is one-to-one in S.

39.3. Explain why the complex plane and the open unit disk cannot be conformally equivalent.

39.4. Show that:

(a). If S' is a conformal image of a domain S, then S is a conformal image of S'.

(b). If S' is a conformal image of a domain S, and if S'' is a conformal image of S', then S'' is a conformal image of S.

(c). Use parts (a) and (b) to prove Corollary 39.1.

39.5. Show that the rational mapping $w = f(z) = P(z)/Q(z)$, where $P(z)$ and $Q(z)$ are polynomials, is conformal at any simple zero of $Q(z)$, and also at $z = \infty$ if the equation $f(z) = f(\infty)$ has no multiple roots.

39.6. Show that the mapping $w = \dfrac{e^z - i}{e^z + i}$ is conformal of the horizontal strip $0 < y < \pi$ onto the disk $|w| < 1$. Furthermore, the x-axis is mapped onto the lower semicircle bounding the disk, and the line $y = \pi$ is mapped onto the upper semicircle.

39.7. Show that the mapping $w = \left(\dfrac{1+z}{1-z}\right)^2$ is conformal of $\{z : |z| < 1,\ \mathrm{Im}(z) > 0\}$ onto $\{w : \mathrm{Im}(w) > 0\}$. Furthermore, $\{z : |z| = 1,\ \mathrm{Im}(z) \geq 0\}$ is mapped onto the negative w-axis, and the segment $-1 < x < 1,\ y = 0$, is mapped onto the positive w-axis.

39.8. Show that the mapping $w = e^{i\alpha}\dfrac{z - z_0}{z\overline{z}_0 - 1}$ is conformal of the unit circle $|z| < 1$ onto $|w| < 1$ so that the given interior point z_0 is transformed into the center of the circle; here, α is an arbitrary parameter.

39.9. Show that the mapping $w = \sin^2 z$ is conformal of the semi-infinite strip $0 < x < \pi/2,\ y > 0$ onto the upper half-plane $\mathrm{Im}(w) > 0$.

39.10. The *Joukowski mapping*

$$w = J(z) = \frac{1}{2}\left(z + \frac{1}{z}\right)$$

occurs when solving a variety of applied problems, particularly in aerodynamics. Show that:

(a). $J(z) = J(1/z)$.

(b). Under the mapping $w = J(z)$, every point of the w-plane except $w = \pm 1$ has exactly two distinct inverse images, z_1 and z_2 satisfying the relation $z_1 z_2 = 1$.

(c). J maps the unit circle $|z| = 1$ onto the real interval $[-1, 1]$.

(d). J maps both the interior and the exterior of the unit circle $|z| = 1$ into the same set in the w-plane.

(e). J maps the circle $|z| = r$ ($r > 0,\ r \neq 1$) onto the ellipse

$$\frac{u^2}{\left[\frac{1}{2}\left(r + \frac{1}{r}\right)\right]^2} + \frac{v^2}{\left[\frac{1}{2}\left(r - \frac{1}{r}\right)\right]^2} = 1,$$

which has foci at ± 1.

(f). J is a one-to-one continuous mapping of both the interior and the exterior of the unit circle $|z| = 1$ onto $\mathbf{C}\backslash[-1, 1]$.

(g). $w = J(z)$ is conformal everywhere except at $z = \pm 1$.

39.11. Let $f(z)$ be analytic on an open set S, and let $z_0 \in S$ and

$f'(z_0) \neq 0$. Show that

$$\frac{2\pi i}{f'(z_0)} = \int_\gamma \frac{1}{f(z) - f(z_0)} dz,$$

where γ is a small circle centered at z_0.

39.12. Let $f(z)$ be analytic on the open disk $B(0, R)$, and $|f(z)| \leq M < \infty$. Show that $|f(z) - f(0)| \leq 2(M/R)r$ for $|z| = r < R$.

39.13. Let $f(z)$ be analytic on the open disk $B(0, R)$, and $|f(z)| \leq M < \infty$. Show that

$$|f'(z_0)| \leq \frac{R}{M} \frac{M^2 - |f(z_0)|^2}{R^2 - |z_0|^2}, \quad z_0 \in B(0, R).$$

39.14 (Schwarz-Pick Lemma). Let $f : B(0,1) \to B(0,1)$ be an analytic function. Furthermore, let $w_1 = f(z_1)$, $w_2 = f(z_2)$, $z_1, z_2 \in B(0,1)$. Show that

$$\left| \frac{w_1 - w_2}{1 - w_1 \overline{w_2}} \right| \leq \left| \frac{z_1 - z_2}{1 - z_1 \overline{z_2}} \right| \quad \text{and} \quad \frac{|dw|}{1 - |w|^2} \leq \frac{|dz|}{1 - |z|^2}.$$

Answers or Hints

39.1. Since the mapping $w = f(z)$ is conformal at z_0 by Theorem 39.1, this follows from (39.1) and the fact that $\arg F(z) = -\arg f(z)$.
39.2. Write $f(z_1) - f(z_1)$ as an integral of f'.
39.3. If they are, then there exists a function $f : \mathbf{C} \to B(0,1)$ that is analytic and bijective. But then $|f(z)| < 1$, so by Theorem 18.6, $f(z)$ is a constant and hence cannot be bijective.
39.4. (a). See Remark 39.2. (b). Follows from the fact that the composition of two conformal maps is conformal. (c). Use Theorem 39.3 and parts (a) and (b).
39.5. Use Theorem 39.1.
39.6. Show that $z \to e^z$ conformally maps $0 < y < \pi$ onto $y > 0$ and takes $y = 0$ (resp. $y = \pi$) onto the positive (resp. negative) x-axis, while $z \to (z - i)/(z + i)$ conformally maps $y > 0$ onto $|z| < 1$ and takes the positive (resp. negative) x-axis onto the lower (resp. upper) semicircle.
39.7. Show that $z \to (1+z)/(1-z)$ conformally maps $\{z : |z| < 1, y > 0\}$ onto the first quadrant and takes $\{z : |z| = 1, y > 0\}$ (resp. the segment $-1 < x < 1, y = 0$) onto the positive y-axis (resp. positive x-axis), while $z \to z^2$ conformally maps the first quadrant onto $y > 0$ and takes the positive y-axis (resp. positive x-axis) onto the negative x-axis (resp. positive x-axis).

39.8. Verify directly.

39.9. Show that $z \to \sin z$ conformally maps $0 < x < \pi/2$, $y > 0$, onto the first quadrant, while $z \to z^2$ conformally maps the first quadrant onto $y > 0$.

39.10. (a). This is clear. (b). Solving $J(z) = w$ gives $z = w \pm \sqrt{w^2 - 1}$. (c). Follows from $J(e^{i\theta}) = \cos \theta$. (d). Follows from (a). (e). Setting $J(re^{i\theta}) = w$ gives $u = [(r+1/r)/2] \cos \theta$, $v = [(r-1/r)/2] \sin \theta$. (f). Follows from (b) and (e). (g). For $z \neq 0$, J is analytic and $J'(z) = (1 - 1/z^2)/2 \neq 0$ except at $z = \pm 1$.

39.11. Since $f'(z_0) \neq 0$, by Theorem 38.2, $f(z)$ is locally one-to-one in a neighborhood S_1 of z_0. Using the analyticity of $f(z)$ at z_0, from (6.2), we have

$$f(z) = f(z_0) + f'(z_0)(z - z_0) + \eta(z)(z - z_0) \tag{39.7}$$

in a small neighborhood S_2 of z_0; here, $\lim_{z \to z_0} \eta(z) = 0$. Take a closed disk $\overline{B}(z_0, r) \subset S_1 \cap S_2$ with γ as its boundary. Then, for any $z \in \gamma$, equation (39.7) holds and $f(z) \neq f(z_0)$. From (39.7), we have

$$\frac{f'(z_0)}{f(z) - f(z_0)} = \frac{1}{z - z_0} - \frac{\eta(z)}{f'(z_0) + \eta(z)} \frac{1}{z - z_0},$$

and hence

$$\frac{1}{2\pi i} \int_\gamma \frac{f'(z_0)}{f(z) - f(z_0)} dz = 1 - \frac{1}{2\pi i} \int_\gamma \frac{\eta(z)}{f'(z_0) + \eta(z)} \frac{dz}{z - z_0}.$$

It suffices to show that $\int_\gamma \frac{\eta(z)}{f'(z_0)+\eta(z)} \frac{dz}{z-z_0} = 0$. Since $\lim_{z \to z_0} \frac{\eta(z)}{f'(z_0)+\eta(z)} = 0$ for any $\epsilon > 0$ there exists a $0 < \delta < r$ such that $z \in \overline{B}(z_0, \delta)$ implies $\left| \frac{\eta(z)}{f'(z_0)+\eta(z)} \right| < \epsilon$. Let γ_δ denote the circle $|z - z_0| = \delta$. Since in view of (39.7) the function $\eta(z)$ is analytic in $\overline{B}(z_0, r) \backslash \{z_0\}$, by Theorem 16.1 it follows that

$$\left| \int_\gamma \frac{\eta(z)}{f'(z_0) + \eta(z)} \frac{dz}{z - z_0} \right| = \left| \int_{\gamma_\delta} \frac{\eta(z)}{f'(z_0) + \eta(z)} \frac{dz}{z - z_0} \right| < \epsilon \frac{1}{\delta} 2\pi\delta = 2\pi\epsilon.$$

Now use the fact that ϵ is arbitrary.

39.12. Let $g(z) = f(z) - f(0)$. Then, $g(0) = 0$ and $|g(z)| \leq |f(z)| + |f(0)| \leq 2M$. Now apply Lemma 39.1.

39.13. Consider $Z = h(z) = R(z - z_0)/(R^2 - \overline{z}_0 z)$, which maps $|z| < R$ analytically onto $|Z| < 1$ with z_0 going into the origin, and $W = g(w) = M(w - w_0)/(M^2 - \overline{w}_0 w)$ ($w_0 = f(z_0)$), which maps $|w| < M$ analytically onto $|W| < 1$ with $g(w_0) = 0$. The function $W = g(f(h^{-1}(Z)))$ satisfies the hypotheses of Lemma 39.1. Hence, $|g(f(h^{-1}(Z)))| \leq |Z|$ or $|g(f(z))| \leq |h(z)|$, which is the same as

$$\left| \frac{f(z) - f(z_0)}{z - z_0} \right| \leq \frac{R}{M} \left| \frac{M^2 - \overline{f(z_0)} f(z)}{R^2 - \overline{z}_0 z} \right|, \quad z \neq z_0.$$

Now let $z \to z_0$.

39.14. Let $\phi(z) = (z + z_1)/(1 + \overline{z}_1 z)$, $\psi(z) = (z - w_1)/(1 - \overline{w}_1 z)$. Then, $(\psi \circ f \circ \phi)(z)$ satisfies the conditions of Lemma 39.1, and hence $|(\psi \circ f \circ \phi)(z)| \leq |z|$. Now let $z = \phi^{-1}(z_2)$. We also have $|(\psi \circ f \circ \phi)'(0)| \leq 1$; i.e., $|\psi'(w_1)f'(z_1)\phi'(0)| \leq 1$.

Lecture 40
Harmonic Functions

In this lecture, we shall employ earlier results to establish some fundamental properties of harmonic functions. The results obtained strengthen our understanding of harmonic functions and are of immense help in solving boundary value problems for the Laplace equation. We begin by proving the following result.

Theorem 40.1. Let $\phi(x, y)$ be a nonconstant harmonic function in a domain S. Then, $\phi(x, y)$ has neither a maximum nor a minimum at any point of S.

Proof. Let $z_0 = x_0 + iy_0$ be an arbitrary point in S, and let $B(z_0, r) \subset S$ be a neighborhood of z_0. As in Lecture 7 (also see Problem 40.2) we construct a function $f(z) = \phi(x, y) + i\theta(x, y)$ that is analytic in $B(z_0, r)$. Clearly, the function $g(z) = e^{f(z)}$ is analytic and nonconstant, and $|g(z)| = e^{\phi(x,y)}$. Now we claim that the function $\phi(x, y)$ cannot have a maximum (minimum) at (x_0, y_0). In fact, if it does, then $|g(z)|$ will have a maximum (minimum) at (x_0, y_0), but this contradicts Theorem 20.1 (Theorem 20.3). ∎

Corollary 40.1. Let $\phi(x, y)$ be a harmonic function in a domain S and continuous on $\overline{S}$. Then, $\phi(x, y)$ attains its maximum and minimum on ∂S.

Corollary 40.2. Let $\phi(x, y)$ be a harmonic function in a domain S, and continuous on $\overline{S}$. If $\phi(x, y) = $ constant for $(x, y) \in \partial S$, then $\phi(x, y) = $ constant for $(x, y) \in \overline{S}$.

Corollary 40.3. Let $S \subset \mathbb{R}^2$ be a bounded domain, and let $g(x, y)$ be a real-valued continuous function on ∂S. Then, the *Dirichlet boundary value problem*

$$\phi_{xx} + \phi_{yy} = 0 \quad \text{in} \quad S \quad \text{and} \quad \phi(x, y) = g(x, y) \quad \text{on} \quad \partial S \quad (40.1)$$

has at most one solution that is twice continuously differentiable in S and continuous in $\overline{S}$.

Problems in which the normal derivative $d\phi/dn = 0$ is known a priori everywhere on the boundary of a domain and where the harmonic function ϕ is sought inside the domain are known as *Neumann problems*.

Theorem 40.2.   Suppose that $w = f(z) = u(x, y) + iv(x, y)$ is a conformal mapping of S into S' and $\phi(u, v)$ is a harmonic function on S'. Then, $\phi \circ f(z) = \phi(u(x, y), v(x, y)) = \psi(x, y)$ is harmonic in S. Thus, if ϕ satisfies $\phi_{uu} + \phi_{vv} = 0$ on S', then $\psi = \phi \circ f$ satisfies $\psi_{xx} + \psi_{yy} = 0$ on S; i.e., the Laplace equation is invariant under a change of variables using a conformal mapping.

Proof.   Let $z_0 \in S$ and $w_0 = f(z_0)$. In view of Problem 40.3, the function ϕ has a harmonic conjugate θ in a disk around w_0. Then, $\phi + i\theta$ is analytic in this disk, and by the composition of analytic functions, $(\phi + i\theta) \circ f$ is analytic at z_0. Hence, in view of Problem 7.20, $\mathrm{Re}[(\phi + i\theta) \circ f] = \mathrm{Re}[\phi \circ f + i(\theta \circ f)] = \phi \circ f$ is harmonic at z_0. Since z_0 was arbitrary, $\phi \circ f$ is analytic in S.   ∎

We also note that, under a change of variables, a level curve $\phi(u, v) = c$ in the uv-plane transforms into the level curve $\phi(u(x, y), v(x, y)) = c$ in the xy-plane. In particular, any portion of the boundary of a region in the uv-plane upon which ϕ has a constant value transforms into a corresponding curve in the xy-plane along which ϕ has the same constant value. Thus, the boundary condition $\phi = c$ in the original problem is also invariant under a change of variables using a conformal mapping. Similarly, if the normal derivative of ϕ vanishes along some curve in the uv-plane, then the normal derivative of ϕ expressed as a function of x and y also vanishes along the corresponding curve in the xy-plane. Hence, the boundary condition $d\phi/dn = 0$ remains unchanged. However, if the boundary condition is not one of these types, then for the transform problem the boundary condition may change significantly.

To see the usefulness of Theorem 40.2 and the remarks above, suppose we are given a domain S in the z-plane. We need to find a harmonic function $\psi(x, y)$ in S that assumes certain conditions on the boundary of S. Suppose we can find a conformal mapping $w = u + iv = f(z)$ that maps S onto a domain S' in the w-plane, and S' has a simpler (familiar) shape than S. Assume that we can find a harmonic function $\phi(u, v)$ in S' that assumes at each point of the boundary of S' the condition that is required by $\psi(x, y)$ at the preimage of that point on the boundary of S. Then, Theorem 40.2 assures that $\psi(x, y) = \phi((u(x, y), v(x, y)))$ will be the required harmonic function in S.

In view of Problem 18.10, the following representation of harmonic functions is immediate.

Theorem 40.3 (Poisson's Integral Formula).   Let $\phi(x, y)$ be harmonic in a domain containing the disk $\overline{B}(0, R)$. Then, for $0 \le r < R$,

$$\phi\left(re^{i\theta}\right) = u(r\cos\theta, r\sin\theta) = \frac{R^2 - r^2}{2\pi} \int_0^{2\pi} \frac{\phi\left(Re^{it}\right)}{R^2 + r^2 - 2Rr\cos(t - \theta)} dt.$$
$$(40.2)$$

Theorem 40.3 has a generalization, which we state in the following result.

Theorem 40.4. Let $\Phi(x, y)$ be defined on the circle $C_R : |z| = R$ and continuous there except for a finite number of jump discontinuities. Then, the function

$$\phi\left(re^{i\theta}\right) = \frac{R^2 - r^2}{2\pi} \int_0^{2\pi} \frac{\Phi\left(Re^{it}\right)}{R^2 + r^2 - 2Rr\cos(t - \theta)} dt \tag{40.3}$$

is harmonic in the disk $B(0, R)$, and as $re^{i\theta}$ approaches any point on C_R where Φ is continuous, $\phi\left(re^{i\theta}\right)$ approaches the value of Φ at that point.

Remark 40.1. Since

$$\frac{R^2 - r^2}{R^2 + r^2 - 2Rr\cos(t - \theta)} = \frac{R^2 - r^2}{R^2} \frac{1}{\left(1 - \frac{r}{R}e^{i(t-\theta)}\right)\left(1 - \frac{r}{R}e^{-i(t-\theta)}\right)}$$

$$= \frac{1}{1 - \frac{r}{R}e^{i(\theta-t)}} + \frac{\frac{r}{R}e^{i(t-\theta)}}{1 - \frac{r}{R}e^{i(t-\theta)}}$$

$$= \sum_{j=0}^{\infty} \left(\frac{r}{R}\right)^j e^{ij(\theta-t)} + \sum_{j=1}^{\infty} \left(\frac{r}{R}\right)^j e^{ij(t-\theta)}$$

$$= 1 + \sum_{j=1}^{\infty} \left(\frac{r}{R}\right)^j \left[e^{ij(\theta-t)} + e^{ij(t-\theta)}\right]$$

$$= 1 + 2\sum_{j=1}^{\infty} \left(\frac{r}{R}\right)^j \cos j(\theta - t)$$

$$= 1 + 2\sum_{j=1}^{\infty} \left(\frac{r}{R}\right)^j \cos j\theta \cos jt$$

$$+ 2\sum_{j=1}^{\infty} \left(\frac{r}{R}\right)^j \sin j\theta \sin jt,$$

from (40.3) it follows that

$$\phi\left(re^{i\theta}\right) = \frac{a_0}{2} + \sum_{j=1}^{\infty} \left(\frac{r}{R}\right)^j (a_j \cos j\theta + b_j \sin j\theta), \tag{40.4}$$

where

$$a_j = \frac{1}{\pi} \int_0^{2\pi} \Phi\left(Re^{it}\right) \cos jt\, dt, \quad j \geq 0, \tag{40.5}$$

and

$$b_j = \frac{1}{\pi} \int_0^{2\pi} \Phi\left(Re^{it}\right) \sin jt\, dt, \quad j \geq 1. \tag{40.6}$$

Thus, the function ϕ can be expanded in a *Fourier series* with the *Fourier coefficients* (40.5) and (40.6).

Remark 40.2. If $w = e^{it}$ and $z = re^{i\theta}$, then it follows that

$$\frac{1 - r^2}{1 + r^2 - 2r\cos(t - \theta)} = \text{Re}\left[\frac{w + z}{w - z}\right] = \frac{1 - |z|^2}{|w - z|^2}. \tag{40.7}$$

Example 40.1. Consider an electrically conducting tube of unit radius $(R = 1)$ that has dielectric material inside and is separated into two halves by means of infinitesimal slits. The top half of the tube $(0 < t < \pi)$ is maintained at an electrical potential of 1 volt, while the bottom half $(\pi < t < 2\pi)$ is at -1 volt. To find the potential ϕ at an arbitrary point (r, θ) inside the tube, we recall that electrostatic potential is a harmonic function. Thus, from (40.3), it follows that

$$\phi(r, \theta) = \frac{1}{2\pi}\int_0^\pi \frac{(1 - r^2)dt}{1 + r^2 - 2r\cos(t - \theta)} - \frac{1}{2\pi}\int_\pi^{2\pi} \frac{(1 - r^2)dt}{1 + r^2 - 2r\cos(t - \theta)}$$

$$= \frac{1}{\pi}\left[2\tan^{-1}\left(\frac{1 + r}{1 - r}\tan\left(\frac{\pi}{2} - \frac{\theta}{2}\right)\right) - \tan^{-1}\left(\frac{1 + r}{1 - r}\tan\left(\pi - \frac{\theta}{2}\right)\right)\right.$$

$$\left. - \tan^{-1}\left(\frac{1 + r}{1 - r}\tan\left(-\frac{\theta}{2}\right)\right)\right].$$

Thus, in particular, at the center, $\phi(0, 0) = 0$.

Once again, for the same problem, (40.5) and (40.6), respectively, give $a_j = 0$, $j \geq 0$ and $b_j = 2(1 - (-1)^j)/(j\pi)$, $j \geq 1$, and hence, in view of (40.4), we have

$$\phi(r, \theta) = \sum_{j=1}^\infty \frac{2r^j}{j\pi}\left(1 - (-1)^j\right)\sin j\theta.$$

Example 40.2. In Theorem 40.4, let $R = 1$ and

$$\Phi(Re^{it}) = \begin{cases} \sin t, & 0 < t < \pi \\ 0, & \pi < t < 2\pi. \end{cases}$$

Then, from (40.3), we have

$$\phi(z) = \frac{1 - r^2}{2\pi}\int_0^\pi \frac{\sin t}{1 + r^2 - 2r\cos(t - \theta)}dt,$$

which in view of (40.7) is the same as

$$\phi(z) = \frac{1}{2\pi}\int_\gamma \frac{w - \overline{w}}{2i}\frac{1 - |z|^2}{|w - z|^2}\frac{dw}{iw},$$

where γ is the upper half unit circle. Since on γ, $\overline{w} = 1/w$, it follows that

$$\phi(z) = -\frac{1}{4\pi} \int_\gamma \frac{\left(w - \frac{1}{w}\right)(1 - |z|^2)}{w(w - z)\left(\frac{1}{w} - \overline{z}\right)} dw$$

$$= -\frac{(1 - |z|^2)}{4\pi} \int_\gamma \frac{(w^2 - 1)}{w(w - z)(1 - w\overline{z})} dw.$$

Thus, from Theorem 31.2, it follows that

$$\phi(z) = -2\pi i \frac{(1 - |z|^2)}{4\pi} \left[\frac{1}{z} + \frac{z^2 - 1}{z(1 - |z|^2)}\right] = \frac{z - \overline{z}}{2i} = r \sin \theta.$$

Once again, for the same problem, (40.5) and (40.6), respectively, give $a_j = 0$, $j \geq 0$ and $b_1 = 1$, $b_j = 0$, $j \geq 2$. Hence, in view of (40.4), we have $\phi(r, \theta) = r \sin \theta$.

Example 40.3. The conformal mapping $w = i(1+z)/(1-z)$ transforms the unit disk $|z| < 1$ onto the upper half-plane $v > 0$, the upper semi-circle is mapped onto the negative u-axis $u < 0$, and the lower semi-circle is mapped onto the positive u-axis $u > 0$. The function

$$\frac{1}{\pi} \log w = \frac{1}{\pi} \text{Log}|w| + \frac{i}{\pi} \arg w, \quad 0 < \arg w < \pi,$$

is analytic in the upper half-plane. The imaginary part of this function; i.e., $(1/\pi) \arg w$, $0 < \arg w < \pi$ is harmonic in the upper half-plane and assumes the value 1 on the negative u-axis and 0 on the positive u-axis. Thus, the function

$$\phi(x, y) = \frac{1}{\pi} \arg \left(i \frac{1 + z}{1 - z}\right)$$

is harmonic in the unit disk satisfying the boundary condition

$$g(\theta) = \begin{cases} 1, & 0 < \theta < \pi \\ 0, & \pi < \theta < 2\pi. \end{cases}$$

Clearly, under the conformal mapping above, the points $z = -1$ and $z = 1$, respectively, correspond to the points $w = 0$ and $w = \infty$, and hence the harmonic function constructed has singularities at $z = \pm 1$.

Once again, for the same problem, (40.5) and (40.6), respectively, give $a_0 = 1$, $a_j = 0$, $j \geq 1$ and $b_j = (1 - (-1)^j)/(j\pi)$, $j \geq 1$, and hence, in view of (40.4), we have

$$\phi(r, \theta) = \frac{1}{2} + \sum_{j=1}^\infty \frac{r^j}{j\pi} \left(1 - (-1)^j\right) \sin j\theta.$$

Problems

40.1. Suppose that $\phi(x,y)$ is a harmonic function in a domain S. Let $\psi = \phi_x - i\phi_y$. Show that ψ is analytic in S. The function ψ is called the *conjugate gradient* of ϕ.

40.2. Suppose that $\phi(x,y)$ is a harmonic function in a simply connected domain S. Show that ϕ has a harmonic conjugate in S given up to an additive constant by

$$\theta(x,y) = \int_{(x_0,y_0)}^{(x,y)} -\phi_y dx + \phi_x dy,$$

where the point (x_0, y_0) is fixed in S and the integral is independent of the path. Furthermore, show that in this problem the connectedness property is crucial.

40.3. Suppose that $\phi(x,y)$ is a harmonic function in a domain S. Show that ϕ admits a harmonic conjugate locally in S.

40.4. Prove that $\phi_{xx} + \phi_{yy} = |f'(z)|^2(\phi_{uu} + \phi_{vv})$, where $w = f(z)$ is analytic and $f'(z) \neq 0$. Hence, deduce Theorem 40.2.

40.5. Let a function $\phi(x,y)$ satisfy *Poisson's equation* $\phi_{xx} + \phi_{yy} = P(x,y)$ in a domain S. If S' is the image of the domain S under the conformal mapping $w = f(z)$, show that $\psi = \phi(x(u,v), y(u,v))$ satisfies another Poisson equation, $\psi_{uu} + \psi_{vv} = |f'(z)|^2 P[x(u,v), y(u,v)]$, in S'.

40.6. Suppose that $\phi(x,y)$ is harmonic and nonnegative in a domain containing the disk $\overline{B}(0,R)$. Establish the *Harnack inequality*

$$\frac{R-r}{R+r}\phi(0) \leq \phi\left(re^{i\theta}\right) \leq \frac{R+r}{R-r}\phi(0).$$

Hence, show that a bounded function that is harmonic in $\mathbf{C}$ is necessarily constant.

40.7. Use (40.4) to find a harmonic function $\phi(x,y)$ in the disk $B(0,a)$ that satisfies $\phi(a,\theta) = g(\theta)$, where

(a). $g(\theta) = \dfrac{1}{2}(1 + \cos\theta), \quad 0 < \theta < 2\pi,$

(b). $g(\theta) = \dfrac{1}{2}(1 + \cos^3\theta), \quad 0 < \theta < 2\pi,$

(c). $g(\theta) = |\theta|, \quad 0 < \theta < 2\pi.$

40.8. Find a harmonic function $\phi(r,\theta)$ in the wedge with three sides $\theta = 0$, $\theta = \beta$, and $r = a$ (see Figure 40.1) and the boundary conditions $\phi(r,0) = 0 = \phi(r,\beta)$, $0 < r < a$ and $\phi(a,\theta) = g(\theta)$, $0 < \theta < \beta$.

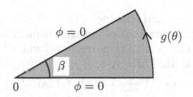

$$\phi = 0$$
$$g(\theta)$$
$$\beta$$
$$0 \qquad \phi = 0$$

Figure 40.1

40.9. Let $g(x)$ be a piecewise continuous and bounded function on $-\infty < x < \infty$. Show that the function defined by

$$\phi(x,y) \;=\; \frac{y}{\pi}\int_{-\infty}^{\infty}\frac{g(t)}{(x-t)^2+y^2}\,dt \tag{40.8}$$

is harmonic in the upper half-plane $y > 0$ and satisfies the boundary condition $\phi(x,0) = g(x)$ at all points of continuity of f. The relation (40.8) is called *Poisson's integral formula for the half-plane*.

40.10. Use Problem 40.9 to find a harmonic function $\phi(x,y)$ in the upper half-plane $y > 0$ that satisfies $\phi(x,0) = g(x)$, where

(a). $g(x) = \begin{cases} 1, & |x| < 1 \\ 0, & |x| > 1, \end{cases}$

(b). $g(x) = \begin{cases} x, & |x| < 1 \\ 0, & |x| > 1, \end{cases}$

(c). $g(x) = \cos x, \quad -\infty < x < \infty.$

Answers and Hints

40.1. It suffices to verify that ϕ_x and $-\phi_y$ satisfy the Cauchy-Riemann conditions.

40.2. See Examples 7.6 and 7.7. Consider the function $\ln|z|$.

40.3. Since S is open, we can find an open disk $B(z_0,r)$ in S. Clearly, this disk is connected.

40.4. Use the chain rule and the Cauchy-Riemann conditions.

40.5. Use Problem 40.4.

40.6. Use $(R-r)^2 \le R^2+r^2-2Rr\cos(t-\theta) \le (R+r)^2$, and note that, for $r = 0$, (40.2) reduces to $\phi(0) = \frac{1}{2\pi}\int_0^{2\pi}\phi\left(Re^{it}\right)dt$.

40.7. (a). $\frac{1}{2}\left(1+\frac{r}{a}\cos\theta\right)$, (b). $\frac{1}{2}\left(1+\frac{3}{4}\frac{r}{a}\cos\theta+\frac{1}{4}\frac{r^3}{a^3}\cos 3\theta\right)$, (c). $\frac{\pi}{2}$ $+\sum_{n=1}^{\infty}\frac{2((-1)^n-1)}{\pi n^2}\frac{r^n}{a^n}\cos n\theta.$

40.8. $\phi(r,\theta) = \sum_{n=1}^{\infty}A_n r^{n\pi/\beta}\sin\frac{n\pi}{\beta}\theta,\ A_n = \frac{2}{\beta}a^{-n\pi/\beta}\int_0^{\beta}g(t)\sin\frac{n\pi}{\beta}t\,dt.$

40.9. Fix $z = x + iy$, $y > 0$, and consider $f(w) = (w - z)/(w - \bar{z})$, $v \geq 0$. Clearly, f maps the open upper half-plane conformally onto $B(0, 1)$. The boundaries $|z| = 1$ and the real axis correspond via $e^{it} = f(\tau) = (\tau - z)/(\tau - \bar{z})$, $t \in [0, 2\pi]$, $\tau \in \mathbb{R}$. This gives $dt = 2y d\tau / |\tau - \bar{z}|^2$.

40.10. (a). $\frac{1}{\pi} \left(\tan^{-1} \frac{y}{x-1} - \tan^{-1} \frac{y}{x+1} \right)$, (b). $\frac{x}{\pi} \left(\tan^{-1} \frac{y}{x-1} - \tan^{-1} \frac{y}{x+1} \right) + \frac{y}{2\pi} \ln \frac{(x-1)^2 + y^2}{(x+1)^2 + y^2}$, (c). $e^{-y} \cos x$.

Lecture 41
The Schwarz-Christoffel Transformation

In this lecture, we shall provide an explicit formula for the derivative of a conformal mapping that maps the upper half-plane onto a given bounded or unbounded polygonal region (boundary contains a finite number of line segments). The integration of this formula (often a formidable task unless done numerically) and then its inversion (another nontrivial task) yields a conformal mapping that maps a polygonal region onto the upper half-plane. Such mappings are often applied in physical problems such as in heat conduction, fluid mechanics, and electrostatics.

Consider the transformation

$$z = g(w) = (w - u_1)^{\alpha/\pi}, \quad 0 < \alpha < 2\pi, \tag{41.1}$$

where u_1 is a point on the real axis of the w-plane. This mapping is a composition of a translation $T(w) = w - u_1$ followed by the real power function $G(w) = w^{\alpha/\pi}$. From Lecture 11, it is clear that under the composite mapping $z = G(T(w)) = (w - u_1)^{\alpha/\pi}$ a ray emanating from u_1 and making an angle θ with the real axis is mapped onto a ray emanating from the origin and making an angle $\alpha\theta/\pi$ with the real axis. The image of the half-plane $v > 0$ is the point $z = 0$ together with the wedge $0 \leq \arg(z) \leq \alpha$. Since $g'(w) = (\alpha/\pi)(w - u_1)^{(\alpha/\pi)-1} \neq 0$ if $w = u + iv$ and $v > 0$, it follows that $z = g(w)$ is a conformal mapping at any point w with $v > 0$. In what follows, we shall use the form of the derivative of the function $g(w)$ to describe a conformal mapping of the upper half-plane $v > 0$ onto an arbitrary polygonal region. For this, we shall first analyze the mapping $h(w)$, which is analytic in the domain $v > 0$ and whose derivative is

$$h'(w) = A(w - u_1)^{(\alpha_1/\pi)-1}(w - u_2)^{(\alpha_2/\pi)-1}, \tag{41.2}$$

where A is a complex number, $u_1 < u_2$ are real, and $0 < \alpha_1, \alpha_2 < 2\pi$. Under the mapping $w = h(z)$, we shall find the images of the intervals $(-\infty, u_1), (u_1, u_2)$, and (u_2, ∞). Since, on the real axis $w = u$, from (41.2) it follows that

$$\mathrm{Arg}(h'(u)) = \mathrm{Arg}(A) + \left(\frac{\alpha_1}{\pi} - 1\right)\mathrm{Arg}(u - u_1) + \left(\frac{\alpha_2}{\pi} - 1\right)\mathrm{Arg}(u - u_2).$$

Thus, if $-\infty < u < u_1$,

$$\mathrm{Arg}(h'(u)) = \mathrm{Arg}(A) + \left(\frac{\alpha_1}{\pi} - 1\right)\pi + \left(\frac{\alpha_2}{\pi} - 1\right)\pi = \mathrm{Arg}(A) + \alpha_1 + \alpha_2 - 2\pi,$$

which is a constant for all u. Hence, the interval $(-\infty, u_1)$ is mapped into a line segment ℓ_1 defined by $z = h(u)$. Similarly, we find that if $u_1 < u < u_2$, then $\text{Arg}(h'(u)) = \text{Arg}(A) + \alpha_2 - \pi$; i.e., the interval (u_1, u_2) is mapped into a line segment ℓ_2 defined by $z = h(u)$, which changes the exterior angle by $\pi - \alpha_1$ from ℓ_1. If $u_2 < u < \infty$, then $\text{Arg}(h'(u)) = \text{Arg}(A)$; i.e., the interval (u_2, ∞) is mapped into a line segment ℓ_3 defined by $z = h(u)$, which changes the exterior angle by $\pi - \alpha_2$ from ℓ_2 (see Figure 41.1).

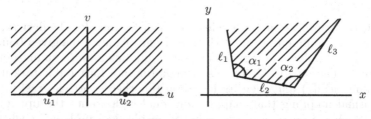

Figure 41.1

Since $h(z)$ is analytic, the image of the half-plane $v \geq 0$ is an unbounded polygonal region (see Figure 41.1). In the following result, we generalize the discussion above for the derivative $f'(w)$ of a function $f(w)$ that maps the half-plane $v \geq 0$ onto a polygonal region with any number of sides.

Theorem 41.1 (The Schwarz-Christoffel Transformation).
Let $f(w)$ be analytic in the domain $v > 0$ and have the derivative

$$f'(w) = A(w - u_1)^{(\alpha_1/\pi)-1}(w - u_2)^{(\alpha_2/\pi)-1} \cdots (w - u_n)^{(\alpha_n/\pi)-1}, \quad (41.3)$$

where $u_1 < u_2 < \cdots < u_n$, $0 < \alpha_j < 2\pi$, $j = 1, 2, \cdots, n$, and A is a complex constant. Then, the upper half-plane $v \geq 0$ is mapped by $z = f(w)$ onto an unbounded polygonal region, say P with interior angles α_j, $j = 1, 2, \cdots, n$.

In integral form, (41.3) can be written as

$$z = f(w) = A \int^{w} (\zeta - u_1)^{(\alpha_1/\pi)-1}(\zeta - u_2)^{(\alpha_2/\pi)-1} \cdots (\zeta - u_n)^{(\alpha_n/\pi)-1} d\zeta + B;$$
$$(41.4)$$

here, in the integral, the lower limit can be chosen arbitrarily. Clearly, from (41.3) and (41.4), it follows that $(u_1, 0), (u_2, 0), \cdots, (u_n, 0)$ are the images in the w-plane of the vertices $z_1, z_2, \cdots, z_n$ of the polygon P in the z-plane. If $w = \infty$ is mapped into one vertex, say z_j then the term containing $(w - u_j)$ is absent in (41.3). If the polygon P is closed, then the last vertex z_n is the same as the initial vertex z_1, and hence instead of n interior angles we have only $n - 1$ interior angles. The size and orientation of the polygon P are determined by the complex constants A and B. This is usually fixed by choosing some of the points u_j on the u-axis. We illustrate some of these remarks in the following examples.

Example 41.1. We shall find the transformation that maps the upper half-plane $v \geq 0$ onto the polygonal region defined by the strip $x \geq 0$, $-1 \leq y \leq 1$ (see Figure 41.2).

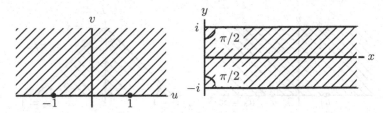

Figure 41.2

Clearly, here we have $\alpha_1 = \alpha_2 = \pi/2$, $z_1 = -i$, and $z_2 = i$. We choose $u_1 = -1$ and $u_2 = 1$, so that (41.4) becomes

$$
\begin{aligned}
z = f(w) &= A \int^w (\zeta + 1)^{-1/2} (\zeta - 1)^{-1/2} d\zeta + B \\
&= -iA \int^w (1 - \zeta^2)^{-1/2} d\zeta + B = -iA \sin^{-1} w + B.
\end{aligned}
$$

Now, since $f(-1) = -i$ and $f(1) = i$, it follows that $A = -2/\pi$, $B = 0$. Hence, the required mapping is

$$
z = i \frac{2}{\pi} \sin^{-1} w.
$$

From this mapping, it is clear that

$$
w = \sin\left(\frac{\pi z}{2i}\right) = -i \sinh\left(\frac{\pi z}{2}\right)
$$

maps the strip $x \geq 0$, $-1 \leq y \leq 1$ onto the upper half-plane $v \geq 0$.

Example 41.2. We shall find the transformation that maps the upper half-plane $v \geq 0$ onto the sector $0 \leq \arg z \leq \alpha\pi$, $0 < \alpha < 2$. For this, we note that the given sector is a polygon with vertices $z_1 = 0$ and $z_2 = \infty$. We choose $u_1 = 0$ and $u_2 = \infty$, so that (41.4) becomes

$$
z = f(w) = A \int^w \zeta^{\alpha-1} d\zeta + B = \frac{A}{\alpha} w^\alpha + B.
$$

Now since $f(0) = 0$ it follows that $B = 0$, and hence $z = f(w) = (A/\alpha)w^\alpha$. To fix the arbitrary constant A, for simplicity we choose $f(1) = 1$, so that $A = \alpha$. Hence, the required mapping is $z = w^\alpha$. From this mapping it is clear that the principal branch of $w = z^{1/\alpha}$ maps the sector $0 \leq \arg z \leq \alpha\pi$, $0 < \alpha < 2$ onto the upper half-plane $v \geq 0$.

Example 41.3. We shall find the transformation that maps the upper half-plane $v \geq 0$ onto the right isosceles triangle. The vertices of the triangle and its images are given in Figure 41.3.

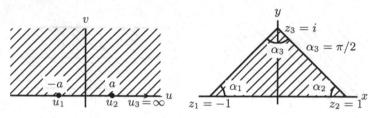

Figure 41.3

Clearly, here we have $\alpha_1 = \alpha_2 = \pi/4$, $\alpha_3 = \pi/2$, $w_1 = -a$, $w_2 = a$, and $w_3 = \infty$, and hence (41.4) can be taken as

$$z = f(w) = A \int_0^w (\zeta+a)^{-3/4}(\zeta+a)^{-3/4}d\zeta + B = A \int_0^w (\zeta^2 - a^2)^{-3/4}d\zeta + B.$$

This integral cannot be evaluated in terms of known functions. It has to be computed numerically for each value of w of interest. Now, since $f(-a) = -1$ and $f(a) = 1$, it follows that

$$-1 = A \int_0^{-a} (\zeta^2 - a^2)^{-3/4}d\zeta + B,$$

$$1 = A \int_0^a (\zeta^2 - a^2)^{-3/4}d\zeta + B,$$

which easily determines

$$A = \frac{1}{\int_0^a (\zeta^2 - a^2)^{-3/4}d\zeta} \quad \text{and} \quad B = 0.$$

Hence, the required mapping is

$$z = \frac{\int_0^w (\zeta^2 - a^2)^{-3/4}d\zeta}{\int_0^a (\zeta^2 - a^2)^{-3/4}d\zeta}.$$

From this the computation of the inverse mapping that transforms the triangle in Figure 41.3 to the upper half-plane $v \geq 0$ is not straightforward.

Example 41.4. We shall find the transformation that maps the upper half-plane $v \geq 0$ onto the rectangle. The vertices of the rectangle and its images are given in Figure 41.4.

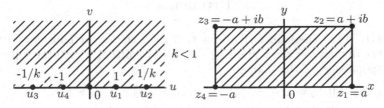

Figure 41.4

Since here $\alpha_1 = \alpha_2 = \alpha_3 = \alpha_4 = \pi/2$, $u_1 = 1$, $u_2 = 1/k$, $u_3 = -1/k$, $u_4 = -1$, (41.4) takes the form

$$z = f(w) = A \int_0^w (\zeta - 1)^{-1/2} \left(\zeta - \frac{1}{k}\right)^{-1/2} \left(\zeta + \frac{1}{k}\right)^{-1/2} (\zeta + 1)^{-1/2} d\zeta + B$$

$$= A \int_0^w \frac{k}{\sqrt{(1 - \zeta^2)(1 - k^2\zeta^2)}} d\zeta + B.$$

We can use the correspondence $w = 0 \to z = 0$ to get $B = 0$. Then,

$$z = C \int_0^w \frac{1}{\sqrt{(1 - \zeta^2)(1 - k^2\zeta^2)}} d\zeta := CF(w, k), \quad C = kA. \quad (41.5)$$

The function $F(w, k)$ is called the *elliptic integral of the first kind*. The correspondence $w = 1 \to z = a$ now yields

$$a = C \int_0^1 \frac{1}{\sqrt{(1 - \zeta^2)(1 - k^2\zeta^2)}} d\zeta = CF(1, k) = CK(k). \quad (41.6)$$

The function $K(k)$ is called the *complete integral of the first kind* and is well-studied in tabular form. Finally, we note that the correspondence $w = 1/k \to a + ib$ leads to

$$a + ib = C \left[\int_0^1 \frac{1}{\sqrt{(1 - \zeta^2)(1 - k^2\zeta^2)}} d\zeta + \int_1^{1/k} \frac{1}{\sqrt{(1 - \zeta^2)(1 - k^2\zeta^2)}} d\zeta \right],$$

which in view of (41.6) is the same as

$$b = C \int_1^{1/k} \frac{1}{\sqrt{(\zeta^2 - 1)(1 - k^2\zeta^2)}} d\zeta. \quad (41.7)$$

For the given values of a and b, the nonlinear equations (41.6) and (41.7) can be solved for the unknowns C and k. The relation (41.5) with these computed values then provides the required transformation.

Problems

41.1. Find the transformations that map the upper half-plane $v \geq 0$ onto the domains in the z-plane as shown in Figure 41.5 with the following correspondences of the points:

(a). $w(u_1 = 0, u_2 = 1, u_3 = \infty) \rightarrow z(z_1 = 0, z_2 = 1, z_3 = \infty)$,

(b). $w(u_1 = 0, u_2 = 1, u_3 = \infty) \rightarrow z(z_1 = 0, z_2 = 1, z_3 = \infty)$,

(c). $w(u_1 = 0, u_2 = 1, u_3 = \infty) \rightarrow z(z_1 = 0, z_2 = \infty, z_3 = \infty)$,

(d). $w(u_1 = 0, u_2 = 1, u_3 = \infty) \rightarrow z(z_1 = 0, z_2 = \infty, z_3 = \infty)$.

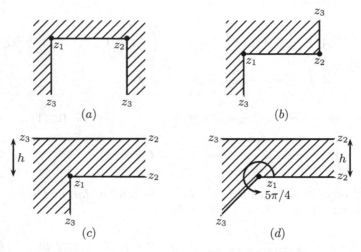

Figure 41.5

Answers or Hints

41.1. (a). $z = (2/\pi)[\sin^{-1}\sqrt{w} - (1 - 2w)\sqrt{w - w^2}]$,

(b). $z = (2/\pi)[\sin^{-1}\sqrt{w} - \sqrt{w - w^2}]$,

(c). $z = (2h/\pi)[\tanh^{-1}\sqrt{w} - \sqrt{w}]$,

(d). $z = (2h/\pi)[\tanh^{-1}w^{1/4} - 2w^{1/4}]$.

Lecture 42
Infinite Products

In this lecture, we shall introduce infinite products of complex numbers and functions and provide necessary and sufficient conditions for their convergence.

Let $\{a_n\}$ be a sequence of complex numbers. The *infinite product*, denoted as $\prod_{j=1}^{\infty} a_j$ is the limit p as $n \to \infty$ of the nth partial product $p_n = \prod_{j=1}^{n} a_j = a_1 a_2 \cdots a_n$ provided this limit exists and is not zero. As in series, we write $\prod_{j=1}^{\infty} a_j = p$ or $\lim_{n \to \infty} \prod_{j=1}^{n} a_j = p$. Now, if $\lim_{n \to \infty} p_n = p \neq 0$, then, since

$$\lim_{n \to \infty} a_n = \lim_{n \to \infty} \frac{\prod_{j=1}^{n} a_j}{\prod_{j=1}^{n-1} a_j} = \frac{p}{p} = 1,$$

for the infinite product $\prod_{j=1}^{\infty} a_j$ to converge it is necessary that $\lim_{n \to \infty} a_n = 1$. Thus, if, for convenience, we consider $a_n = 1 + b_n$, then the infinite product

$$\prod_{j=1}^{\infty} (1 + b_j) \tag{42.1}$$

diverges if $\lim_{n \to \infty} b_n \neq 0$ or some $b_n = -1$. The following example suggests that $\lim_{n \to \infty} b_n = 0$ is not sufficient for the convergence of (42.1).

Example 42.1. Both the infinite products, $\prod_{j=2}^{\infty}(1 - 1/j)$ and $\prod_{j=1}^{\infty}(1 + 1/j)$, diverge. In fact, respectively, we have

$$p_n = \frac{1}{2} \frac{2}{3} \frac{3}{4} \cdots \frac{n-1}{n} \frac{n}{n+1} = \frac{1}{n+1} \to 0 \quad \text{as} \quad n \to \infty$$

and

$$p_n = \frac{2}{1} \frac{3}{2} \frac{4}{3} \cdots \frac{n+1}{n} = n+1 \to \infty \quad \text{as} \quad n \to \infty.$$

The following results that we state, leaving the proofs as exercises, provide necessary and/or sufficient conditions for the convergence of the infinite product (42.1).

Theorem 42.1. The infinite product (42.1) converges if and only if the infinite series

$$\sum_{j=1}^{\infty} \text{Log}(1 + b_j) \tag{42.2}$$

converges.

Theorem 42.2. The infinite product (42.1) converges if and only if for every $\epsilon > 0$ there exists an integer $N = N(\epsilon)$ such that

$$|(1 + b_{n+1})(1 + b_{n+2}) \cdots (1 + b_{n+m}) - 1| < \epsilon \qquad (42.3)$$

whenever $n \geq N$ and $m \geq 1$.

Theorem 42.3. If the series $\sum_{j=1}^{\infty} b_j$ and $\sum_{j=1}^{\infty} b_j^2$ converge, then the infinite product (42.1) converges.

The infinite product (42.1) is said to be *absolutely convergent* if the infinite product

$$\prod_{j=1}^{\infty}(1 + |b_j|) \qquad (42.4)$$

converges. A convergent infinite product that is not absolutely convergent is called *conditionally convergent*.

Theorem 42.4. The infinite product (42.1) is absolutely convergent if and only if the series $\sum_{j=1}^{\infty} b_j$ is absolutely convergent.

Theorem 42.5. If (42.4) converges, then (42.1) converges.

Example 42.2. For the infinite product $\prod_{j=2}^{\infty}\left(1 - (-1)^j/j\right)$, we have

$$p_n = \frac{1}{2}\frac{4}{3}\frac{3}{4} \cdots \frac{n - (-1)^n}{n} \frac{(n+1) - (-1)^{n+1}}{n+1} = \begin{cases} \dfrac{1}{2} & \text{if } n \text{ is odd} \\ \dfrac{1}{2}\dfrac{n+2}{n+1} & \text{if } n \text{ even,} \end{cases}$$

and hence $\lim_{n\to\infty} p_n = 1/2$. However, in view of Example 42.1, the infinite product $\prod_{j=1}^{\infty}(1 + 1/j)$ diverges, and hence the converse of Theorem 42.5 is not true.

Now we shall consider infinite products of complex functions. Let $\{b_n(z)\}$ be a sequence of complex functions defined on a set S. An infinite product

$$\prod_{j=1}^{\infty}(1 + b_j(z)) \qquad (42.5)$$

is said to be *uniformly convergent* on S if the sequence of partial products $p_n(z) = \prod_{j=1}^{n}(1 + b_j(z))$ is uniformly convergent on S to a function $p(z)$ that is different from zero for all $z \in S$.

Theorem 42.6. Suppose that the series $\sum_{j=1}^{\infty} |b_j(z)|$ converges uniformly on every compact subset of S. Then, (42.5) converges uniformly to $p(z)$ on every compact subset of S.

Proof. Let S_1 be a compact subset of S. Since $\sum_{j=1}^{\infty} |b_j(z)|$ converges uniformly on S_1, it is uniformly bounded. Thus, for all $z \in S_1$, it follows that

$$|p_n(z)| \leq \prod_{j=1}^{n}(1 + |b_j(z)|) \leq \prod_{j=1}^{n} e^{|b_j(z)|} = \exp\left(\sum_{j=1}^{n} |b_j(z)|\right) \leq M.$$

Let $\epsilon > 0$ be arbitrary. Then, there exists an integer N sufficiently large such that $\sum_{j=m+1}^{n} |b_j(z)| < \epsilon$ for every $n > m \geq N$ and all $z \in S_1$. Now an easy argument for all $z \in S_1$ gives

$$
\begin{aligned}
|p_n(z) - p_m(z)| &= |p_m(z)| \left| \prod_{j=m+1}^{n} (1 + b_j(z)) - 1 \right| \\
&\leq |p_m(z)| \left[\prod_{j=m+1}^{n} (1 + |b_j(z)|) - 1 \right] \\
&\leq |p_m(z)| \left[\exp\left(\sum_{j=m+1}^{n} |b_j(z)| \right) - 1 \right] \\
&\leq M\left(e^{\epsilon} - 1\right).
\end{aligned}
$$

But this shows that the sequence $\{p_n(z)\}$ converges uniformly on S_1. ∎

Corollary 42.1. Let $\{M_n\}$ be a sequence of positive constants, and let $|b_n(z)| \leq M_n$ for all $z \in S$. Then, if $\sum_{j=1}^{\infty} M_j$ converges, (42.5) converges uniformly on S.

Corollary 42.2. If in Theorem 42.6 the functions $b_n(z)$ are analytic on S, then the limiting function $p(z)$ is analytic on S.

Example 42.3. For $|z| < 1$, uniformly we have

$$
\begin{aligned}
\prod_{j=0}^{\infty} \left(1 + z^{2^j}\right) &= (1+z)(1+z^2)(1+z^4)\cdots \\
&= \frac{1}{1-z}(1-z^2)(1+z^2)(1+z^4)\cdots \\
&= \frac{1}{1-z} \lim_{n \to \infty} \left(1 - z^{2^n}\right) = \frac{1}{1-z}.
\end{aligned}
$$

We also note that this infinite product diverges for $|z| \geq 1$.

Example 42.4. Since $(1 + z/\sqrt{j})(1 - z/\sqrt{j}) = (1 - z^2/j)$, $j \geq 2$ and $(1 + zi/\sqrt{j})(1 - zi/\sqrt{j}) = (1 + z^2/j)$, $j \geq 1$ in view of Example 42.1, both the infinite products

$$\prod_{j=2}^{\infty}\left(1+\frac{z}{\sqrt{j}}\right)\left(1-\frac{z}{\sqrt{j}}\right) \quad \text{and} \quad \prod_{j=1}^{\infty}\left(1+\frac{zi}{\sqrt{j}}\right)\left(1-\frac{zi}{\sqrt{j}}\right)$$

diverge at $z = \pm 1$.

Example 42.5. Consider the zeros of the function $\sin z$; i.e., $z = j\pi$, $j = 0, \pm 1, \pm 2, \cdots$. The infinite product

$$z\prod_{j=1}^{\infty}(z+j\pi)(z-j\pi) = z\prod_{j=1}^{\infty}(z^2 - j^2\pi^2) = z\prod_{j=1}^{\infty}(1+(z^2 - j^2\pi^2 - 1))$$

does not converge at any point $z \neq j\pi$, as the necessary condition $z^2 - j^2\pi^2 - 1 \to 0$ as $j \to \infty$ is violated.

Remark 42.1. If the conditions of Corollary 42.2 are satisfied, then it can be shown that

$$\frac{p'(z)}{p(z)} = \sum_{j=1}^{\infty}\frac{b_j'(z)}{1+b_j(z)}.$$

Corollary 42.3. Let the conditions of Theorem 42.6 be satisfied. Then, if $\{n_1, n_2, \cdots, n_j, \cdots\}$ is any permutation of the positive integers, then $p(z) = \prod_{j=1}^{\infty}(1+b_{n_j}(z))$ on every compact subset of S.

Proof. Let S_1, M, ϵ, and N be as in Theorem 42.6. Choose K so large that $\{1, 2, \cdots, N\} \subset \{n_1, n_2, \cdots, n_K\}$. Furthermore, let $q_k(z) = \prod_{j=1}^{k}(1+b_{n_j}(z))$. Then, for all $z \in S_1$, we have

$$|q_K(z) - p_N(z)| = |p_N(z)|\left|\prod_j(1+b_{n_j}(z)) - 1\right|,$$

where $j \in \{n_1, n_2, \cdots, n_K\}\backslash\{1, 2, \cdots, N\}$. Thus, as in Theorem 42.6, we have $|q_K(z) - p_N(z)| \leq M(e^\epsilon - 1)$, which shows that $\{q_k(z)\}$ converges uniformly on S_1. ∎

Corollary 42.4. Let the conditions of Theorem 42.6 be satisfied. Then, $p(z_0) = 0$ for some z_0 in a compact subset of S if and only if $b_j(z_0) = -1$ for some j.

Proof. Let S_1, $0 < \epsilon < 1/2$ and N be as in Theorem 42.6. Then, for every $n > m \geq N$ and all $z \in S_1$, we have $|p_n(z) - p_m(z)| \leq |p_m(z)|(e^\epsilon - 1) \leq 2|p_m(z)|\epsilon$. Thus, it follows that

$$|p_n(z)| \geq |p_m(z)| - |p_n(z) - p_m(z)| \geq |p_m(z)| - 2|p_m(z)|\epsilon = (1-2\epsilon)|p_m(z)|.$$

Hence, as $n \to \infty$, $|p(z)| \geq (1 - 2\epsilon)|p_m(z)|$, where $(1 - 2\epsilon) > 0$. This implies that if $|p_m(z)| > 0$ for each m and $z \in S_1$, then $|p(z)| > 0$. The converse is obvious. ∎

Corollary 42.5. If, for each j, $0 \leq b_j < 1$, then $\prod_{j=1}^{\infty}(1 - b_j) > 0$ if and only if $\sum_{j=1}^{\infty} b_j < \infty$.

Proof. Let $q_n = \prod_{j=1}^{n}(1 - b_j)$. Then, $q_1 \geq q_2 \geq \cdots \geq 0$, and therefore $\lim_{n \to \infty} q_n = q$ exists. If $\sum_{j=1}^{\infty} b_j < \infty$, then $q > 0$ by Corollary 42.4 since each factor $1 - b_j > 0$. Conversely, $0 \leq q \leq q_n \leq \prod_{j=1}^{n}(1 - b_j) \leq \exp\left(-\sum_{j=1}^{n} b_j\right)$, and hence if $\sum_{j=1}^{\infty} b_j = \infty$, then $\exp\left(-\sum_{j=1}^{\infty} b_j\right) = 0$, which implies that $q = 0$. ∎

Problems

42.1. Evaluate the following infinite products:

(a). $\displaystyle\prod_{j=2}^{\infty}\left(1 - \frac{1}{j^2}\right)$, (b). $\displaystyle\prod_{j=1}^{\infty}\left(1 + \frac{1}{j(j + 2)}\right)$, (c). $\displaystyle\prod_{j=1}^{\infty}\left(1 - \frac{2}{(j+1)(j+2)}\right)$.

42.2. Show that the infinite product $\prod_{j=1}^{\infty}\left(1 + e^{jz}\right)$ diverges for $\text{Re}(z) \geq 0$ and converges absolutely for $\text{Re}(z) < 0$.

42.3. Show that the infinite product $\prod_{j=1}^{\infty}\left(1 + z^j/j\right)$ is uniformly and absolutely convergent for $|z| \leq r$, where $0 < r < 1$.

42.4. Show that the infinite product $\prod_{j=1}^{\infty}\left(1 + 1/j^z\right)$ is uniformly and absolutely convergent in the half-plane $\text{Re}(z) \geq 1 + \epsilon$, where $\epsilon > 0$. Use the principal value of j^z.

42.5. Show that the infinite product $\prod_{j=1}^{\infty}\left(1 - z^2/j^2\right)$ is uniformly and absolutely convergent in any closed and bounded region, where z is not a positive or negative integer.

42.6. Show that the infinite product

$$\left\{\left(1 - \frac{z}{1}\right)e^z\right\}\left\{\left(1 + \frac{z}{1}\right)e^{-z}\right\}\left\{\left(1 - \frac{z}{2}\right)e^{z/2}\right\}\left\{\left(1 + \frac{z}{2}\right)e^{-z/2}\right\}\cdots$$

is uniformly and absolutely convergent in any closed and bounded region, where z is not a positive or negative integer.

42.7 (Blaschke Product). Let the sequence $\{z_n\}$ be such that $0 < |z_n| < 1$ for all n and $\sum_{j=1}^{\infty}(1 - |z_j|)$ converges. Show that the infinite product

$$f(z) = \prod_{j=1}^{\infty} \frac{z_j - z}{1 - \overline{z}_j z} \frac{|z_j|}{z_j}$$

converges uniformly on the disk $\overline{B}(0,r)$, $0 < r < 1$. Hence, $f(z)$ is analytic on $B(0,1)$, z_n, $n = 1, 2, \cdots$ arc its zeros, $f(z)$ has no other zeros, and $|f(z)| \leq 1$.

Answers and Hints

42.1. (a). $1/2$, (b). 2, (c). $1/3$.

42.2. For $\text{Re}(z) \geq 0$ use the necessary condition for convergence, and for $\text{Re}(z) < 0$ use Corollary 42.1.

42.3. $|b_n(z)| = |z^n|/n \leq r^n$. Now use Corollary 42.1.

42.4. Use Corollary 42.1.

42.5. Use Corollary 42.1.

42.6. Since $\left(1 - \frac{z}{n}\right) e^{z/n} = 1 + b_n(z)$, where $b_n(z) = -\sum_{j=2}^{\infty} \frac{(j-1)}{j!} \left(\frac{z}{n}\right)^j$, we have $|b_n(z)| \leq \sum_{j=2}^{\infty} \frac{1}{(j-2)!} \left(\frac{R}{n}\right)^j = \left(\frac{R}{n}\right)^2 e^{R/n} \leq e \left(\frac{R}{n}\right)^2$ when $n > R$. Similarly, we have $\left(1 + \frac{z}{n}\right) e^{-z/n} = 1 + c_n(z)$, where $|c_n(z)| \leq e \left(\frac{R}{n}\right)^2$ when $n > R$. Now use Corollary 42.1.

42.7. Follows from Corollary 42.1 since

$$b_j(z) := \frac{z_j - z}{1 - \overline{z}_j z} \frac{|z_j|}{z_j} - 1 = -\frac{1 + z\frac{|z_j|}{z_j}}{1 - z\overline{z}_j}(1 - |z_j|)$$

satisfies

$$|b_j(z)| \leq \frac{1 + |z|}{1 - |z||z_j|}(1 - |z_j|) \leq \frac{1 + r}{1 - r}(1 - |z_j|) =: M_j \quad \text{for all} \quad z \in \overline{B}(0,r)$$

and $\sum_{j=1}^{\infty} M_j$ converges.

Lecture 43
Weierstrass's Factorization Theorem

In this lecture, we shall provide representations of entire functions as finite/infinite products involving their finite/infinite zeros. We begin with the following simple cases.

Theorem 43.1. If an entire function $f(z)$ has no zeros, then $f(z)$ is of the form $f(z) = e^{g(z)}$, where $g(z)$ is an entire function.

Proof. Clearly, the function

$$h(z) = \frac{f'(z)}{f(z)} = \frac{d}{dz} \text{Log} f(z)$$

is entire, and hence

$$\int_0^z h(\xi) d\xi = \text{Log} f(\xi) \Big|_0^z = \text{Log} f(z) - \text{Log} f(0).$$

Thus, $f(z) = e^{g(z)}$, where $g(z) = \int_0^z h(\xi) d\xi + \text{Log} f(0)$. ∎

Theorem 43.2. Let $z_1, \cdots, z_m$ be the distinct zeros of an entire function $f(z)$, where z_j is of order k_j, $j = 1, \cdots, m$. Then, $f(z)$ is of the form

$$f(z) = (z - z_1)^{k_1} \cdots (z - z_m)^{k_m} e^{g(z)}, \tag{43.1}$$

where $g(z)$ is an entire function.

Proof. It suffices to note that the function

$$F(z) = \frac{f(z)}{(z - z_1)^{k_1} \cdots (z - z_m)^{k_m}}$$

is entire and has no zeros. The result now follows from Theorem 43.1. ∎

Example 43.1. In Theorem 43.2, if $g(z)$ is a constant, then $f(z)$ is a polynomial. Otherwise, it is an entire transcendental function.

Remark 43.1. If $f_1(z)$ and $f_2(z)$ are two entire functions having the same zeros with the same multiplicities, then $f_1(z) = f_2(z)e^{g(z)}$, where $g(z)$

is an entire function. Conversely, if $g(z)$ is an entire function, then $f(z)e^{g(z)}$ has the same zeros as $f(z)$, counting multiplicities.

Now let us assume that the entire function $f(z)$ has an infinite number of zeros, say $z_1, z_2, \cdots$, which have been ordered in increasing absolute value; i.e., $|z_1| \leq |z_2| \leq \cdots$. Clearly, $|z_n| \to \infty$, for otherwise the function $f(z)$ will be identically zero (see Corollary 26.3). In view of Examples 42.4 and 42.5, the function $f(z)$ cannot be represented as $A \prod_{j=1}^{\infty} (1 - z/z_j)$ or $A \prod_{j=1}^{\infty} (z - z_j)$. In fact, to find a proper representation of such a function, we need *Weierstrass's elementary functions*:

$$E_0(z) = (1-z), \quad E_\ell(z) = (1-z)\exp(z+(z^2/2)+\cdots+(z^\ell/\ell)), \quad \ell = 1, 2, \cdots$$

and the following lemmas.

Lemma 43.1. Each elementary function $E_\ell(z)$, $\ell = 0, 1, 2, \cdots$ is an entire function having a simple zero at $z = 1$. Furthermore,

(i). $E_\ell'(z) = -z^\ell \exp(z + (z^2/2) + \cdots + (z^\ell/\ell))$,

(ii). if $E_\ell(z) = \sum_{j=0}^{\infty} a_j z^j$ is the power series expansion of $E_\ell(z)$ about $z = 0$, then $a_0 = 1$, $a_1 = a_2 = \cdots = a_\ell = 0$ and $a_j \leq 0$ for $j > \ell$, and

(iii). if $|z| \leq 1$, then $|E_\ell(z) - 1| \leq |z|^{\ell+1}$.

Proof. (i). Follows by a direct verification.

(ii). That $a_0 = 1$ is obvious. Since $E_\ell'(z)$ has a zero of multiplicity ℓ at 0, and since term-by-term differentiation is permissible, in $E_\ell(z) = \sum_{j=0}^{\infty} a_j z^j$, it follows that $a_1 = a_2 = \cdots = a_\ell = 0$. Next, in the expansion of $-E_\ell'(z) = z^\ell \exp(z + (z^2/2) + \cdots + (z^\ell/\ell))$, the coefficient of each z^j is a nonnegative real number. Hence, the coefficients in the expansion of $E_\ell(z)$ must be nonpositive.

(iii). From part (ii), we have

$$|E_\ell(z) - 1| \leq \left| \sum_{j=\ell+1}^{\infty} a_j z^j \right| \leq \sum_{j=\ell+1}^{\infty} |a_j||z|^j \leq |z|^{\ell+1} \sum_{j=\ell+1}^{\infty} (-a_j)|z|^{j-\ell-1}.$$

Now, since $E_\ell(1) = 0 = 1 + \sum_{j=\ell+1}^{\infty} a_j$ or $-\sum_{j=\ell+1}^{\infty} a_j = 1$, for $|z| \leq 1$ it follows that

$$|E_\ell(z) - 1| \leq |z|^{\ell+1} \left(\sum_{j=\ell+1}^{\infty} -a_j \right) \cdot 1 \leq |z|^{\ell+1}. \quad \blacksquare$$

Corollary 43.1. For any nonzero z_j, $|E_\ell(z/z_j) - 1| \leq |z/z_j|^{\ell+1}$ for $|z| \leq |z_j|$.

Lemma 43.2. Let $\{z_n\}$ be a sequence of complex numbers such that $z_n \neq 0$, $n = 1, 2, \cdots$ and $|z_n| \to \infty$ as $n \to \infty$. Then, there exists a sequence of nonnegative integers $\{\ell_n\}$ such that the series $\sum_{j=1}^{\infty} |z/z_j|^{\ell_j + 1}$ is uniformly convergent on every closed disk $\overline{D}(0, r)$, $r < \infty$.

Proof. Let $\ell_n = n - 1$, $n = 1, 2, \cdots$. Since $|z_n| \to \infty$ as $n \to \infty$, there exists N large enough so that $|z_n| \geq 2r$ for $n \geq N$. For all these n and for $|z| \leq r$, we have

$$\left| \frac{z}{z_n} \right|^{\ell_n + 1} = \left| \frac{z}{z_n} \right|^n \leq \left(\frac{r}{2r} \right)^n = \frac{1}{2^n}.$$

The result now follows from Theorem 22.4. ∎

The main result of this lecture is the following theorem.

Theorem 43.3 (Weierstrass's Factorization Theorem). Let $\{z_n\}$ be a sequence of complex numbers such that $z_n \neq 0$, $n = 1, 2, \cdots$ and $|z_n| \to \infty$ as $n \to \infty$, and let m be a nonnegative integer. Furthermore, let $\{\ell_n\}$ be a sequence of nonnegative integers such that the series $\sum_{j=1}^{\infty} |z/z_j|^{\ell_j + 1}$ converges uniformly on compact subsets of the complex plane (in view of Lemma 43.2, such a choice is always possible). Then,

$$f(z) = z^m \prod_{j=1}^{\infty} E_{\ell_j} \left(\frac{z}{z_j} \right) \tag{43.2}$$

represents an entire function. This function has a zero at $z = 0$ of multiplicity m and at each z_j, $j = 1, 2, \cdots$ of multiplicity k_j; here, k_j is the number of times z_j occurs in the sequence $\{z_n\}$.

Proof. Since $E_{\ell_n}(z/z_n) = 1 + (E_{\ell_n}(z/z_n) - 1)$, in view of Theorem 42.6, (43.2) converges uniformly and absolutely provided the series $\sum_{j=1}^{\infty} (E_{\ell_j}(z/z_j) - 1)$ converges uniformly and absolutely on every disk of finite radius. Let $|z| \leq r$. Then, there exists N large enough so that $|z/z_n| \leq 1$ for $n \geq N$. But, then in view of Corollary 43.1, we have $|E_{\ell_n}(z/z_n) - 1| \leq |z/z_n|^{\ell_n + 1}$. Since ℓ_n can be chosen as in Lemma 43.2, it follows that the series $\sum_{j=1}^{\infty} |z/z_j|^{\ell_j + 1}$ converges uniformly on the closed disk $\overline{B}(0, r)$, which in turn implies that the series $\sum_{j=1}^{\infty} (E_{\ell_j}(z/z_j) - 1)$ converges uniformly and absolutely on $\overline{B}(0, r)$. Finally, since $r > 0$ is arbitrary, in (43.2) the limit function $f(z)$ is entire and has the prescribed zeros. ∎

Remark 43.2. Since the sequence of positive integers $\{\ell_n\}$ can be chosen differently, the representation (43.2) is not unique.

Remark 43.3. The general entire function with simple zeros at $z = z_n$, $n = 1, 2, \cdots$ and a zero of multiplicity m at $z = 0$ is given by $f(z)e^{g(z)}$ where $f(z)$ is as in (43.2) and $g(z)$ is an arbitrary entire function.

Example 43.2.   The function $\sin \pi z$ is entire and has simple zeros at $n = 0, \pm 1, \pm 2, \cdots$. In Theorem 43.3, we let $\ell_n = 1$ for each n. Since $\sum_{j=1}^{n} |z/n|^2$ converges uniformly on every disk of finite radius, the function

$$P(z) \;=\; z \prod_{j=1}^{\infty} \left(1 - \frac{z}{j} \right) \left(1 + \frac{z}{j} \right) \;=\; z \prod_{j=1}^{\infty} \left(1 - \frac{z^2}{j^2} \right)$$

is entire and has simple zeros at nonzero integers. Thus, in view of Remark 43.3, it follows that

$$\sin \pi z \;=\; e^{g(z)} \, z \prod_{j=1}^{\infty} \left(1 - \frac{z^2}{j^2} \right), \tag{43.3}$$

where the entire function $g(z)$ has to be determined. For this, differentiating both sides of (43.3) and using Remark 42.1, we obtain

$$\pi \cos \pi z \;=\; \left(g'(z) + \frac{1}{z} + \sum_{j=1}^{\infty} \frac{-2z}{j^2 - z^2} \right) \sin \pi z,$$

which is the same as

$$\pi \cot \pi z \;=\; g'(z) + \frac{1}{z} + \sum_{j=1}^{\infty} \left(\frac{1}{z - j} + \frac{1}{z + j} \right)$$

for all z such that $\sin \pi z \neq 0$; i.e., for all $z \neq n$. Hence, we have

$$g''(z) \;=\; \sum_{j=-\infty}^{\infty} \frac{1}{(z - j)^2} - \frac{\pi^2}{\sin^2 \pi z}.$$

Now we observe that the function $g''(z)$ is periodic of period 1. Also, since $|\sin \pi z|^2 = \sin^2 \pi x + \sinh^2 \pi y$ it follows that for $0 \leq x \leq 1$, $|y| \geq 1$, the function $g''(z)$ converges uniformly to zero as $y \to \infty$. This ensures that $|g''(z)|$ is bounded in the strip $0 \leq x \leq 1$, $|y| \geq 0$. But then, by periodicity, $|g''(z)|$ is bounded on the whole complex plane. By Theorem 18.6, $g''(z)$ must be a constant, and since the limit is zero, the constant must also be zero; i.e., $g''(z) = 0$. This leads to the interesting identity

$$\sum_{j=-\infty}^{\infty} \frac{1}{(z - j)^2} \;=\; \frac{\pi^2}{\sin^2 \pi z}. \tag{43.4}$$

Thus, $g'(z)$ must be a constant. But, since $g'(z) = -g'(-z)$, this constant also must be zero. Hence, we also have the identity

$$\pi \cot \pi z \;=\; \frac{1}{z} + \sum_{j=1}^{\infty} \left(\frac{1}{z - j} + \frac{1}{z + j} \right). \tag{43.5}$$

Therefore, $g(z)$ must be a constant, say $e^{g(z)} = c$, and thus (43.3) can be written as

$$\pi \frac{\sin \pi z}{\pi z} = c \prod_{j=1}^{\infty} \left(1 - \frac{z^2}{j^2}\right). \tag{43.6}$$

Taking $z \to 0$ on both sides of (43.6) gives $c = \pi$. Thus, we have the infinite product expansion

$$\sin \pi z = \pi z \prod_{j=1}^{\infty} \left(1 - \frac{z^2}{j^2}\right). \tag{43.7}$$

Remark 43.4. For $z = 1/2$, (43.7) reduces to Wallis's formula

$$\frac{\pi}{2} = \prod_{j=1}^{\infty} \frac{(2j)^2}{(2j+1)(2j-1)} = \frac{2 \cdot 2 \cdot 4 \cdot 4 \cdot 6 \cdot 6 \cdot 8 \cdot 8 \cdots}{1 \cdot 3 \cdot 3 \cdot 5 \cdot 5 \cdot 7 \cdot 7 \cdot 9 \cdots}. \tag{43.8}$$

Remark 43.5. From (43.7), it follows that

$$\sinh \pi z = \pi z \prod_{j=1}^{\infty} \left(1 + \frac{z^2}{j^2}\right). \tag{43.9}$$

Remark 43.6. Since

$$\sum_{j=-2N-1}^{2N+1} \frac{(-1)^j}{z-j} = \sum_{j=-N}^{N} \frac{1}{z-2j} - \sum_{j=-N-1}^{N} \frac{1}{z-1-2j}$$

$$= \frac{1}{z} + \sum_{j=1}^{N} \left(\frac{1}{z-2j} + \frac{1}{z+2j}\right) - \frac{1}{z+1+2N}$$

$$- \frac{1}{z-1} - \sum_{j=1}^{N} \left(\frac{1}{z-1-2j} + \frac{1}{z-1+2j}\right)$$

as $N \to \infty$, from (43.5), we find

$$\sum_{j=-\infty}^{\infty} \frac{(-1)^j}{z-j} = \frac{\pi}{2} \cot \frac{\pi z}{2} - \frac{\pi}{2} \cot \frac{\pi(z-1)}{2},$$

which immediately gives the series

$$\frac{\pi}{\sin \pi z} = \sum_{j=-\infty}^{\infty} \frac{(-1)^j}{z-j}. \tag{43.10}$$

Remark 43.7. In (43.10), replace z by $z - 1/2$ to get

$$\frac{\pi}{\cos \pi z} = \sum_{j=-\infty}^{\infty} \frac{2(-1)^{j+1}}{2z-1-2j}, \tag{43.11}$$

which for $z = 0$ gives the Gregory-Leibniz series

$$\frac{\pi}{4} = \sum_{j=0}^{\infty} \frac{(-1)^j}{2j+1} = 1 - \frac{1}{3} + \frac{1}{5} - \frac{1}{7} + \frac{1}{9} - \cdots. \tag{43.12}$$

Remark 43.8. From the expansions above the following series are immediate

$$\frac{1}{\sinh z} = \frac{1}{z} - \frac{2z}{z^2 + \pi^2} + \frac{2z}{z^2 + 4\pi^2} - \frac{2z}{z^2 + 9\pi^2} + \cdots, \tag{43.13}$$

$$\frac{1}{\cosh z} = \frac{\pi}{(\pi/2)^2 + z^2} - \frac{3\pi}{(3\pi/2)^2 + z^2} + \frac{5\pi}{(5\pi/2)^2 + z^2} - \cdots, \tag{43.14}$$

and

$$\tanh z = \frac{2z}{(\pi/2)^2 + z^2} + \frac{2z}{(3\pi/2)^2 + z^2} + \frac{2z}{(5\pi/2)^2 + z^2} + \cdots. \tag{43.15}$$

Lecture 44
Mittag-Leffler Theorem

In this lecture, we shall construct a meromorphic function in the entire complex plane with preassigned poles and the corresponding principal parts.

Recall that a function is meromorphic in a region if it is analytic except for isolated singularities which are at most poles. In what follows, if necessary, we shall redefine the function at its removable singularities, so that it has only poles. Clearly, a meromorphic function in the entire complex plane can have at most a finite number of poles in every bounded subset. Otherwise, there will be a finite limit point of poles, and this limit point will be an essential singularity. If z_0 is a pole of order m of a function $f(z)$, then in a neighborhood of z_0, $f(z) = G(z) + g(z)$, where $g(z)$ is analytic at z_0 and

$$G(z) = \frac{a_{-m}}{(z - z_0)^m} + \frac{a_{-m+1}}{(z - z_0)^{m-1}} + \cdots + \frac{a_{-1}}{(z - z_0)}$$

is the principal part of $f(z)$ at z_0 (observe that $G(z)$ is a rational function). Thus, with a pole, the principal part of the function at the pole is always associated. In this lecture for a given sequence of complex numbers $\{z_n\}$ and a sequence of functions $\{G_n(z)\}$ of the type $G(z)$ above, we shall construct a meromorphic function $f(z)$ that has poles at z_n, $n = 1, 2, \cdots$ with the corresponding principal parts $G_n(z)$. The construction of such a function $f(z)$ is trivial if the number of points z_n is finite, say N. In fact, then $f(z) = \sum_{j=1}^{N} G_j(z) + g(z)$, where $g(z)$ is analytic, is the required function. When the number of points z_n is infinite, we have the following theorem.

Theorem 44.1 (Mittag-Leffler Theorem). Let $\{z_n\}$ be a sequence of complex numbers such that $0 < |z_1| \le |z_2| \le \cdots$ and, $\lim_{n \to \infty} |z_n| = \infty$, and for each n, let $G_n(z)$ be the rational function given by

$$G_n(z) = \frac{a^n_{-m(n)}}{(z - z_n)^{m(n)}} + \frac{a^n_{-m(n)+1}}{(z - z_n)^{m(n)-1}} + \cdots + \frac{a^n_{-1}}{(z - z_n)}.$$

Then, there exist polynomials $h_n(z)$, $n = 1, 2, \cdots$ such that

$$f(z) = \sum_{j=1}^{\infty} (G_j(z) - h_j(z)) \tag{44.1}$$

defines a meromorphic function with poles at z_n, $n = 1, 2, \cdots$ and principal parts $G_n(z)$, respectively.

Proof.   Consider the power series expansion of the principal part $G_n(z)$ in the neighborhood of the point $z = 0$; i.e.,

$$G_n(z) = a_0^{(n)} + a_1^{(n)}z + \cdots + a_\ell^{(n)}z^\ell + \cdots. \tag{44.2}$$

Clearly, this series is convergent in the open disk $B(0, |z_n|)$, and its convergence is uniform in every closed disk of radius strictly less than $|z_n|$; in particular, in the open disk $B(0, |z_n|/2)$. Let $h_n(z) = a_0^{(n)} + a_1^{(n)}z + \cdots + a_{N(n)}^{(n)}z^{N(n)}$, where $N(n)$ is so large that

$$|G_n(z) - h_n(z)| = \left| \sum_{k=N(n)+1}^{\infty} a_k^{(n)}z^k \right| \leq \frac{1}{2^n}, \quad z \in B(0, |z_n|/2). \tag{44.3}$$

Now consider a disk $B(0, r)$, $r > 0$, and let $n_0 = n_0(r)$ be the smallest integer such that $|z_n| > 2r$ for all $n > n_0$. Since $B(0, r) \subset B(0, |z_n|/2)$ for all $n > n_0$, $B(0, r)$ contains none of the points $z_{n_0+1}, z_{n_0+2}, \cdots$. Thus, (44.3) holds for all $n > n_0$ and $z \in B(0, r)$. Therefore, by Theorem 22.4, the series

$$\sum_{j=n_0+1}^{\infty} [G_j(z) - h_j(z)]$$

converges uniformly on $B(0, r)$ and hence represents an analytic function $Q_{n_0}(z)$, on $B(0, r)$. Thus, if $f(z) = \sum_{j=1}^{\infty}[G_j(z) - h_j(z)]$, we have the representation

$$f(z) = f_{n_0}(z) + Q_{n_0}(z), \quad z \in B(0, r),$$

where $Q_{n_0}(z)$ is analytic on $B(0, r)$, and the partial sum

$$f_{n_0}(z) = \sum_{j=1}^{n_0}[G_j(z) - h_j(z)]$$

is a rational function whose poles in $B(0, r)$ are those points of the sequence $\{z_n\}$ that lie in $B(0, r)$. Moreover, the principal part of $f_{n_0}(z)$, and hence of $f(z)$, at any point $z_n \in B(0, r)$ is just $G_n(z)$. Finally, since r can be chosen arbitrarily large, the result follows.   ■

Remark 44.1.   In Theorem 44.1, if $z_0 = 0$ is one of the points of the sequence $\{z_n\}$ with the given rational function $G_0(z)$, then in (44.1) we need to add a corresponding term.

Remark 44.2.   If $f_1(z)$ and $f_2(z)$ are meromorphic functions with the same poles and the same corresponding principal parts, then their difference $f_1(z) - f_2(z)$ is analytic in the entire plane, and hence equals to an entire function. Conversely, the addition of an entire function to a given meromorphic function does not alter its poles and the corresponding principal part. Hence, if $g(z)$ is an entire function, then (44.1) can be replaced by

$$f(z) = \sum_{j=1}^{\infty}(G_j(z) - h_j(z)) + g(z). \tag{44.4}$$

This representation is called a *partial fraction expansion* of $f(z)$.

Example 44.1. Consider the problem of finding a meromorphic function with simple poles at each integer n and with the principal part $G_n(z) = 1/(z-n)$ at n. The series $\sum_{j=-\infty}^{\infty} G_j(z) = \sum_{j=-\infty}^{\infty} 1/(z-j)$ does not converge uniformly for noninteger z in $B(0,r)$ for any $r > 0$. However, for any $n \neq 0$,

$$\left|\frac{1}{z-n} + \frac{1}{n}\right| = \left|\frac{z}{n(z-n)}\right| = \frac{1}{n^2}\frac{|z|}{|(z/n)-1|} \leq \frac{|z|}{n^2}$$

for large n. Hence, for noninteger z with $|z| \leq r$, the series

$$\sum_{j \neq 0}\left(\frac{1}{z-j} + \frac{1}{j}\right)$$

converges uniformly for each $r > 0$. Hence, in Theorem 44.1, letting $h_n(z) = 1/n$ for each nonzero integer n, we have that

$$\frac{1}{z} + \sum_{j \neq 0}\left(\frac{1}{z-j} + \frac{1}{j}\right) = \frac{1}{z} + \sum_{j=1}^{\infty}\left(\frac{1}{z-j} + \frac{1}{j} + \frac{1}{z+j} - \frac{1}{j}\right)$$

$$= \frac{1}{z} + \sum_{j=1}^{\infty}\left(\frac{2z}{z^2 - j^2}\right)$$

is a meromorphic function with simple poles at integers and the principal part $1/(z-n)$ at each integer n. Thus, in view of (43.5), the desired meromorphic function is $\pi\cot\pi z$.

Example 44.2. Consider the problem of constructing a meromorphic function with double poles at each integer n and with the principal part $G_n(z) = 1/(z-n)^2$ at n. The series $\sum_{j=-\infty}^{\infty} G_j(z) = \sum_{j=-\infty}^{\infty} 1/(z-j)^2$ converges uniformly and absolutely for noninteger z in $B(0,r)$ for any $r > 0$. Hence, in Theorem 44.1, letting $h_n(z) = 0$ for each integer n, we find the desired meromorphic function $\sum_{j=-\infty}^{\infty} 1/(z-j)^2$, which in view of (43.4) is the same as $\pi^2/\sin^2\pi z$.

Example 44.3. Let $\{z_n\}$ be a sequence of complex numbers such that $0 < |z_1| < |z_2| < \cdots$ and $\lim_{n\to\infty}|z_n| = \infty$, and let $\{A_n\}$ be a sequence of arbitrary complex numbers. We shall find an entire function $f(z)$ such that

$$f(z_n) = A_n, \quad n = 1, 2, \cdots. \tag{44.5}$$

This is a simple *interpolation problem*. For this, first we follow Theorem 43.3 to construct an entire function $\phi(z)$ with simple zeros at the points $z_1, z_2, \cdots,$

$$\phi(z) = \prod_{j=1}^{\infty} \left(1 - \frac{z}{z_j}\right) \exp\left(\frac{z}{z_j} + \frac{z^2}{2z_j^2} + \cdots + \frac{z^j}{jz_j^j}\right).$$

Then, we compute the derivative $\phi'(z)$ at every point z_n, which gives a sequence of nonzero complex numbers $\{\phi'(z_n)\}$. Next, we use Theorem 44.1 to find a meromorphic function $\psi(z)$ with simple poles at the points $z_1, z_2, \cdots$ and the corresponding principal parts

$$\frac{A_n}{\phi'(z_n)(z - z_n)}, \quad n = 1, 2, \cdots;$$

i.e.,

$$\psi(z) = \sum_{j=1}^{\infty} \left[\frac{A_j}{\phi'(z_j)(z - z_j)} - h_j(z)\right],$$

where $h_j(z)$ are suitably chosen polynomials. Then, the desired function is $f(z) = \phi(z)\psi(z)$. For this, clearly $f(z)$ is entire, and since

$$f(z_n) = \lim_{z \to z_n} \phi(z)\psi(z) = \lim_{z \to z_n} \left[\frac{\phi(z) - \phi(z_n)}{z - z_n}\psi(z)(z - z_n)\right] = \frac{\phi'(z_n)A_n}{\phi'(z_n)},$$

it satisfies the conditions (44.5).

Finally, we state the following result, which provides an explicit partial fraction expansion of a meromorphic function $f(z)$.

Theorem 44.2. Let $f(z)$ be a meromorphic function, and let $\{\gamma_n\}$ be a sequence of closed rectifiable Jordan curves with the following properties:

1. The origin lies inside each curve γ_n, $n = 1, 2, \cdots$.
2. None of the curves passes through poles of $f(z)$.
3. $\gamma_n \subset I(\gamma_{n+1})$, $n = 1, 2, \cdots$.
4. If r_n is the distance from the origin to γ_n, then $\lim_{n\to\infty} r_n = \infty$.

Let the poles of $f(z)$ be ordered in such a way that γ_n contains the first $m_n + 1$ poles $z_0 = 0, z_1, \cdots, z_{m_n}$, with principal parts $G_0(z), G_1(z), \cdots,$ $G_{m_n}(z)$, so that $m_n < m_{n+1}$, $n = 1, 2, \cdots$. Moreover, suppose that, for some integer $p \geq -1$,

$$\limsup_{n\to\infty} \int_{\gamma_n} \frac{|f(\xi)|}{|\xi|^{p+1}} ds < \infty, \tag{44.6}$$

where $ds = |d\xi|$. Then, at an arbitrary regular point of $f(z)$,

$$f(z) = \lim_{n \to \infty} \sum_{j=0}^{m_n} G_j(z), \quad p = -1, \tag{44.7}$$

and

$$f(z) = \lim_{n \to \infty} \sum_{j=0}^{m_n} [G_j(z) - h_j(z)], \quad p > -1, \tag{44.8}$$

where $h_n(z)$ are polynomials of degree at most p. The convergence in both (44.7) and (44.8) is uniform on every compact set containing no poles of $f(z)$.

Example 44.4. We shall use Theorem 44.2 to expand the function $\sec z$ in partial fractions. For this, we choose γ_n to be the square with center at the origin and sides of length $2n\pi$ parallel to the coordinate axis. Clearly, for $z = \pm n\pi + iy$ and $z = x \pm in\pi$, respectively, we have

$$|\sec z| = \frac{1}{\cosh y} \quad \text{and} \quad |\sec z| < \frac{1}{\sinh n\pi},$$

and hence

$$\int_{\gamma_n} |\sec \xi| ds < 2 \int_{-n\pi}^{n\pi} \frac{dy}{\cosh y} + 4n\pi \frac{1}{\sinh n\pi},$$

which shows that (44.6) holds with $p = -1$. Inside γ_n, the function $\sec z$ has simple poles at the points $z = (j - 1/2)\pi$, $-n + 1 \leq j \leq n$. Furthermore, since, in view of (31.2),

$$R\left[\sec z, \left(j - \frac{1}{2}\right)\pi\right] = \lim_{z \to (j-1/2)\pi} \frac{(z - 1/2)\pi}{\cos z} = -\frac{1}{\sin(j - 1/2)\pi} = (-1)^n,$$

the principal part of $\sec z$ at the pole $z = (j - 1/2)\pi$ is $G_j(z) = (-1)^j/[z - (j-1/2)\pi]$. Furthermore, since $z = 0$ is not a pole of $\sec z$, we have $G_0(z) = 0$. Hence, (44.7) gives

$$\sec z = \lim_{n \to \infty} \sum_{j=-n+1}^{n} \frac{(-1)^j}{z - (j - 1/2)\pi} = \sum_{j=1}^{\infty} (-1)^j \frac{(2j - 1)\pi}{z^2 - (j - 1/2)^2 \pi^2},$$

which converges uniformly on every compact set containing none of the points $z = (j - 1/2)\pi$, $j = 0, \pm 1, \pm 2, \cdots$.

Lecture 45
Periodic Functions

Recall from Lecture 8 that a complex number $\omega \neq 0$ is a *period* of a function $f(z)$ if $f(z + \omega) = f(z)$ for all z. For example, e^z has the period $2\pi i$, and $\sin z$ and $\cos z$ have the period 2π. If ω_1 and ω_2 are periods of $f(z)$, then

$$f(z + \omega_1 + \omega_2) = f(z + \omega_1) = f(z);$$

i.e., $\omega_1 + \omega_2$ is also a period. In particular, if ω is a period, then $n\omega$ is also a period, where n is any integer. The following results for periodic functions are important:

R1. Let $f(z)$ be a meromorphic periodic function with period ω. Then, ω is also a period of all derivatives of $f(z)$. In fact, differentiating n times the relation $f(z + \omega) = f(z)$, we get $f^{(n)}(z + \omega) = f^{(n)}(z)$.

R2. Let $f(z)$ be a meromorphic periodic function with period ω, and let there exist a meromorphic function $F(z)$ such that $F'(z) = f(z)$. Then, there exists a constant $c = c(\omega)$ such that $F(z + \omega) - F(z) = c$. Indeed, since $F'(z + \omega) - F'(z) = f(z + \omega) - f(z) = 0$, the function $F(z + \omega) - F(z)$ must be a constant. Furthermore, $F(z + 2\omega) = F(z + \omega + \omega) = F(z + \omega) + c = F(z) + 2c$. Also, replacing z by $z - \omega$, the relation $F(z + \omega) - F(z) = c$ can be written as $F(z - \omega) - F(z) = -c$. Hence, we have $F(z + n\omega) = F(z) + nc$ for all integers n by induction.

The function $f(z)$ is said to be *simply periodic* if all its periods are of the form $n\omega$. The functions e^z, $\sin z$, and $\cos z$ are simply periodic. The function $f(z)$ is called *doubly periodic* if all of its periods are of the form $n_1\omega_1 + n_2\omega_2$, where n_1 and n_2 are arbitrary integers and ω_2/ω_1 is not real. Here ω_1 and ω_2 are called fundamental periods for $f(z)$.

A pair of fundamental periods is not unique. If (ω_1', ω_2') is also a pair of fundamental periods, then there are integers a, b, c, d such that

$$\begin{pmatrix} \omega_1' \\ \omega_2' \end{pmatrix} = \begin{pmatrix} a & b \\ c & d \end{pmatrix} \begin{pmatrix} \omega_1 \\ \omega_2 \end{pmatrix}.$$

Since a, b, c, d are real, then

$$\begin{pmatrix} \overline{\omega}_1' \\ \overline{\omega}_2' \end{pmatrix} = \begin{pmatrix} a & b \\ c & d \end{pmatrix} \begin{pmatrix} \overline{\omega}_1 \\ \overline{\omega}_2 \end{pmatrix},$$

so

$$\begin{pmatrix} \omega_1' & \overline{\omega}_1' \\ \omega_2' & \overline{\omega}_2' \end{pmatrix} = \begin{pmatrix} a & b \\ c & d \end{pmatrix} \begin{pmatrix} \omega_1 & \overline{\omega}_1 \\ \omega_2 & \overline{\omega}_2 \end{pmatrix}.$$

Similarly, there are also integers a', b', c', d' such that

$$\begin{pmatrix} \omega_1 & \overline{\omega}_1 \\ \omega_2 & \overline{\omega}_2 \end{pmatrix} = \begin{pmatrix} a' & b' \\ c' & d' \end{pmatrix} \begin{pmatrix} \omega_1' & \overline{\omega}_1' \\ \omega_2' & \overline{\omega}_2' \end{pmatrix},$$

so

$$\begin{pmatrix} \omega_1 & \overline{\omega}_1 \\ \omega_2 & \overline{\omega}_2 \end{pmatrix} = \begin{pmatrix} a' & b' \\ c' & d' \end{pmatrix} \begin{pmatrix} a & b \\ c & d \end{pmatrix} \begin{pmatrix} \omega_1 & \overline{\omega}_1 \\ \omega_2 & \overline{\omega}_2 \end{pmatrix}.$$

Since

$$\begin{vmatrix} \omega_1 & \overline{\omega}_1 \\ \omega_2 & \overline{\omega}_2 \end{vmatrix} = \omega_1 \overline{\omega}_2 - \overline{\omega}_1 \omega_2 = -2i \operatorname{Im} \overline{\omega}_1 \omega_2 = -2i |\omega_1|^2 \operatorname{Im} \frac{\omega_2}{\omega_1} \neq 0,$$

this implies that

$$\begin{vmatrix} a' & b' \\ c' & d' \end{vmatrix} \begin{vmatrix} a & b \\ c & d \end{vmatrix} = 1,$$

and hence

$$\begin{vmatrix} a & b \\ c & d \end{vmatrix} = \begin{vmatrix} a' & b' \\ c' & d' \end{vmatrix} = \pm 1$$

since each determinant is an integer.

A doubly periodic and meromorphic function is called an elliptic function. Elliptic functions have been studied extensively, particularly because of their importance in algebra and number theory.

Example 45.1. Weierstrass's elliptic function with periods ω_1 and ω_2 is defined by

$$\wp(z; \omega_1, \omega_2) = \frac{1}{z^2} + \sum_{(n_1, n_2) \neq (0,0)} \left\{ \frac{1}{(z - n_1\omega_1 - n_2\omega_2)^2} - \frac{1}{(n_1\omega_1 + n_2\omega_2)^2} \right\}.$$

Alternatively, $\wp(z)$ may be defined in terms of $\tau = \omega_2/\omega_1$ as

$$\wp(z; \tau) = \frac{1}{z^2} + \sum_{(n_1, n_2) \neq (0,0)} \left\{ \frac{1}{(z - n_1 - n_2\tau)^2} - \frac{1}{(n_1 + n_2\tau)^2} \right\}.$$

These two definitions are related by

$$\wp(z; \tau) = \wp(z; 1, \tau)$$

and

$$\wp(z; \omega_1, \omega_2) = \frac{\wp(z/\omega_1; \omega_2/\omega_1)}{\omega_1^2}.$$

Let $f(z)$ be an elliptic function with fundamental periods ω_1, ω_2. Any parallelogram P with vertices of the form $z_0, z_0 + \omega_1, z_0 + \omega_2$, and $z_0 + \omega_1 + \omega_2$, for some $z_0 \in \mathbf{C}$, is called a fundamental parallelogram for $f(z)$. First we prove the following elementary result.

Theorem 45.1.   If $f(z)$ is a nonconstant elliptic function and P is a fundamental parallelogram for $f(z)$, then there is a pole of $f(z)$ in P.

Proof.   If not, then $f(z)$ is bounded on P. But then it is bounded everywhere by the periodicity, and hence constant by Liouville's Theorem.   ∎

An immediate consequence of Theorem 45.1 is the following corollary.

Corollary 45.1.   If $f(z)$ is an entire elliptic function, then $f(z)$ is a constant.

Since the poles of $f(z)$ are isolated, there are only a finite number of them in the compact set P, so the point z_0 may be chosen such that there are no poles on the boundary of P. Then, we have the following result.

Theorem 45.2.   If $f(z)$ is an elliptic function and P is a fundamental parallelogram for $f(z)$ with no poles of $f(z)$ on its boundary, then

$$\int_{\partial P} f(z)dz = 0.$$

Proof.   Since f has periods ω_1, ω_2, the integrals over opposite sides of P cancel each other out, and hence the contour integral vanishes.   ∎

Combining this theorem with the residue theorem now gives the following corollary.

Corollary 45.2.   The sum of the residues of $f(z)$ in P is zero.

Thus, $f(z)$ cannot have a single simple pole in P; it must have at least two simple poles or a higher-order pole.

Since the zeros of $f(z)$ are the poles of $1/f(z)$, which is also an elliptic function with P as a fundamental parallelogram, $f(z)$ also cannot have a single simple zero in P; it must have at least two simple zeros or a zero of higher multiplicity. In particular, $f(z)$ cannot attain any complex value only once in P, counting multiplicities. Indeed, if $f(z)$ attains $c \in \mathbf{C}$ only once in P, then $f(z) - c$ is an elliptic function with P as a fundamental parallelogram having a single simple zero there.

Now let $f(z)$ be a nonconstant elliptic function, let $a_1, \cdots, a_j$ be the distinct poles of $f(z)$ in the fundamental parallelogram P, and let $m_1, \cdots, m_j$ be their respective multiplicities. The number

$$m = \sum_{l=1}^{j} m_l$$

(i.e., the total number of poles of $f(z)$ in P counting multiplicities) is called the order of $f(z)$. Since $f(z)$ must have at least two simple poles or a higher-order pole, $m \geq 2$.

Example 45.2. The function $\wp(z)$ of Example 45.1 has only a single pole of order 2 in any fundamental parallelogram, and hence it is of order 2. Its derivative

$$\wp'(z; \omega_1, \omega_2) = -\frac{2}{z^3} - \sum_{(n_1, n_2) \neq (0,0)} \frac{2}{(z - n_1\omega_1 - n_2\omega_2)^3}$$

has order 3.

Note that P may be further chosen so that there are also no zeros of $f(z)$ on its boundary. Let $b_1, \cdots, b_k$ be the distinct zeros in P, with respective multiplicities $n_1, \cdots, n_k$. We have the following result.

Theorem 45.3. If $f(z)$ is a nonconstant elliptic function and P is a fundamental parallelogram for $f(z)$ with no poles or zeros of $f(z)$ on its boundary, then

$$\sum_{l=1}^{k} n_l = m.$$

Proof. We have

$$f(z) = \frac{(z - b_1)^{n_1} \cdots (z - b_k)^{n_k}}{(z - a_1)^{m_1} \cdots (z - a_j)^{m_j}} g(z), \quad z \in P$$

for some analytic function $g(z)$ with no zeros in P. A simple calculation shows that

$$\frac{f'(z)}{f(z)} = -\sum_{l=1}^{j} \frac{m_l}{z - a_l} + \sum_{l=1}^{k} \frac{n_l}{z - b_l} + \frac{g'(z)}{g(z)}.$$

Applying the Residue Theorem and noting that $g'(z)/g(z)$ is analytic in P gives

$$\frac{1}{2\pi i} \int_{\partial P} \frac{f'(z)}{f(z)} \, dz = -\sum_{l=1}^{j} m_l + \sum_{l=1}^{k} n_l.$$

On the other hand, $f'(z)/f(z)$ is also an elliptic function with P as a fundamental parallelogram, so this integral is zero by Theorem 45.2. ∎

Since $f(z)$ attains the complex value c precisely at the zeros of the elliptic function $f(z) - c$, for which P is again a fundamental parallelogram, we have the following corollary.

Corollary 45.3. The function $f(z)$ attains every complex value m-times in P, counting multiplicities.

Example 45.3. By Example 45.2 and Corollary 45.3, the function $\wp'(z)$ has three zeros in the fundamental parallelogram P with vertices $0, \omega_1, \omega_2$, and $\omega_1 + \omega_2$. Let us determine these zeros. Since $\wp'(z)$ is an odd periodic function

$$\wp'(z) \;=\; -\wp'(-z) \;=\; -\wp'(-z + \omega)$$

for any period ω, and taking $z = \omega/2$ gives

$$\wp'(\omega/2) \;=\; -\wp'(\omega/2),$$

which implies

$$\wp'(\omega/2) \;=\; 0.$$

In particular, $\omega_1/2, \omega_2/2, (\omega_1 + \omega_2)/2$ are the zeros of $\wp'(z)$ in P.

We end this lecture by deriving a differential equation satisfied by the function $\wp(z)$. Referring to the example above, $\wp'(z)$ has simple zeros at $\omega_1/2, \omega_2/2$, and $(\omega_1 + \omega_2)/2$ and a pole of order 3 at $z = 0$, so the function

$$f(z) \;=\; \wp'^2(z)$$

has zeros of order 2 at $\omega_1/2, \omega_2/2, (\omega_1 + \omega_2)/2$ and a pole of order 6 at $z = 0$. Let

$$e_1 \;=\; \wp(\omega_1/2), \qquad e_2 \;=\; \wp(\omega_2/2), \qquad e_3 \;=\; \wp((\omega_1 + \omega_2)/2),$$

and consider the function

$$g(z) \;=\; (\wp(z) - e_1)(\wp(z) - e_2)(\wp(z) - e_3).$$

Since $\omega_1/2$ and $\omega_2/2, (\omega_1 + \omega_2)/2$ are simple zeros of $\wp'(z)$, they are double zeros of $\wp(z) - e_1, \wp(z) - e_2$, and $\wp(z) - e_3$, respectively, so $g(z)$ also has zeros of order 2 at $\omega_1/2, \omega_2/2$, and $(\omega_1 + \omega_2)/2$ and a pole of order 6 at $z = 0$. Then, $f(z)/g(z)$ is an entire elliptic function, and hence constant by Corollary 45.2, so

$$f(z) \;=\; A g(z)$$

for some $A \in \mathbf{C}$. Comparing the coefficients of $1/z^6$ on the two sides shows that $A = 4$, so $\wp(z)$ satisfies the differential equation

$$\wp'^2(z) \;=\; 4(\wp(z) - e_1)(\wp(z) - e_2)(\wp(z) - e_3).$$

Lecture 46
The Riemann Zeta Function

The Riemann zeta function is one of the most important functions of classical mathematics, with a variety of applications in analytic number theory. In this lecture, we shall study some of its elementary properties.

Recall that for any positive real number r and a complex number z, $r^z = e^{z \mathrm{Log}\, r}$ is a well-defined complex number, and if $z = x + iy$, then

$$\left| r^z \right| = \left| e^{(x+iy)\mathrm{Log}\, r} \right| = \left| e^{x\mathrm{Log}\, r} \right| \left| e^{iy\mathrm{Log}\, r} \right| = r^x = r^{\mathrm{Re}(z)}.$$

Let D_1 be the open half-plane $D_1 = \{z \in \mathbf{C} : \mathrm{Re}(z) > 1\}$. The *Riemann zeta function* is defined on D_1 by the equation

$$\zeta(z) = \sum_{j=1}^{\infty} \frac{1}{j^z}. \tag{46.1}$$

Analyticity property. Note that, for any $\delta > 0$ and all $z \in D_1$ such that $\mathrm{Re}(z) \geq 1 + \delta$,

$$\sum_{j=1}^{\infty} \left| \frac{1}{j^z} \right| \leq \sum_{j=1}^{\infty} \frac{1}{j^{1+\delta}} < \infty,$$

i.e., the series in (46.1) converges uniformly and absolutely for z with $\mathrm{Re}(z) \geq 1 + \delta$. In particular, the series converges uniformly on compact subsets of D_1. Now, since for each j the function $1/j^z = e^{-z\mathrm{Log}\, j}$ is entire, it follows that the function $\zeta(z)$ is analytic on D_1. Moreover, in view of Corollary 22.4, for any $k \geq 1$ and all $z \in D_1$, we have

$$\zeta^{(k)}(z) = \sum_{j=1}^{\infty} (-1)^k \frac{(\mathrm{Log}\, j)^k}{j^z}. \tag{46.2}$$

Connection with prime numbers. Note that

$$\left(1 - \frac{1}{2^z}\right)\zeta(z) = 1 + \frac{1}{3^z} + \frac{1}{5^z} + \frac{1}{7^z} + \cdots$$

as all even terms cancel each other, and also

$$\left(1 - \frac{1}{3^z}\right)\left(1 - \frac{1}{2^z}\right)\zeta(z) = 1 + \frac{1}{5^z} + \frac{1}{7^z} + \frac{1}{11^z} + \cdots.$$

This time all terms having 3 as a factor in the denominator cancel each other. Proceeding in this way, if $p_1, p_2, \cdots, p_k$ are the first k prime numbers, then we have

$$\prod_{j=1}^{k} \left(1 - \frac{1}{p_j^z}\right) \zeta(z) = 1 + \sum_{m} \frac{1}{m^z},$$

where m consists of those positive integers that do not have $p_1, p_2, \cdots, p_k$ as a factor. Thus, letting $k \to \infty$, which is permissable, we obtain

$$\prod_{j=1}^{\infty} \left(1 - \frac{1}{p_j^z}\right) \zeta(z) = 1,$$

which is the same as

$$\frac{1}{\zeta(z)} = \prod_{j=1}^{\infty} \left(1 - \frac{1}{p_j^z}\right), \tag{46.3}$$

where the product on the right is over all the primes and converges for $\text{Re}(z) > 1$.

Analytic continuation. This is accomplished step by step by extending the function first to the domain $D_0 = \{z \in \mathbf{C} : \text{Re}(z) > 0\}$, next to the domain $D_{-1} = \{z \in \mathbf{C} : \text{Re}(z) > -1\}$, and so on, adding a vertical strip to the domain at each step. Observe that, for each n, the function

$$g_n(z) = \frac{1}{n^{z-1}} - \frac{1}{(n+1)^{z-1}}$$

is entire, with a zero at 1. Hence, for each n, the function $\phi_n(z)$ defined by

$$\phi_n(z) = \frac{1}{z-1} \left(\frac{1}{n^{z-1}} - \frac{1}{(n+1)^{z-1}}\right) = \int_0^1 \frac{dt}{(n+t)^z} = \int_n^{n+1} \frac{dt}{t^z} \tag{46.4}$$

is also entire. Similarly, for each n, the function $f_n(z)$ defined by the difference of two entire functions,

$$f_n(z) = \frac{1}{(n+1)^z} - \phi_n(z) = \frac{1}{(n+1)^z} - \int_0^1 \frac{dt}{(n+t)^z} = -z \int_0^1 \frac{t}{(n+t)^{z+1}} dt, \tag{46.5}$$

is also entire. Moreover, for all z, we also have the estimate

$$|f_n(z)| \leq |-z| \int_0^1 \frac{t}{|(n+t)^{z+1}|} dt \leq |z| \int_0^1 \frac{1}{(n+t)^{\text{Re}(z)+1}},$$

and hence, if z is such that $\text{Re}(z) > \delta$ and $|z| < M$ for some $\delta > 0$ and $M > 0$, we have

$$|f_n(z)| \leq M \frac{1}{n^{1+\delta}},$$

which implies that $\sum_{j=1}^{\infty} f_j(z)$ converges uniformly on compact sets of D_0, and therefore defines an analytic function on D_0. On the other hand, using (46.5), we have, for $z \in D_1$,

$$
\begin{aligned}
\zeta(z) &= 1 + \frac{1}{2^z} + \frac{1}{3^z} + \cdots = 1 + \sum_{j=1}^{\infty} \frac{1}{(j+1)^z} \\
&= 1 + \sum_{j=1}^{\infty} \left(\int_0^1 \frac{dt}{(j+t)^z} + \frac{1}{(j+1)^z} - \int_0^1 \frac{dt}{(j+t)^z} \right) \\
&= 1 + \sum_{j=1}^{\infty} \int_0^1 \frac{dt}{(j+t)^z} - z \sum_{j=1}^{\infty} \int_0^1 \frac{t}{(j+t)^{z+1}} dt \\
&= 1 + \int_1^{\infty} \frac{dt}{t^z} - z \sum_{j=1}^{\infty} \int_0^1 \frac{t}{(j+t)^{z+1}} dt,
\end{aligned}
$$

and hence

$$
\zeta(z) = 1 + \frac{1}{z-1} - z \sum_{j=1}^{\infty} \int_0^1 \frac{t}{(j+t)^{z+1}} dt. \tag{46.6}
$$

Thus, for $z \in D_1$, $\zeta(z)$ is given by (46.6). But, in (46.6) the last term is analytic on D_0, while $1 + 1/(z-1)$ is meromorphic on the entire complex plane with a simple pole at 1. Hence, the function defined by the right-hand side of (46.6) is a meromorphic function on D_0 with a simple pole at 1 and coincides with the definition of the Riemann zeta function on D_1. Therefore, we can define the function $\zeta(z)$ on D_0 by (46.6). This extends $\zeta(z)$ from D_1 to D_0.

We can similarly extend the function $\zeta(z)$ to D_{-1}. Note that

$$
-z \int_0^1 \frac{t}{(n+t)^{z+1}} dt = -\frac{z}{2(n+1)^{z+1}} - \frac{z(z+1)}{2} \int_0^1 \frac{t^2}{(n+t)^{z+2}} dt. \tag{46.7}
$$

Both the terms on the right-hand side of (46.7) define entire functions. Moreover, the series

$$
-\sum_{j=1}^{\infty} \frac{z(z+1)}{2} \int_0^1 \frac{t^2}{(j+t)^{z+2}} dt
$$

now converges uniformly for z such that $|z| < M$ and $\mathrm{Re}(z) > \delta > -1$ for any $\delta > -1$ and $M > 0$ and thus converges uniformly on compact subsets of D_{-1}, and therefore defines an analytic function there. Now, from (46.6) and (46.7), we have

$$
\zeta(z) = 1 + \frac{1}{z-1} - \frac{z}{2}[\zeta(z+1) - 1] - \frac{z(z+1)}{2} \sum_{j=1}^{\infty} \int_0^1 \frac{t^2}{(j+t)^{z+2}} dt. \tag{46.8}
$$

which is analytic everywhere on D_{-1} except at 1, where it has a simple pole. Note that $[\zeta(z+1) - 1]$ has a simple pole at $z = 0$, and hence $(z/2)[\zeta(z+1) - 1]$ is analytic at $z = 0$.

The extension of the domain of $\zeta(z)$ from D_{-1} to D_{-2} follows similarly. Indeed, integration by parts first gives

$$\int_0^1 \frac{t^2}{(n+t)^{z+2}} dt = \frac{1}{3(n+1)^{z+2}} + \frac{(z+2)}{3} \int_0^1 \frac{t^3}{(n+t)^{z+3}} dt,$$

and then its substitution in (46.8) leads to

$$\zeta(z) = 1 + \frac{1}{z-1} - \frac{z}{2!}[\zeta(z+1) - 1] - \frac{z(z+1)}{3!}[\zeta(z+2) - 1]$$
$$- \frac{z(z+1)(z+2)}{3!} \sum_{j=1}^{\infty} \int_0^1 \frac{t^3}{(j+t)^{z+3}} dt, \quad z \in D_{-2}.$$

The extension of the Riemann zeta function $\zeta(z)$ over the entire complex plane follows analogously. It is a meromorphic function with a single pole at $z = 1$. The residue of $\zeta(z)$ at 1 is one.

Relation between gamma and zeta functions. In the definition of gamma function in (33.17), we change t to jt, where j is a positive integer, to get

$$\Gamma(z) = j^z \int_0^{\infty} e^{-jt} t^{z-1} dt, \quad \mathrm{Re}(z) > 0.$$

Hence, it follows that

$$\Gamma(z) \sum_{j=1}^{\infty} \frac{1}{j^z} = \Gamma(z)\zeta(z) = \sum_{j=1}^{\infty} \int_0^{\infty} e^{-jt} t^{z-1} dt, \quad \mathrm{Re}(z) > 1.$$

It can be shown that the order of summation and integration on the right-hand side can be interchanged, so that, for $\mathrm{Re}(z) > 1$, we have

$$\Gamma(z)\zeta(z) = \int_0^{\infty} \left(\sum_{j=1}^{\infty} e^{-jt} \right) t^{z-1} dt = \int_0^{\infty} \frac{t^{z-1}}{e^t - 1} dt$$

and hence

$$\zeta(z) = \frac{1}{\Gamma(z)} \int_0^{\infty} \frac{t^{z-1}}{e^t - 1} dt, \quad \mathrm{Re}(z) > 1. \tag{46.9}$$

Using this relation between the gamma and zeta functions, it is possible to extend $\zeta(z)$ first to D_0 and then to D_{-1}, and so on, to the whole plane.

Riemann hypothesis. The series (46.1) for real z was introduced by Euler in 1787. In Example 36.1, we have seen that $\zeta(2) = \pi^2/6$ and

$\zeta(4) = \pi^4/90$. In fact, $\zeta(2n)$ can be calculated for all positive integers n. From elementary calculus, we know that $\zeta(1)$ diverges. The study of $\zeta(2n+1)$ is significantly more difficult. In 1979, Apéry proved that $\zeta(3)$ is irrational, but no similar results are known for other odd integers. Riemann in 1859 used the zeta function over the complex numbers for the first time to study the distribution of the prime numbers. This function has zeros on the real axis only at the negative integers $-2, -4, -6, \cdots$. One of the most important open problems in mathematics today is the *Riemann hypothesis*, which is a conjecture made by Riemann, who stated that the analytic continuation of the zeta function has infinitely many nonreal roots and that all these roots lie on the critical line $x = 1/2$. This conjecture was on the list of 23 problems David Hilbert proposed in 1900 as being most worthy of solution in the coming century. However, it has defeated the finest minds in the world. In recent years computers have been brought to bear on the conjecture, and by 2005 it had been verified that the first 100 billion zeros were on the critical line-but it does not constitute a proof.

Lecture 47
Bieberbach's Conjecture

Let S be the class of functions that are analytic and one-to-one in the unit disk $B(0,1)$ and are normalized by the conditions $f(0) = 0$ and $f'(0) = 1$. The class S has many interesting properties. A function $f \in S$ in terms of Maclaurin's series can be expressed as

$$f(z) = z + a_2 z^2 + a_3 z^3 + \cdots. \tag{47.1}$$

In 1916, Ludwig Bieberbach (1886-1982), a German mathematician (remembered as a notorious uniform-wearing Nazi and vicious anti-Semite, who sought to eliminate Jews from the profession of German mathematics) conjectured that in (47.1) the coefficients $|a_n| \leq n$, $n \geq 2$. This conjecture attracted the attention of several distinguished mathematicians. The proof for the case $n = 2$ was known to Bieberbach. In 1923, K. Löwner used a differential equation to treat the case $n = 3$, whereas in 1925 Littlewood proved that $|a_n| \leq en$ for all n, showing that the Bieberbach conjecture is true up to a factor of e. Several authors later reduced the constant in the inequality below e. Variational methods were employed in the 1930s, which led to the conjecture established for $n = 4$ in 1955 by Garabedian and Schiffer and for $n = 6$ by Pederson in 1968 and Ozawa in 1969. The case $n = 5$ was settled by Pederson and Schiffer in 1972. From time to time, proofs of other special cases were announced, but they have not been substantiated. Finally, twelve years later, Louis de Branges in 1984 proved the general case. As expected, his proof is not simple, it ran to over 350 pages. At one point, even a computer was used to validate the work; however, the proof itself does not rely on a machine. In recent years, the proof of Bieberbach's conjecture (also now known as de Branges's Theorem) has been shortened considerably, but still it is outside the scope of our book. In this lecture, we shall prove the conjecture for the case $n = 2$, and for the general case the reader can refer to [27]. Our proof requires the following result.

Theorem 47.1 (Bieberbach's Area Theorem). Let $g \in \mathcal{T}$, where $\mathcal{T}$ is the class of functions that have Laurent's expansion

$$w = g(z) = z + c_0 + \frac{c_1}{z} + \frac{c_2}{z^2} + \cdots, \tag{47.2}$$

which are analytic and one-to-one in $\{z : |z| > 1\}$ except for a pole at infinity. Then,

$$\sum_{j=1}^{\infty} j |c_j|^2 \leq 1. \tag{47.3}$$

Proof. We assume that $c_0 = 0$; otherwise, we can replace $w - c_0$ by w. Since the function g is one-to-one, it maps the circle $z = re^{i\theta}$, $0 \leq \theta \leq 2\pi$, $r > 1$ onto a closed contour γ. Clearly, γ can be written as $w(\theta) = u(\theta) + iv(\theta)$, $0 \leq \theta \leq 2\pi$, where $w = g(z)$ is as in (47.2) with $c_0 = 0$. The area $A(\Omega)$, where Ω is the region enclosed by γ, is given by

$$A(\Omega) = \int_0^{2\pi} u(\theta)v'(\theta)d\theta, \tag{47.4}$$

where $v'(\theta)$ is the derivative of $v(\theta)$ with respect to θ. Now, from (47.2), we have

$$2u(\theta) = w(\theta) + \overline{w}(\theta) = re^{i\theta} + re^{-i\theta} + \sum_{j=1}^{\infty} \frac{c_j e^{-ij\theta} + \overline{c}_j e^{ij\theta}}{r^j}$$

and

$$2iv'(\theta) = w'(\theta) - \overline{w}'(\theta) = i\left[re^{i\theta} + re^{-i\theta} - \sum_{j=1}^{\infty} \frac{jc_j e^{-ij\theta} + j\overline{c}_j e^{ij\theta}}{r^j}\right].$$

Thus, from (47.4), we obtain

$$A(\Omega) = \pi r^2 - \pi \sum_{j=1}^{\infty} j|c_j|^2 r^{-2j}.$$

Since the area is always positive, it follows that

$$\sum_{j=1}^{\infty} j|c_j|^2 r^{-2j} \leq r^2. \tag{47.5}$$

The inequality (47.3) now follows by letting $r \to 1$ in (47.5). ∎

The main result of this lecture is the following theorem.

Theorem 47.2 (Bieberbach's Conjecture for $n = 2$). Let $f \in \mathcal{S}$ be as in (47.1). Then, $|a_2| \leq 2$.

Proof. The proof requires the following three steps.

Step 1. $F(z) = [f(1/z)]^{-1} \in \mathcal{T}$. For this, first we shall show that $F(z)$ is one-to-one. Assume that $[f(1/z_1)]^{-1} = [f(1/z_2)]^{-1}$, where $|z_1| > 1$ and $|z_2| > 1$. Then, we have

$$f\left(\frac{1}{z_1}\right) = f\left(\frac{1}{z_2}\right), \quad \left|\frac{1}{z_1}\right| > 1 \quad \text{and} \quad \left|\frac{1}{z_2}\right| > 1.$$

The one-to-one property of $[f(1/z)]^{-1}$ for $|z| > 1$ now follows from the one-to-one property of $f(z)$ for $|z| < 1$.

Next, we shall demonstrate that $F(z)$ is analytic for $|z| > 1$. If we can show that $f(1/z) \neq 0$ for $|z| > 1$, then from the analyticity of $f(z)$ on $B(0,1)$ it will follow that $F(z)$ is analytic for $|z| > 1$. Let z_0 be such that $0 < |1/z_0| < 1$ and $f(1/z_0) = 0$, but then $f(0) = f(1/z_0) = 0$, which contradicts the one-to-one property of $f(z)$ on $B(0,1)$.

Step 2. If $f(z) \in \mathcal{S}$, then $F_1(z) = z\sqrt{f(z^2)/z^2} \in \mathcal{S}$. For this, since

$$F_1(z) = z(1 + a_2 z^2 + a_3 z^4 + a_4 z^6 + \cdots)^{1/2}, \tag{47.6}$$

the function $F_1(z)$ is analytic with $F_1(0) = 0$ and $F_1'(0) = 1$. Moreover, $F_1(z) \neq 0$ for $z \neq 0$. To show that $F_1(z)$ is one-to-one on $B(0,1)$, assume that $F_1(z_1) = F_1(z_2)$. But this implies that $f(z_1^2) = f(z_2^2)$, and hence, in view of the one-to-one property of $f(z)$, we must have $z_1^2 = z_2^2$; i.e., $z_1 = \pm z_2$. Now notice that $F_1(-z) = -F_1(z)$; i.e., $F_1(z)$ is an odd function. Hence, $z_1 = -z_2$ gives $F_1(z_1) = F_1(-z_2) = -F_1(z_2)$; i.e., then $F_1(z_1) \neq F_1(z_2)$. Therefore, $z_1 = z_2$ is the only solution of $F_1(z_1) = F_1(z_2)$.

Step 3. From (47.6), it follows that $F_1'''(0) = 3a_2$, and hence

$$F_1(z) = z + \frac{1}{2}a_2 z^3 + \cdots.$$

Since, by Step 1, $[F_1(1/z)]^{-1} \in \mathcal{T}$, we have

$$\frac{1}{F_1\left(\dfrac{1}{z}\right)} = \frac{1}{\dfrac{1}{z}\left[1 + \dfrac{1}{2}a_2\dfrac{1}{z^2} + \cdots\right]} = z - \frac{1}{2}a_2\frac{1}{z} + \cdots,$$

which in view of Theorem 47.1 implies that

$$2\left|\frac{1}{2}a_2\right|^2 \leq 1,$$

and hence certainly we must have $|a_2| \leq 2$. ∎

Remark 47.1.   The inequality $|a_2| \leq 2$ in Theorem 47.2 is the best possible. For this, we consider the function

$$w = h(z) = \frac{z}{(1 - \lambda z)^2} = z + 2\lambda z^2 + 3\lambda^2 z^3 + \cdots,$$

where $|\lambda| = 1$. We claim that this function is one-to-one on $\mathcal{S}$. In fact, we have

$$\frac{z_1}{(1 - \lambda z_1)^2} = \frac{z_2}{(1 - \lambda z_2)^2}$$

if and only if

$$(z_1 - z_2)(1 - \lambda^2 z_1 z_2) = 0.$$

However, since $1 - \lambda^2 z_1 z_2 \neq 0$ as $|z_1| < 1$ and $|z_2| < 1$, it follows that $z_1 = z_2$. It is also clear that the function $h(z)$ is analytic in $B(0,1)$. Thus, $h(z)$ is the desired function that belongs to S for which $|a_n| = n$. This function is known as *Koebe's function*. It can be shown that this is the only function for which $|a_n| = n$. Furthermore, for $\lambda = 1$, $w = h(z)$ maps $B(0,1)$ onto the entire w-plane with a branch cut $-\infty < u < -1/4$, $v = 0$.

We conclude this lecture by proving the following interesting result.

Theorem 47.3 (Koebe's Covering Theorem). If $f \in S$ and $f(z) \neq c$ for $|z| < 1$, then $|c| \geq 1/4$; i.e., if the point c is not in the image of $f(z)$, $z \in B(0,1)$, then the distance between c and the origin is at least $1/4$.

Proof. Since $f(z) \neq c$, from (47.1), it follows that the function

$$g(z) = \frac{cf(z)}{c - f(z)} = z + \left(a_2 + \frac{1}{c}\right)z^2 + \cdots$$

is analytic and one-to-one on $B(0,1)$. Thus, from Theorem 47.2, it follows that

$$\left|\frac{1}{c}\right| - |a_2| \leq \left|a_2 + \frac{1}{c}\right| \leq 2,$$

which in view of $|a_2| \leq 2$ gives

$$\left|\frac{1}{c}\right| \leq 2 + |a_2| \leq 4,$$

i.e., $|c| \geq 1/4$. ∎

Remark 47.2.   Theorem 47.3 shows that for each $f \in S$ the image $w = f(z)$, $z \in B(0,1)$ in the w-plane contains the disk $\{w : |w| < 1/4\}$.

Remark 47.3.   There is a beautiful regularity theorem of Hayman, which he established in 1953. The limit $\lim_{n \to \infty} |a_n|/n$ exists for every $f \in S$ and is smaller than 1 unless f is a Koebe's function.

Lecture 48
The Riemann Surfaces

A Riemann surface is an ingenious construct for visualizing a multi-valued function. We treat all branches of a multi-valued function as a single-valued function on a domain that consists of many sheets of the z-plane. These sheets are then glued together so that in moving from one sheet to another we pass continuously from one branch of the multi-valued function to another. This glued structure of sheets is called a Riemann surface for the multi-valued function. For example, in a multi-story car park, floors can be thought of as sheets of the z-plane, that are glued by the ramps on which cars can go from one level to another. Riemann surfaces have proved to be of inestimable value, especially in the study of algebraic functions. Although there is much literature on the subject, in this lecture we shall construct Riemann surfaces for some simple functions.

Riemann surface for $w = z^{1/2}$. Recall from Lecture 9, that this function has two branches, represented by the single-valued functions $f_1(z)$ and $f_2(z)$ (see Figure 48.1). The respective domains S_1 and S_2 of these functions are obtained by cutting the z-plane along the negative x-axis. The range R_1 for $f_1(z)$ consists of the right half-plane and the positive v-axis; the range R_2 for $f_2(z)$ consists of the left half-plane and the negative v-axis. The sets R_1 and R_2 are glued together along the positive and negative v-axes to form the w-plane with the origin deleted.

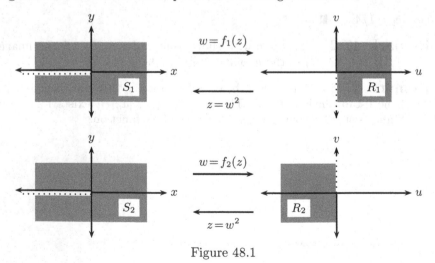

Figure 48.1

Now we place these two copies of the cut z-plane S_1 and S_2 directly on each other so that the upper sheet S_1 maps to the right w half-plane R_1, and the lower sheet S_2 maps to the left w half-plane R_2. The dashed edge of S_1 is glued to the black edge of S_2, and the dashed edge of S_2 is glued to the black edge of S_1. The surface R thus obtained is a Riemann surface for the mapping $w = z^{1/2}$ (for portions of S_1 and S_2 see Figure 48.2). Although in a physical sense this gluing procedure is impossible, on R the function $w = z^{1/2}$ is single-valued, and continuous for all $z \neq 0$.

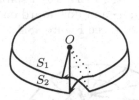

Figure 48.2

Riemann surface for $w = z^{1/n}$.

Recall that this function has n branches, represented by the single-valued functions

$$f_k(z) = |z|^{1/n}\left(\cos\frac{\arg z + 2k\pi}{n} + i\sin\frac{\arg z + 2k\pi}{n}\right), \quad k = 0, 1, \cdots, n-1.$$

The respective domains $S_0, S_1, \cdots, S_{n-1}$ of these functions are obtained by cutting the z-plane along the negative x-axis. Now we place these n copies of the cut z-plane directly on each other. Let s_k^+ and s_k^- denote the black and dashed edges of S_k. We glue s_0^- to s_1^+, s_1^- to s_2^+, $\cdots, s_{n-2}^-$ to s_{n-1}^+, and finally s_{n-1}^- to s_0^+. This results in an n-sheeted Riemann surface R (see Figure 48.3 for $n = 4$). All n sheets meet at the branch point $z = 0$. On R, the function $w = z^{1/n}$ is single-valued, and continuous for all $z \neq 0$.

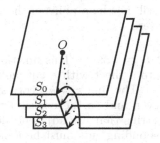

Figure 48.3

Riemann surface for $w = \log z$. Recall that

$$\log z \;=\; \text{Log}\,|z| + i \arg z \;=\; \text{Log}\,|z| + i\,\text{Arg}\,z + 2k\pi i, \quad k = 0, \pm 1, \pm 2, \cdots$$

is a countably many-valued function. Thus, its Riemann surface requires infinitely many copies S_k, $k = \cdots, -n, \cdots, -1, 0, 1, \cdots, n, \cdots$ of the z-plane cut along the negative x-axis. As earlier, we place these infinite copies of the cut z-plane directly on each other and let s_k^+ and s_k^- denote the black and dashed edges of S_k. For each integer k, we glue s_k^+ to s_{k+1}^-. The Riemann surface for the function $w = \log z$ looks like a spiral staircase that extends upward on the sheets $S_1, S_2, \cdots$ and downward on the sheets $S_{-1}, S_{-2}, \cdots$ (for portions of S_1, S_0, and S_{-1}, see Figure 48.4).

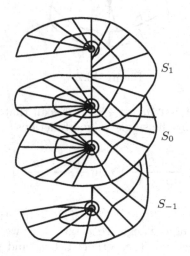

S_1

S_0

S_{-1}

Figure 48.4

Riemann surface for $w = e^z$. This function maps each parallel strip $(k-1)2\pi < y < k \cdot 2\pi$ onto a sheet with a cut along the positive axis. The sheets are attached to each other so that they form an endless screw in the w-plane. The origin will not be a point of the Riemann surface due to the fact that e^z is never zero.

Riemann surface for $w = \cos z$. This function maps each parallel strip $(k-1)\pi < x < k\pi$ onto a sheet with a cut along the real axis from $-\infty$ to -1 and from 1 to ∞. The line $z = k\pi$ corresponds to both edges of the positive cut if k is even and to the edges of the negative cut if k is odd. If we consider two strips that are adjacent along the line $z = k\pi$, then the edges of the corresponding cuts must be joined crosswise so as to generate a simple branch point at $w = \pm 1$. The resulting Riemann surface has infinitely many simple branch points over $w = 1$ and $w = -1$ that alternatingly connect the odd and even sheets. Figure 48.5 represents a cross section of the surface in the case that the cuts are chosen parallel to each other. Here, any two points on the same level are joined by an arc that does not intersect any of the cuts.

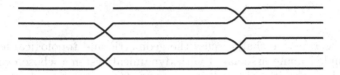

Figure 48.5

Riemann surface for $w = \sin z$. Since $w = \sin z = \cos(\pi/2 - z)$, the Riemann surface for $w = \sin z$ is the same as for $w = \cos z$.

Riemann surface for $w = \tan z$. The Riemann surface has infinitely many sheets with two logarithmic branch points above $w = \pm i$. It is constructed by joining an infinite number of w-planes with a cut $u = 0$, $|v| \le 1$, corresponding to the vertical strips $k\pi < x < (k+1)\pi$.

Riemann surface for $w = \cot z$. Since $\cot z = \tan(\pi/2 - z)$, the Riemann surface for $w = \cot z$ is the same as for $w = \tan z$.

Riemann surface for $w = \cosh z$. Since $\cosh z = \cos iz$, the Riemann surface for $w = \cosh z$ is the same as for $w = \cos z$.

Riemann surface for $w = \sinh z$. Since $\sinh z = -i \sin iz$, the Riemann surface for $w = \sinh z$ is obtained by rotating the surface for $w = \sin z$ through $\pi/2$ about the origin.

Riemann surface for $w = \tanh z$. Since $\tanh z = -i \tan iz$, the Riemann surface for $w = \tanh z$ is obtained by rotating the surface for $w = \tan z$ through $\pi/2$ about the origin.

Riemann surface for $w = \coth z$. Since $\coth z = i \cot iz$, the Riemann surface for $w = \coth z$ is the same as for $w = \tanh z$.

Lecture 49
Julia and Mandelbrot Sets

In this lecture, we shall discuss the geometric and topological features of the complex plane associated with dynamical systems whose evolution is governed by the iterative scheme $z_{n+1} = f(z_n)$, $z_0 = p$ where $f(z)$ is a complex valued function and $p \in \mathbf{C}$. Such systems occur in physical, engineering, medical, and aesthetic problems, especially those exhibiting chaotic behavior.

The word *fractal* is related to the words fractured and fraction, meaning broken or not whole. This term was coined in 1975 by Benoit Mandelbrot to describe objects that in a certain sense can have dimensions that are not whole numbers; however, the fractal phenomenon had been noticed earlier, in 1918, by the French mathematicians Gaston Julia and Pierre Fatou when exploring iterations of the complex functions; i.e., analyzing the complex sequence $\{z_n\}$ generated by the iterating scheme

$$z_{n+1} = f(z_n), \quad z_0 = p \in \mathbf{C}. \tag{49.1}$$

The findings of Julia and Fatou did not receive much attention because during that period computer graphics were not available. With the recent advancement of computers, the study of complex iterative sequences has attracted many prominent mathematicians and computer scientists. We shall discuss some of their interesting results for the quadratic function $f_c(z) = z^2 + c$ and show graphically how the iterates for such a simple function produce startling pictures. For this, we need the following definitions.

Let $f(z)$ be an analytic function in $\mathbf{C}$. A point z_0 is called an *attracting point (repelling point)* for the function $f(z)$ if $|f'(z_0)| < 1$ ($|f'(z_0)| > 1$). A *q-cycle* for $f(z)$ is a set $Q = \{z_0, z_1, \cdots, z_{q-1}\}$ of q complex numbers such that $z_{n+1} = f(z_n)$, $n = 0, 1, \cdots, q - 2$, and $f(z_{q-1}) = z_0$. If $q = 1$, then z_0 is a *fixed point* of $f(z)$. Each $z_n \in Q$ is called a *periodic point*. In fact, once $z_n = f(z_{n-1})$ reaches a periodic point, thereafter it cycles indefinitely through the points of the cycle. The q-cycle is said to be *attracting (repelling)* if $|F_q'(z_0)| < 1$ ($|F_q'(z_0)| > 1$), where F_q is the composition of f with itself n times, e.g., when $q = 2$, then $F_2(z) = (f \circ f)(z) = f(f(z))$. From the chain rule, it immediately follows that the derivative of $F_q(z)$ takes the same value at each point of the q-cycle. The significance of an attracting point is explained in the following result.

Theorem 49.1. If z^* is an attracting fixed point of $f(z)$, then there exists a disk $B(z^*, r)$ such that $f(z)$ draws any point $z \in B(z^*, r)$ toward

z^*, in the sense that $|f(z) - z^*| < |z - z^*|$. Furthermore, the sequence $\{z_n\}$ generated by (49.1) with any $z_0 = z \in B(z^*, r)$ converges to z^*.

Proof. Since $f(z)$ is analytic, for every $\epsilon > 0$ there exists some $r > 0$ such that if $z \in B(z^*, r)$, then

$$\left| \frac{f(z) - f(z^*)}{z - z^*} - f'(z^*) \right| < \epsilon.$$

Since $|f'(z^*)| < 1$, we can let $\epsilon = 1 - |f'(z^*)|$ to obtain

$$\left| \frac{f(z) - f(z^*)}{z - z^*} \right| - |f'(z^*)| \leq \left| \frac{f(z) - f(z^*)}{z - z^*} - f'(z^*) \right| < 1 - |f'(z^*)|,$$

which immediately gives $|f(z) - z^*| < |z - z^*|$. Hence, there exists a $c \in (0, 1)$ such that for all $z \in B(z^*, r)$, $|f(z) - z^*| \leq c|z - z^*|$. Thus, in particular, if $z_0 \in B(z^*, r)$, then $|z_1 - z^*| = |f(z_0) - z^*| \leq c|z_0 - z^*| \leq cr$; i.e., $z_1 \in B(z^*, cr)$. Furthermore, $|z_2 - z^*| = |f(z_1) - z^*| \leq c|z_1 - z^*| \leq c^2 r$; i.e., $z_2 \in B(x^*, c^2 r)$. Now, an easy induction gives $z_n \in B(z^*, c^n r)$, and this in turn implies that $z_n \to z_0$. ∎

Example 49.1. For the function $f_c(z) = z^2 + c$, z^* is a fixed point if and only if $f_c(z^*) = z^*$, which gives $z_1^* = (1 + \sqrt{1 - 4c})/2$ and $z_2^* = (1 - \sqrt{1 - 4c})/2$, where the square root designates the principal square root function. Since $f'(z) = 2z$, the fixed point z_1^* (z_2^*) is attracting if and only if $|1 + \sqrt{1 - 4c}| < 1$ ($|1 - \sqrt{1 - 4c}| < 1$). In particular, for the function $f(z) = z^2$ the fixed points are $z_1^* = 1$ and $z_2^* = 0$. The fixed point 1 is repelling, whereas the fixed point 0 is attracting. In fact, the iterative scheme (49.1) gives $z_n = p^{2n}$, and hence, if $|p| < 1$, then $z_n \to 0$; if $|p > 1$, then $|z_n| \to \infty$; and if $|p| = 1$, the sequence $\{z_n\}$ either oscillates around the unit circle or converges to 1. Thus, the unit circle divides the complex plane into two regions separated by the unit circle. A starting value p in one region results in z_n being attracted to 0 and in the other region results in repulsion.

The nature of the sequence $\{z_{n,c}(p)\}$ generated by (49.1) for the function $f_c(z) = z^2 + c$ depends critically on the choice of c. Thus, for a fixed c, we define the sets

$$E_c = \{p : |z_{n,c}(p)| \to \infty\} \quad (\text{escape set})$$
$$K_c = \mathbf{C} \backslash E_c \quad (\text{keep set}).$$

The following properties of these sets are known:

P1. $z \in E_c$ (K_c) if and only if $-z \in E_c$ (K_c).

P2. $z \in E_c$ (K_c) implies that $f_c(z) \in E_c$ (K_c).

P3. $K_c \subseteq \overline{B}(0, r_c)$, where r_c is the nonnegative root (if it exists) of the equation $x^2 + c = x$.

P4. E_c is open and connected.

P5. K_c is closed and simply connected.

P6. K_c is connected if and only if $0 \in K_c$ (*Fatou-Julia Theorem*). In particular, the sets K_0 and K_i are connected.

P7. If the point z_0 is a periodic attractor for $f_c(z)$, then it is an interior point of K_c.

P8. If the point z_0 is a periodic repeller for $f_c(z)$, then it is on the boundary ∂K_c of K_c.

The boundary ∂K_c of K_c is known as the *Julia set* for the function $f_c(z)$, and the set $K_c \cup \partial K_c$ is called the *filled-in Julia set*. For an assigned value of c, the Julia set of $f_c(z)$ can be viewed as a curve that divides the complex plane into two regions. From Example 49.1, it is clear that the Julia set for $f_0(z)$ is the unit circle $|z| = 1$. It turns out that K_c is a nice simple set only when $c = 0$ or $c = -2$. Mandelbrot discovered that for every other value of c the Julia set of $f_c(z)$ is a fractal. In fact, it exhibits a complicated structure under any degree of magnification and describes an object whose dimensionality might not be a whole number. It may fragment into a multitude of tiny flecks (called Fatou dusts), with K_c having no interior points at all.

The *Mandelbrot set* denoted as M is defined as

$$M = \{c : z_{n,c}(0) \text{ does not tend to infinity}\}.$$

Clearly, $c \in M$ if and only if $0 \in K_c$. The Fatou-Julia Theorem characterizes M in terms of K_c as

$$M = \{c : K_c \text{ is connected}\}.$$

Theorem 49.2. $\overline{B}(0, 1/4) \subseteq M \subseteq \overline{B}(0, 2)$.

Proof. To prove $\overline{B}(0, 1/4) \subseteq M$ by inductive arguments, we shall show that $|z_{n,c}(0)| \leq 1/2$, $n \geq 1$. Clearly, if $|c| \leq 1/4$, then $|z_{1,c}(0)| = |c| \leq 1/4$. Now, assuming that $|z_{n,c}(0)| \leq 1/2$, we have $|z_{n+1,c}(0)| = |z_{n,c}(0) + c| \leq |z_{n,c}(0)|^2 + |c| \leq 1/4 + 1/4 = 1/2$. To prove $M \subseteq \overline{B}(0, 2)$, we shall show that, if $|c| > 2$, then $c \notin M$. Clearly, $|z_{1,c}(0)| = |c| > 2$, and

$$
\begin{aligned}
|z_{2,c}(0)| &= |z_{1,c}^2(0) + c| = |z_{1,c}(0)||z_{1,c}(0) + c/z_{1,c}(0)| \\
&\geq |z_{1,c}(0)|(|z_{1,c}(0)| - |c|/|z_{1,c}(0)|) = |c|(|c| - 1),
\end{aligned}
$$

which also implies that $|z_{2,c}(0)| > |c|$. Next, we have

$$|z_{3,c}(0)| \geq |z_{2,c}(0)|(|z_{2,c}(0)| - |c|/|z_{2,c}(0)|) > |z_{2,c}(0)|(|c| - 1) \geq |c|(|c| - 1)^2.$$

Continuing in this way, we find

$$|z_{n,c}(0)| > |c|(|c| - 1)^{n-1},$$

which in view of $|c| > 2$ implies that $|z_{n,c}(0)| \to \infty$. ∎

For $|c| \le 2$, if we encounter an iterate $z_{n,c}(0)$ such that $|z_{n,c}(0)| > 2$, then as above it follows that

$$|z_{n+m,c}(0)| > |z_{n,c}(0)|(|z_{n,c}(0)| - 1)^m,$$

which immediately implies that $|z_{n,c}(0)| \to \infty$, and this means that $c \notin M$. It is believed that if we reach $z_{1000,c}(0)$ such that $|z_{1000,c}(0)| \le 2$, then there is very little probability that the sequence will diverge to infinity. We can then, with great safety, say that $c \in M$.

The elements of the Mandelbrot set M for values of c in the range $-2 \le \operatorname{Re} c \le 1,\ -1.5 \le \operatorname{Im} c \le 1.5$ are plotted in Figure 49.1.

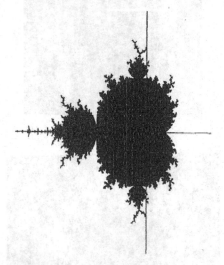

Figure 49.1

The following properties of the set M are known:

Q1. If c is any real number greater than $1/4$, then $c \notin M$.

Q2. M is a closed subset of $\overline{B}(0,2)$, and hence compact.

Q3. M is symmetric about the real axis, which it intersects in the interval $[-2, 1/4]$.

Q4. M is simply connected.

The set M is not self-similar, although it may look that way. There are subtle variations in its infinite complexity. The boundary of the set M is its most fascinating aspect. A magnification of a portion of the boundary of M in Figure 49.2 reveals its fractal nature and the presence of infinitely many hairlike branching filaments. The connectedness of M relies on the existence of these filaments. A further magnification is shown in Figure 49.3. These pictures justify the statement of Hubbard that the Mandelbrot set M is the most complicated object in mathematics.

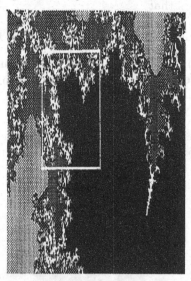

Figure 49.2

Figure 49.3

Finally, we remark that the connection between chaotic systems and fractals has been explored in many recent books.

Lecture 50
History of Complex Numbers

The problem of complex numbers dates back to the 1st century, when Heron of Alexandria (about 75 AD) attempted to find the volume of a frustum of a pyramid, which required computing the square root of $81 - 144$ (though negative numbers were not conceived in the Hellenistic world). We also have the following quotation from Bhaskara Acharya (working in 486 AD), a Hindu mathematician: "The square of a positive number, also that of a negative number, is positive: and the square root of a positive number is two-fold, positive and negative; there is no square root of a negative number, for a negative number is not square." Later, around 850 AD, another Hindu mathematician, Mahavira Acharya, wrote: "As in the nature of things, a negative (quantity) is not a square (quantity), it has therefore no square root." In 1545, the Italian mathematician, physician, gambler, and philosopher Girolamo Cardano (1501-76) published his *Ars Magna* (The Great Art), in which he described algebraic methods for solving cubic and quartic equations. This book was a great event in mathematics. In fact, it was the first major achievement in algebra in 3000 years, after the Babylonians showed how to solve quadratic equations. Cardano also dealt with quadratics in his book. One of the problems that he called "manifestly impossible" is the following: Divide 10 into two parts whose product is 40; i.e., find the solution of $x + y = 10$, $xy = 40$, or, equivalently, the solution of the quadratic equation $40 - x(10 - x) = x^2 - 10x + 40 = 0$, which has the roots $5 + \sqrt{-15}$ and $5 - \sqrt{-15}$. Cardano formally multiplied $5 + \sqrt{-15}$ by $5 - \sqrt{-15}$ and obtained 40; however, to calculations he said "putting aside the mental tortures involved." He did not pursue the matter but concluded that the result was "as subtle as it is useless." This event was historic since it was the first time the square root of a negative number had been explicitly written down. For the cubic equation $x^3 = ax + b$, the so-called Cardano formula is

$$x = \sqrt[3]{\frac{b}{2} + \sqrt{\left(\frac{b}{2}\right)^2 - \left(\frac{a}{3}\right)^3}} + \sqrt[3]{\frac{b}{2} - \sqrt{\left(\frac{b}{2}\right)^2 - \left(\frac{a}{3}\right)^3}}.$$

When applied to the historic example $x^3 = 15x + 4$, the formula yields

$$x = \sqrt[3]{2 + \sqrt{-121}} + \sqrt[3]{2 - \sqrt{-121}}.$$

Although Cardano claimed that his general formula for the solution of the cubic equation was inapplicable in this case (because of the appearance of

$\sqrt{-121}$), square roots of negative numbers could no longer be so lightly dismissed. Whereas for the quadratic equation (e.g., $x^2 + 1 = 0$) one could say that no solution exists, for the cubic $x^3 = 15x + 4$ a real solution, namely $x = 4$, does exist; in fact, the two other solutions, $-2 \pm \sqrt{3}$, are also real. It now remained to reconcile the formal and "meaningless" solution $x = \sqrt[3]{2 + \sqrt{-121}} + \sqrt[3]{2 - \sqrt{-121}}$ of $x^3 = 15x + 4$, found by using Cardano's formula, with the solution $x = 4$, found by inspection. The task was undertaken by the hydraulic engineer Rafael Bombelli (1526-73) about thirty years after the publication of Cardano's work.

Bombelli had the "wild thought" that since the radicals $2 + \sqrt{-121}$ and $2 - \sqrt{-121}$ differ only in sign, the same might be true of their cube roots. Thus, he let

$$\sqrt[3]{2 + \sqrt{-121}} = a + \sqrt{-b} \quad \text{and} \quad \sqrt[3]{2 - \sqrt{-121}} = a - \sqrt{-b}$$

and proceeded to solve for a and b by manipulating these expressions according to the established rules for real variables. He deduced that $a = 2$ and $b = 1$ and thereby showed that, indeed,

$$\sqrt[3]{2 + \sqrt{-121}} + \sqrt[3]{2 - \sqrt{-121}} = (2 + \sqrt{-1}) + (2 - \sqrt{-1}) = 4.$$

Bombelli had thus given meaning to the "meaningless." This event signaled the birth of complex numbers. A breakthrough was achieved by thinking the unthinkable and daring to present it in public. Thus, the complex numbers forced themselves in connection with the solutions of cubic equations rather than the quadratic equations.

To formalize his discovery, Bombelli developed a calculus of operations with complex numbers. His rules, in our symbolism, are $(-i)(-i) = -1$ and

$$(\pm 1)i = \pm i, \quad (+i)(+i) = -1, \quad (-i)(+i) = +1,$$
$$(\pm 1)(-i) = \mp i, \quad (+i)(-i) = +1.$$

He also considered examples involving addition and multiplication of complex numbers, such as $8i + (-5i) = +3i$ and

$$\left(\sqrt[3]{4 + \sqrt{2i}} \right) \left(\sqrt[3]{3 + \sqrt{8i}} \right) = \sqrt[3]{8 + 11\sqrt{2i}}.$$

Bombelli thus laid the foundation stone of the theory of complex numbers. However, his work was only the beginning of the saga of complex numbers. Although his book *l'Algebra* was widely read, complex numbers were shrouded in mystery, little understood, and often entirely ignored. In fact, for complex numbers Simon Stevin (1548-1620) in 1585 remarked that "there is enough legitimate matter, even infinitely much, to exercise oneself without occupying oneself and wasting time on uncertainties." John

Wallis (1616-1703) had pondered and puzzled over the meaning of imaginary numbers in geometry. He wrote, "These Imaginary Quantities (as they are commonly called) arising from the Supposed Root of a Negative Square (when they happen) are reputed to imply that the Case proposed is Impossible." Gottfried Wilhelm von Leibniz (1646-1716) made the following statement in 1702: "The imaginary numbers are a fine and wonderful refuge of the Divine Sprit, almost an amphibian between being and non-being." Christiaan Huygens (1629-95) a prominent Dutch mathematician, astronomer, physicist, horologist, and writer of early science fiction, was just as puzzled as Leibniz. In reply to a query he wrote to Leibniz: "One would never have believed that $\sqrt{1 + \sqrt{-3}} + \sqrt{1 - \sqrt{-3}} = \sqrt{6}$ and there is something hidden in this which is incomprehensible to us." Leonhard Euler (1707-83) was candidly astonished by the remarkable fact that expressions such as $\sqrt{-1}$, $\sqrt{-2}$, etc., are neither nothing, nor greater than nothing, nor less than nothing, which necessarily constitutes them imaginary or impossible. In fact, he was confused by the absurdity $\sqrt{(-4)(-9)} = \sqrt{36} = 6 \neq \sqrt{-4}\sqrt{-9} = (2i)(3i) = 6i^2 = -6$.

Similar doubts concerning the meaning and legitimacy of complex numbers persisted for two and a half centuries. Nevertheless, during the same period complex numbers were extensively used and a considerable amount of theoretical work was done by such distinguished mathematicians as René Descartes (1596-1650) (who coined the term *imaginary number*, before him these numbers were called *sophisticated* or *subtle*), and Euler (who was the first to designate $\sqrt{-1}$ by i); Abraham de Moivre (1667-1754) in 1730 noted that the complicated identities relating trigonometric functions of an integer multiple of an angle to powers of trigonometric functions of that angle could be simply reexpressed by the well-known formula $(\cos\theta + i\sin\theta)^n - \cos n\theta + i\sin n\theta$ and many others. Complex numbers also found applications in map projection by Johann Heinrich Lambert (1728-77) and by Jean le Rond d'Alembert (1717-83) in hydrodynamics.

The desire for a logically satisfactory explanation of complex numbers became manifest in the latter part of the 18th century, on philosophical, if not on utilitarian grounds. With the advent of the Age of Reason, when mathematics was held up as a model to be followed not only in the natural sciences but also in philosophy as well as political and social thought, the inadequacy of a rational explanation of complex numbers was disturbing. By 1831, the great German mathematician Karl Friedrich Gauss (1777-1855) had overcome his scruples concerning complex numbers (the phrase complex numbers is due to him) and, in connection with a work on number theory, published his results on the geometric representation of complex numbers as points in the plane. However, from Gauss's diary, which was left among his papers, it is clear that he was already in possession of this interpretation by 1797. Through this representation, Gauss clarified the "true metaphysics of imaginary numbers" and bestowed on them complete franchise in mathematics. Similar representations by the Norwegian surveyor Casper Wessel (1745-1818) in 1797 and by the Swiss clerk Jean-Robert Argand (1768-1822) in 1806 went largely unnoticed. The concept *modulus* of complex numbers is due to Argand, and absolute value, for modulus, is due to Karl Theodor Wilhelm Weierstrass (1815-97). The Cartesian coordinate system called the complex plane or Argand diagram is also named after the same Argand. Mention should also be made of an excellent little treatise by C.V. Mourey (1828), in which the foundations for the theory of directional numbers are scientifically laid. The general acceptance of the theory is not a little due to the labors of Augustin Louis Cauchy (1789-1857) and Niels Henrik Abel (1802-29), especially the latter, who was the first to boldly use complex numbers, with a success that is well-known.

Geometric applications of complex numbers appeared in several memoirs of prominent mathematicians such as August Ferdinand Möbius (1790-1868), George Peacock (1791-1858), Giusto Bellavitis (1803-80), Augustus De Morgan (1806-71), Ernst Kummer (1810-93), and Leopold Kronecker (1823-91). In the next three decades, further development took place. Especially, in 1833 William Rowan Hamilton (1805-65) gave an essentially rigorous algebraic definition of complex numbers as pairs of real numbers. However, a lack of confidence in them persisted; for example, the English mathematician and astronomer George Airy (1801-92) declared: "I have not the smallest confidence in any result which is essentially obtained by the use of imaginary symbols." The English logician George Boole (1815-64) in 1854 called $\sqrt{-1}$ an "uninterpretable symbol." The German mathematician Leopold Kronecker believed that mathematics should deal only with whole numbers and with a finite number of operations, and is credited with saying: "God made the natural numbers; all else is the work of man." He felt that irrational, imaginary, and all other numbers excluding the positive integers were man's work and therefore unreliable. However, the French mathematician Jacques Salomon Hadamard (1865-1963) said the shortest path between two truths in the real domain passes through the complex domain. By the latter part of the 19th century, all vestiges of mystery and distrust of complex numbers could be said to have disappeared, although some resistance continued among a few textbook writers well into the 20th century. Nowadays, complex numbers are viewed in the following different ways:

1. points or vectors in the plane;

2. ordered pairs of real numbers;

3. operators (i.e., rotations of vectors in the plane);

4. numbers of the form $a + bi$, with a and b real numbers;

5. polynomials with real coefficients modulo $x^2 + 1$;

6. matrices of the form $\begin{bmatrix} a & b \\ -b & a \end{bmatrix}$, with a and b real numbers;

7. an algebraically closed complete field (a field is an algebraic structure that has the four operations of arithmetic).

The foregoing descriptions of complex numbers are not the end of the story. Various developments in the 19th and 20th centuries enabled us to gain a deeper insight into the role of complex numbers in mathematics (algebra, analysis, geometry, and the most fundamental work of Peter Gustav Lejeune Dirichlet (1805-59) in number theory); engineering (stresses and strains on beams, resonance phenomena in structures as different as tall buildings and suspension bridges, control theory, signal analysis, quantum mechanics, fluid dynamics, electric circuits, aircraft wings, and electromagnetic waves); and physics (relativity, fractals, and the Schrödinger equation).

Although scholars who employ complex numbers in their work today do not think of them as mysterious, these quantities still have an aura for the mathematically naive. For example, the famous 20th-century French intellectual and psychoanalyst Jacques Lacan (1901-81) saw a sexual meaning in $\sqrt{-1}$.

References for Further Reading

[1] M.J. Ablowitz and A.S. Fokas, *Complex Variables: Introduction and Applications*, Cambridge University Press, 2nd ed., Cambridge, 2003.

[2] L.V. Ahlfors and L. Sario, *Riemann Surfaces*, Princeton University Press, Princeton, New Jersey, 1960.

[3] L.V. Ahlfors, *Complex Analysis*, 3rd ed., McGraw-Hill, New York, 1979.

[4] R.B. Ash, *Complex Variables*, Academic Press, New York, 1971.

[5] R.P. Boas, *Entire Functions*, Academic Press, New York, 1954.

[6] R.P. Boas, *Invitation to Complex Analysis*, Random House, New York, 1987.

[7] J.B. Conway, *Functions of One Complex Variable*, Springer-Verlag, New York, Vol. 1, 1973.

[8] B.R. Gelbaum, *Problems in Real and Complex Analysis*, Springer-Verlag, New York, 1992.

[9] W.K. Hayman, *Meromorphic Functions*, Clarendon Press, Oxford, 1964.

[10] E. Hille, *Analytic Function Theory*, Vols. 1 and 2, 2nd ed., Chelsea, New York, 1973.

[11] S. Krantz, *Handbook of Complex Variables*, Birkhäuser, Boston, 1999.

[12] S. Lang, *Complex Analysis*, 2nd ed., Springer-Verlag, New York, 1985.

[13] N. Levinson and R.M. Redheffer, *Complex Variables*, Holden-Day, San Francisco, 1970.

[14] P.T. Mocanu and S.S. Miller, *Differential Subordinations: Theory and Applications*, Marcel Dekker, New York, 2000.

[15] P.J. Nahin, *An Imaginary Tale: The Story of $\sqrt{-1}$*, Princeton University Press, Princeton, New Jersey, 1998.

[16] R. Narasimhan, *Complex Analysis in One Variable*, Birkhäuser, Boston, 1984.

[17] Z. Nehari, *Conformal Mapping*, Dover, New York, 1975.

[18] R. Nevanlinna and V. Paatero, *Introduction to Complex Analysis*, Addison-Wesley, Reading, Massachusetts, 1969.

[19] R.N. Pederson, The Jordan curve theorem for piecewise smooth curves, *American Mathematical Monthly*, **76**(1969), 605-610.

[20] R. Remmert, *Theory of Complex Functions*, translated by R.B. Burckel, Springer-Verlag, New York, 1991.

[21] W. Rudin, *Real and Complex Analysis*, 3rd ed., McGraw-Hill, New York, 1987.

[22] S. Saks and A. Zygmund, *Analytic Functions*, Elsevier, New York, 1971.

[23] R.A. Silverman, *Complex Analysis with Applications*, Prentice-Hall, Englewood Cliffs, New Jersey, 1974.

[24] F. Smithies, *Cauchy and the Creation of Complex Function Theory*, Cambridge University Press, Cambridge, 1997.

[25] E.M. Stein and R. Shakarchi, *Complex Analysis*, Princeton University Press, Princeton, New Jersey, 2003.

[26] I. Stewart and D. Tall, *Complex Analysis*, Cambridge University Press, Cambridge, 1983.

[27] N. Steinmetz, de Branges' proof of the Bieberbach conjecture, in *International Series of Numerical Mathematics*, **80**(1987), 3-16.

Index